MOLLUSCA AND RADIATA

ENCYCLOPAEDIA OF LIFE SCIENCE SERIES

MOLLUSCA AND RADIATA

BARON CUVIER

ANMOL PUBLICATIONS PVT. LTD.
NEW DELHI-110 002 (INDIA)

ANMOL PUBLICATIONS PVT. LTD.
4374/4B, Ansari Road, Daryaganj
New Delhi - 110 002

Mollusca and Radiata
First Edition, 1998
ISBN 81-261-0115-6

PRINTED IN INDIA

Published by J.L. Kumar for Anmol Publications Pvt. Ltd., New Delhi and Printed at Tarun Offset Press, Delhi.

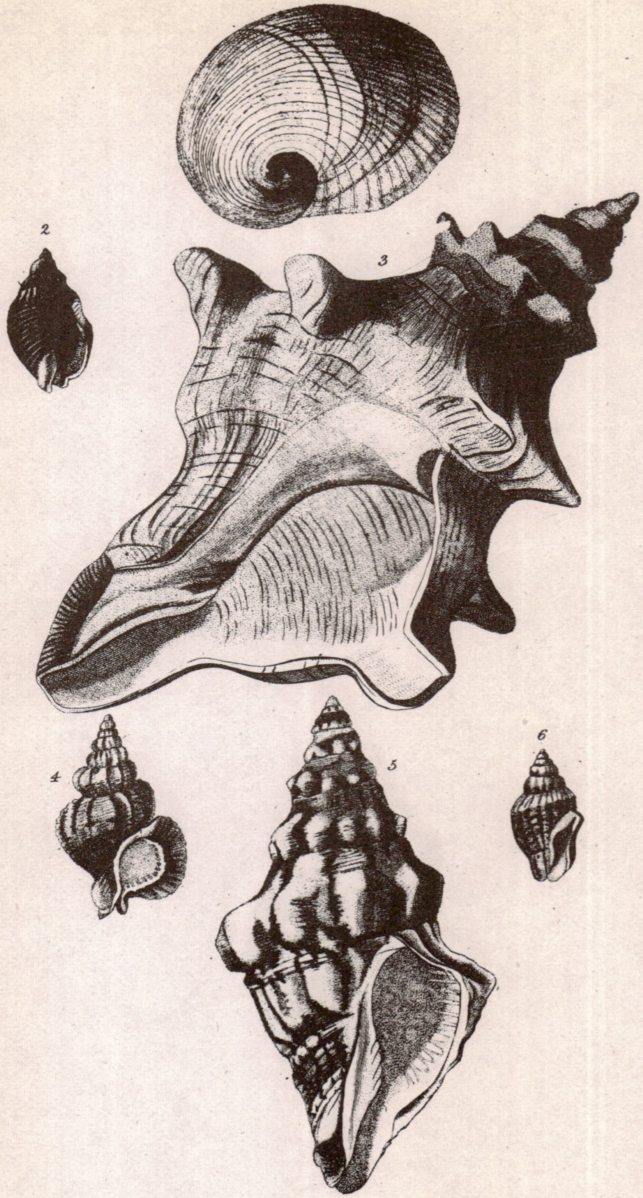

1 *Cryptostoma Javanica.* 4 *Triton Nassoides.*

2 *Cyllene Owenii.* 5 *Turbenella ceratus.*

3 *Fasciolaria trapezium var.* 6 *Columbella suturalis.*

LIST OF PLATES.

MOLLUSCA.

LIST OF PLATES.

ZOOPHYTES.

ANIMALIA MOLLUSCA.

THE mollusca have neither an articulated skeleton, nor a vertebral canal. Their nervous system is not united in a spinal marrow, but merely in a certain number of medullary masses, dispersed in different parts of the body, the chief of which, termed the brain, is situated transversely on the œsophagus, and envelopes it with a nervous collar. Their organs of motion and sensation have not the same uniformity of number and position as in the vertebrata, and the irregularity is still more striking in the viscera, particularly as respects the position of the heart and respiratory organs, and even as regards the structure of the latter; for some of these respire elastic air, and others salt or fresh water. Their external organs, however, and those of locomotion, are generally arranged symmetrically on the two sides of an axis.

The circulation of the mollusca is always double; that is, their pulmonary circulation describes a separate and distinct circle. This function is at least always aided by a fleshy ventricle, situated between the veins of the lungs and the arteries of the body, and not, as in fishes, between the veins of the body and the arteries of the lungs. It is then an aortic ventricle. The cephalopoda alone are provided with a pulmonary ventricle, which is even divided into two. The aortic ventricle is also divided in some genera as in *Arca* and *Lingula ;* at other times, as in other bivalves, its auricle only is divided. When there is more than one ventricle they are not united in a single mass, as in warm-blooded animals, but often sufficiently remote the one from the other, and then one might say there are many hearts.

The blood of the mollusca is white, or bluish, and it appears to contain a smaller proportionate quantity of fibrine than that of the vertebrata. There are reasons for believing that their veins fulfil the functions of absorbent vessels.

Their muscles are attached to various points of their skin, forming there tissues, which are more or less complex and dense. Their motions consist of contractions in different directions, which produce inflexions and prolongations, or relaxations, of their various parts, by means of which they creep, swim, and seize upon objects, just as the form of these parts may permit; but as the limbs are not supported by articulated and solid levers, they cannot proceed rapidly, or by leaps.

The irritability of most of them is extremely great, and remains for a long time after they are divided. Their skin is naked, very sensible, and usually covered with a humour that oozes from its pores. No particular organ of smell has been discovered in them, though they enjoy that sense; it may possibly reside in the entire skin, for it greatly resembles a pituitary membrane. All the acephala, brachiopoda, cirrhopoda, and part of the gasteropoda, and pteropoda, are destitute of eyes. The cephalopoda, on the contrary, have them at least as complicated as those of the warm-blooded animals; they are the only ones in which the organ of hearing has been discovered, and whose brain is enclosed within a particular cartilaginous box.

Nearly all the mollusca have a development of the skin, which covers their body, and which bears more or less resemblance to a *mantle*; it is often, however, narrowed into a simple disk, formed into a pipe, hollowed into a sac, or extended and divided in the form of fins.

The *naked mollusca* are those in which the mantle is simply membranous or fleshy: most frequently, however, it forms in its thickness one or several laminæ, of a substance more or less hard, deposited in layers, and increasing in extent, as well as in thickness, because the recent layers always outedge the old ones.

When this substance remains concealed in the thickness of the mantle, it is still customary to call the animals naked mollusca. Most generally, however, it becomes so much developed, that the contracted animal finds shelter beneath it. It is then termed a *shell*, and the animal is said to be *testaceous*. The epidermis which covers it is thin, and sometimes desiccated.

Until my labours on the subject were published, the testacea were made a particular order; but there are so many insensible transitions from the naked mollusca to the testacea, and their natural divisions form such groups with each other, that this distinction can no longer be admitted. Besides, there are several of the testacea which are not mollusca.

The variety in the form, colour, surface, substance, and brilliancy of shells, is infinite. Most of them are calcareous, some are horny; but they always consist of matters deposited in layers, or exuded from the skin under the epidermis, like the mucous covering, the nails, horns, scales, and even teeth. The tissue of shells differs according to the mode of this disposition, which is either in parallel laminæ, or crowded vertical filaments.

All the modes of mastication and deglutition are found in the mollusca. Their stomachs are sometimes simple, sometimes multiple, frequently provided with a peculiar armature, and their intestines are variously prolonged. They most generally have salivary glands, and always a large liver, but neither pancreas, nor mesentery ; several have secretions which are peculiar to them.

They also present examples of all the modes of generation. Several of them possess the faculty of self-impregnation ; others, although hermaphrodites, have need of a reciprocal intercourse. Many have the sexes separated. Some are viviparous, others oviparous ; the eggs of the latter are sometimes enveloped with a shell more or less hard, sometimes with a simple viscosity. These varieties of the digestive and generative processes are found in the same order, and sometimes in the same family. The mollusca in general appear to be animals that are but slightly developed, possessed of but little industry, and which are only preserved by their fecundity and vital tenacity.

DIVISION OF THE MOLLUSCA INTO SIX CLASSES.

The general form of the body of the mollusca being in proportion to the complication of their internal organization, indicates their natural division.

The body of some resembles a sac, open in front, containing the branchiæ, whence issues a well-developed head, crowned with long and strong fleshy productions, by means of which they crawl, and seize various objects. These we term the CEPHALOPODA.

That of others is closed ; the appendages of the head are either wanting, or are extremely reduced ; the principal organs of locomotion are two wings or membranous fins, situated on the sides of the neck, and which frequently support the branchial tissue. They constitute the PTEROPODA.

Others, again, crawl by means of a fleshy disk on their belly, sometimes, though rarely, compressed into a fin, and have almost always a distinct head before. We call these the GASTEROPODA.

A fourth class is composed of those where the mouth remains hidden in the bottom of the mantle, which also encloses the branchiæ and viscera, and is open either throughout its length, at both ends, or at one extremity only. Such are our ACEPHALA.

A fifth comprises those which, also inclosed in a mantle, and without an apparent head, have fleshy or membranaceous arms, furnished with cilia of the same nature. We term these BRACHIOPODA.

Finally, there are some which, although similar to the other mollusca in the mantle, branchiæ, &c., differ from them in numerous horny and articulated limbs, and in a nervous system more nearly allied to that of the Articulata. They will constitute our last class, or that of the CIRRHOPODA.

FIRST CLASS OF MOLLUSCA.

CEPHALOPODA.

THEIR mantle unites under the body, forming a muscular sac, which envelopes all the viscera. In several its sides are extended into fleshy fins. The head projects from the opening of the sac; it is rounded, furnished with two large eyes, and crowned with arms or feet, conical, fleshy, more or less elongated, capable of being flexed in every direction, and extremely vigorous, the surface of which is armed with suckers or *cups*, by means of which they adhere with great tenacity to every body they embrace. These feet are their instruments of prehension, swimming, and walking. They swim with the head backwards, and crawl in all directions with the head beneath and the body above.

A fleshy funnel, placed at the opening of the sac, before the neck, affords a passage to the excretions.

The cephalopoda have two branchiæ within the sac, one on each side, resembling a highly complicated fern leaf; the great vena cava, having arrived between them, divides into two branches, which pour their contents into two fleshy ventricles, each of which is placed at the base of the gill on its own side, and propels the blood into it.

The two branchial veins communicate with a third ventricle, situated near the bottom of the sac, which, by means of various arteries, distributes the blood to every part of the body.

Respiration is effected by the water which flows into the sac, and issues through the funnel. It appears that it can even penetrate into two cavities of the peritoneum, traversed by the vena cava in their passage to the branchiæ, and act upon the venous blood by means of a glandular apparatus attached to those veins.

Between the base of the feet we find the mouth, armed with two stout horny jaws, resembling the beak of a parrot.

Between the jaws is a tongue, bristling with horny points; the œsophagus swells into a crop, and then communicates with a gizzard as fleshy as that of a bird, to which succeeds a third membranous and spiral stomach, which receives the bile from the two ducts of the very large liver. The intestine is simple and short; the rectum terminates in the funnel.

These animals are remarkable for a peculiar and intensely black excretion, with which they darken the surrounding water when they wish to conceal themselves. It is produced by a gland, and held in reserve by a sac, variously situated, according to the species.

Their brain, which is contained in a cartilaginous cavity of the head, gives off a cord on each side, which produces a large ganglion in each orbit, whence are derived innumerable optic filaments. The eye consists of several membranes, and is covered by the skin which becomes diaphanous in that particular spot, sometimes forming folds, which supply the want of eye-lids. The ear is merely a slight cavity on each side near the brain, without semi-circular canals, or an external meatus, where a membranous sac is suspended, which contains a little stone.

The skin of these animals, of the octopi particularly, changes colour in places, by spots, with a rapidity which greatly surpasses that of the chameleon.

The sexes are separated. The ovary of the female is in the bottom of the sac; two oviducts take up the ova and pass them out through large glands which envelope them in a viscid matter, and collect them into clusters. The testis of the male, placed like the ovary, communicates with a vas deferens, which terminates in a fleshy penis, situated on the left of the anus. A bladder and prostate terminate there likewise. There is reason to believe that fecundation is effected by sprinkling, as is the case with most fishes. In the spawning season, the bladder contains a multitude of little filiform bodies, which, by means of a peculiar mechanism, are ruptured the moment they reach the water, where they move about with great rapidity, and diffuse a humour with which they are filled.

These animals are voracious and cruel; possessed both of agility and numerous modes of seizing their prey, they destroy immense quantities of fish and crustacea. Their flesh is eaten; their *ink* is employed in painting, and the Indian or China ink is supposed to be made from it.

The cephalopoda comprise but a single order, which is divided into genera, according to the nature of the shell.

Those which have no external shell, according to Linnæus, formed but the single genus,

SEPIA, *Lin.*,

Which is now divided as follows:

OCTOPUS, *Lam.* *Polypus* of the ancients.

But two small conical granules, of a horny substance, on the two sides, of the thickness of the back; the sac, having no fins, resembles an oval purse; eight feet, all of which are about equal, very large in proportion to the body, and united at base by a membrane; they are employed by the animal in swimming, crawling, and seizing its prey. The length and strength of these limbs render them fearful weapons, which they twine round animals; in this way it has even destroyed men while bathing. The eyes are small in proportion, and the skin contracts over them so as entirely to cover them at the will of the animal. The receptacle of the ink is sunk in the liver; the glands of the oviducts are small. Some of them,

POLYPUS, *Aristotle*,

Have the suckers alternating in two rows along each foot. The common species, *Sepia octopodia*, Lin., with a slightly rough skin, arms six times the length of its body, and furnished with one hundred and twenty pairs of cups, infests the coast of Europe in summer, and destroys immense numbers of fishes and crustacea.

The seas of hot climates produce another, *Sepia rugosa*, Bosc., Seb., III. ii. 2, 3, whose body is rougher, arms something longer than the body, and furnished with ninety pairs of cups. It is from this species that some authors suppose the Indian ink is procured. Others, again,

ELEDON, *Aristotle*,

Have but a single row of cups along each foot. One of them, the *Poulpe musqué*, Lam., Mém. de la Soc. d'Hist. Nat., 4to. pl. ii.; Rondelet, 516; is found in the Mediterranean, and is remarkable for its musky odour.

ARGONAUTA, *Lin.*

Octopi with two rows of cups; the pair of feet which are nearest to the back being dilated at the extremity into a broad membrane. The two cartilaginous granules of the common octopus are wanting; but these mollusca are always found in a very thin shell, symmetrically fluted and spirally convoluted; the last whorl so large that it bears some resemblance to a galley, of which the spire should be the poop. The animal makes a consequent use of it, and in calm weather whole fleets of them may be observed navigating the surface of the ocean, employing six of their tentacula as oars, and elevating the two membranous ones by way of a sail. If the sea becomes rough, or they perceive any danger, the argonaut withdraws all its arms, concentrates itself in its shell, and descends to the bottom. The body of the animal does not penetrate to the bottom of the spires of the shell, and it appears that it does not adhere to it, at least there is no muscular attachment, a circumstance which has induced some authors to believe that its residence there is that of a parasite, like the *Pagurus Bernhardus*, for instance. As it is always found in the same shell, however, and as no other animal is ever seen there, although it is very common, and so formed as to show itself frequently on the surface, and as the germ of it is visible even in the ovum of the argonaut, this opinion must be considered as highly problematical, to say nothing more of it.

The ancients were well acquainted with this singular animal and its manœuvres. It is their *Nautilus* and their *Pompilus.*—Pliny, IX. c. xxix.

Several species are known, closely resembling each other, both in the animal and the shell, which were united by Linnæus under the name of *Argonauta argo*, or the paper nautilus.

Bellerophon, *Montf.*

Certain fossil shells, so called, the animal of which is supposed to have been analagous to the argonauts. They are spirally and symmetrically convoluted, without septa, but thick, and not fluted; the last whorl proportionably shorter.

Loligo, *Lam.*

An ensiform lamina of horn in the back, in lieu of a shell; the sac has two fins, and besides the eight feet promiscuously loaded with little cups on short pedicles, the head is furnished with two much longer arms, provided with cups near the end only, which is widened. The animal uses these latter to keep itself immoveable, as if at anchor. The receptacle of the colouring matter is lodged in the liver, and the glands of the oviducts are very large. The eggs are deposited closely together in narrow garlands, and in two rows.

They are now sub-divided according to the number and armature of the feet, and the form of the fins.

Loligopsis, *Lam.*,

Or the Calmarets, should have but eight feet, as in Octopus. They are only known, however, by drawings of but little authority.

In the true loligo the long arms are furnished with cups like the other tentacula, and the fins are placed near the point of the sac. Three species are found in the European seas.

L. vulgaris; Sepia loligo; L. Rondel, 506; Salv. 169. Fins forming a rhomb at the bottom of the sac.

L. Sagittata, Lam.; Seb. III. iv. Fins forming a triangle

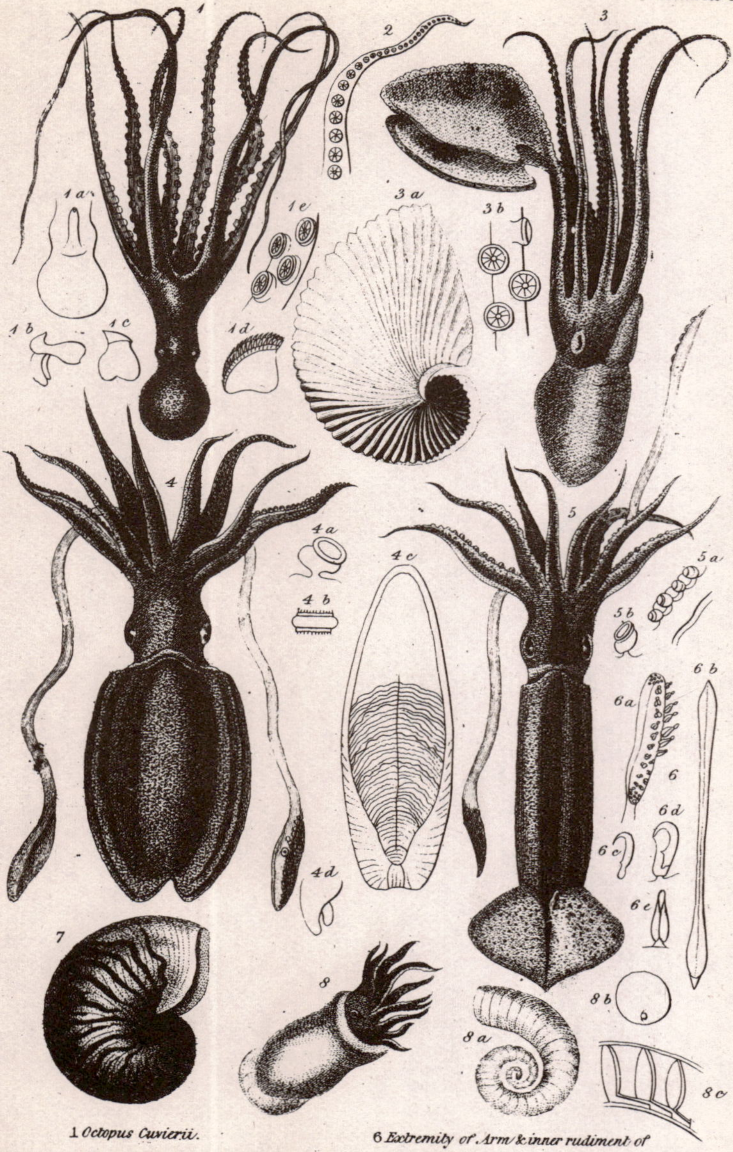

1 *Octopus Cuvierii.*

2 *Portion of Arm of Eledone moschatus.*

3 *Argonauta argo.* 4 *Sepia officinalis.*

5 *Loligo Brogniartii.*

6 *Extremity of Arm & inner rudiment of Onychoteuthis angulata.*

7 *Nautilus pompilius.*

8 *Spirula australis.*

at the bottom of the sac; arms shorter than the body, and loaded with cups for about half their length.

L. media; Sep. media; L. Rondel, 508. Fins forming an ellipsis at the bottom of the sac, which terminates in a sharp point.

ONYKIA, *Lesueur.* ONYCHOTHEUTHIS, *Lichtenst.*

The long arms furnished with cups, terminating in hooks; in other respects the form is the same.

SEPIOLA, *Cuv.*

The rounded fins attached to the sides of the sac, and not to its point; one species. *S. vulgaris; S. sepiola;* L. Rondel, 519; inhabits European seas. The sac is short and obtuse, and the fins small and circular. It seldom exceeds three inches in length, and its horny lamina is as slender, and sharp as a stilet.

CHONDROSEPIA, *Leukard.* SEPIOTHEUTES, *Blainv.*,

The whole margin of the sac, on each side, bordered with the fins, as in sepia; but the shell horny, as in loligo.

SEPIA, *Lam.*

The Sepiæ, properly so called, have the two long arms of a loligo, and a fleshy fin extending along the whole length of each side of the sac. The shell is oval, thick, convex, and composed of numerous and parallel calcareous laminæ, united by thousands of little hollow columns, running perpendicularly from one to the other. This structure, rendering it friable, causes it to be employed for polishing various kinds of work; it is also given to birds in aviaries, for the purpose of whetting their beaks.

The ink-pouch of the sepiæ is detached from the liver, and situated more deeply in the abdomen. The glands of the oviducts are enormous. The eggs are produced attached to each other in branching clusters, resembling those of grapes, and are commonly termed *sea grapes*.

The species most commonly found in the seas of Europe, *Sepia officinalis*, L.; Rondel, 498; Seb. III. iii., attains the length of a foot or more. Its skin is smooth, whitish, and dotted with red.

The Indian Ocean produces another, *Sepia tuberculata*, Lam. Soc. d'Hist. Nat. 4to. pl. i. f. 1.

NAUTILUS, *Lin.*

In this genus Linnæus united all spiral, symmetrical, and chambered shells, that is to say, such as are divided by septa into several cavities; their inhabitants he supposed to be cephalopoda. One of them really does belong to a cephalopod that strongly resembles a sepia, but it has shorter arms; it forms the genus

SPIRULA, *Lam.*

In the hind part of the body, which is that of a sepia, is an interior shell, which, although very different from the bone of that animal as to figure, differs but little in its formation. A correct idea of the latter may be obtained by imagining the successive laminæ, instead of remaining parallel and approximated, to be concave towards the body, more distant, increasing little in breadth, and forming an angle between them, thus producing an elongated cone, spirally convoluted in one plane, and divided transversely into chambers. Such is the shell of the spirula, which has additional chambers, consisting of a single hollow column that occupies the internal side of each chamber, continuing its tube with those of the other chambers to the very extremity of the shell. This column is termed the *siphon*. The turns of the spire do not come into contact.

But a single species, *Nautilus spirula*, L.; List. 550-2, is known. The

NAUTILUS, properly so called,

Has a shell which differs from the spirula in the sudden crossing of the laminæ, and in the last turns of the spire, which not only touch the preceding ones but envelope them. The siphon occupies the centre of each septum.

N. pompilius, L.; List. 551, the most common species; it is very large, formed internally of a beautiful mother-of-pearl, and covered externally with a white crust, varied with fawn-coloured streaks or bands.

The animal, according to Rumph, is partly contained within the last cell, has the sac, eyes, parrot beak, and funnel of the other cephalopoda; but its mouth, instead of having their large feet and arms, is surrounded by several circles of numerous small tentacula without cups. A ligament arising from the back traverses the whole siphon and fastens it there. It is probable that the epidermis is extended over the outside of the shell, though we may presume it is very thin over the parts that are coloured.

Individuals are sometimes found, *Naut. pompilius*, *β*, Gmel.; List. 552; AMMONIE, Montf. 74.; in which the last whorl does not envelope and conceal the others, but where all of them, though in contact, are exposed, a circumstance which approximates them to the ammonites; they so closely resemble the common species, however, in all the rest of the shell, that it is scarcely possible to believe them to be any thing more than a variety of it.

Fossil nautili are found of a large or moderate size, and much more various, as to form, than those now taken in the ocean.

Chambered shells are also found among fossils, furnished with simple septa and a siphon, the body of which, at first arched, or even spirally convoluted, remains straight in the more recent parts; they are the LITUUS of Breyn, in which the whorls are sometimes contiguous and sometimes distinct. The HORTOLES of Montfort.

In others, the orthoceratites, it is altogether straight; it is not improbable that the animal resembled that of the nautilus, or of the spirula. The

BELEMNITES

Probably belong to this family; but it is impossible to ascertain the fact, as they are only found among fossils. Every thing, however, proves them to have been internal shells, thin, and double; that is, composed of two cones united at base, the inner one much shorter than the other, and divided into chambers by parallel septa, which are concave on the side next to the base. A siphon extends from the summit of the external cone to that of the internal one, and continues thence, sometimes along the margin of the septa, and sometimes through their centre. The interval between the two testaceous cones is filled with a solid substance, here composed of radiating fibres, and there of self-involving conical layers, the base of each being on the margin of one of the septa of the inner cone. In one specimen we only find this hard portion, and in another we also find the nuclei of the chambers of the inner cone, or what are termed the alveoli. Most commonly these nuclei and the chambers themselves have left no further traces than some projecting circles on the inside of the internal cone. In other specimens, again, we find more or fewer of the nuclei, and still in piles, but detached from the double conical sheath that enveloped them.

Of all the fossils the Belemnites are the most abundant, particularly in chalk and compact limestone.

M. de Blainville divides them according as the interior cone or chambered part penetrates to a greater or less depth, as the edges of the exterior cone have or have not a little cleft, as the external surface is marked on one side by a longitudinal furrow, or by two or more furrows towards the summit, or, finally, as that surface is smooth and without furrows.

Bodies very similar to the Belemnites, but without a cavity, and with rather a prominent base, form the genus ACTINOCAMAX of Miller. It is also upon conjecture of a similar nature that reposes the classification of the

AMMONITES, *Brug.*,

Or the Cornua-ammonis, for they no longer exist, except among fossils. They are distinguished from the nautili by their septa, which, instead of being plane or simply concave, are angular, and sometimes undulated, but most frequently crimped on the edge like the leaf of an acanthus. The smallness of their last cell seems to indicate that, like the spirula, they were internal shells. They are very abundant in the strata of secondary mountains, where they are found varying from the size of a lentil to that of a coach-wheel. Their subdivisions are based upon the variation of their volutes and siphons.

The name of AMMONITES, *Lam.* (SIMPLEGADES, *Montf.*, 82), is particularly restricted to those species in which all the whorls are visible. Their siphon is near the margin.

They have lately been divided into the AMMONITES, Planites of Haan, where the edge of the septa is foliaceous, and into the CERATITES of Haan, where it is simply angular and undulated.

Those in which the last whorl envelopes all the others form the ORBITULITES, *Lam.*, or the GLOBITES and the GONIALITES of Haan, or the PELAGUSES, *Montf.* 62, in all of which the siphon is situated as in the preceding ones.

The SCAPHITES, *Sowerb.*, are those in which the whorls are contiguous and in the same plane, the last one excepted, which is detached, and reflexed on itself.

Some, BACULITES, *Lam.*, are entirely straight, without any

spiral portion whatever. Some of them are round, and others compressed. The last sometimes have a lateral siphon.

There are some with their first cells arched. The HAMITES, *Sowerb.*, are arcuated.

Finally, those which vary most from the usual form of this family are the TURRILITES, *Montf.* 118, where the whorls, so far from running in the same plane, suddenly descend, giving to the shell that form of an obelisk which is called *turreted.*

It is also thought, and from similar considerations, that we should refer to the cephalopoda, and consider as internal shells the

CAMERINES, *Brug.* NUMMULITES, *Lam.*,

Commonly called *nummulites*, *lenticular stones*, &c., which are only found among fossils, and present, externally, a lenticular figure, without any apparent opening, and a spiral cavity internally, divided by septa into numerous small chambers, but without a siphon. They are amongst the most universally diffused of all fossils, forming, *per se*, entire chains of calcareous hills and immense bodies of building stone.

The most common, and those which attain the greatest size, form a complete disk, and have only a single range of cells in each whorl. Some very small species are also found in certain seas.

The margin of other small species (the SIDEROLITHES, *Lam.*), both fossil and living, are bristled with points, which give them a stellated appearance.

The labours and researches, conducted with infinite patience by Bianchi (or Janus Plancus), Soldani, Fichtel, Moll, and D'Orbigny, have ascertained an astonishing number of these chambered shells without a siphon, like the nummulites, that are extremely small and frequently microscopical, both in the sea, among the sand, fuci, &c., and in a fossil state in the sand formations of various countries.

They vary in a remarkable degree as to their general form, the number and relative position of the chambers, &c. One or two species, in which alone the animals have been observed, appeared to present a small oblong body, crowned by numerous red tentacula, which, added to the septa of the shell, have caused them to be placed immediately after the cephalopoda, like the genera just mentioned. This arrangement, however, requires to be confirmed by more numerous observations, before we can consider it as final.

Such of these species as were known in the time of Linnæus and Gmelin, were placed by those naturalists among the nautili. M. D'Orbigny, who has exceeded every other person in attention to this subject, forms them into an order which he calls FORAMINIFERA, on account of the only communication between the cells being by means of holes, and divides them into families according to the manner in which the cells are disposed. When the cells are simple and spirally arranged, they constitute his HELICOSTEGA, which are again subdivided. If the whorls are enveloped, as is particularly the case in the nummulites, they become his HELICOSTEGA NAUTILOIDA.

If the whorls do not envelope each other, they are the HELICOSTEGA AMMONOIDA.

If the whorls are elevated, as in most of the univalves, they are the HELICOSTEGA TURBINOIDA.

Simple cells may also be strung upon a single straight or slightly curved axis, constituting the family of the STYCOSTEGA.

Or they may be arranged in two alternate series, when they become the ENALLOSTEGA.

Or a few of them may be collected and united as in a pellet, forming the AGATHISTEGA.

Finally, in the entomostega, the cells are not simple, as in the other families, but are sub-divided by transverse septa in such a way that a section of the shell exhibits a sort of trellis.

SECOND CLASS OF MOLLUSCA.

THE PTEROPODA

Swim like the cephalopoda in the waters of the sea, but cannot fix themselves, nor creep there, for want of feet. Their organs of locomotion consist only of fins, placed like wings at the two sides of the mouth. But small species are known, and few in number, all hermaphrodites.

CLIO, *Lin.* CLIONE, *Pall.*,

Have the body oblong, membranaceous, without a mantle; the head formed of two rounded tubes, from which issue small tentacula; two small fleshy lips and a ligula on the front of the mouth; and the fins provided with a vascular net-work, which takes the place of gills. The anus and the orifice of generation are under the right gill. Some authors attribute eyes to these animals.

The mass of the viscera does not nearly fill the exterior envelope; the stomach is broad, the intestine short, the liver considerable.

The most celebrated species (*Clio Borealis*, L.) swarms in the seas of the north, and constitutes, from its abundance, plenty of food for the whales, though each individual is scarcely an inch long.

Bruguiéres has observed one larger, and not less abundant, in the Indian Ocean. It is distinguished by its rose colour, by its emarginated tail, and its body divided by furrows into six lobes.—Encyc. Meth. pl. des Mollusc. pl. lxxv. f. 1, 2.

It appears that we must likewise place here,

The CYMBULIA of *Péron*,

Which have a cartilaginous or gelatinous envelope, in the form of a boat, or rather of a hoof, bristling with some little points in longitudinal series. The animal has two great wings, with a vascular tissue, answering as both gills and fins, and between them, on the open side, a third smaller lobe, with three points. The mouth, with two small tentacula, is between the wings, towards the closed side of the shell, and above, two small eyes, and the orifice of generation, from which issues a penis, in the form of a small proboscis. From their transparency, the heart, brain, and viscera, can be distinguished through the integuments.

PNEUMODERMON, *Cuv.*,

Begins to be somewhat removed from *Clio:* the body is oval, without mantle, and without shells, the gills attached to the surface, and formed of small leaflets, ranged in two or three lines, disposed like an H on the side opposite to the head. The fins are small; the mouth, furnished with two small lips and two bundles of numerous tentacula, terminated each by a sucker, has a small lobe or fleshy tentaculum underneath.

The known species (*Pneumodermon Peronii*, Cuv., Ann. du Mus. IV. pl. lix., and Péron, ib. XV. pl. ii.), has been taken in the ocean by Péron. It is scarcely more than an inch long.

LIMACINA, *Cuv.*,

Should, according to the description of Fabricius, have close relations with pneumodermon; but their body is terminated by a spiral tail, and is lodged in a very thin shell, of one whorl and a half, umbilicated on one side, and flatted at the other. The animal makes use of its shell as a boat, and of its wings as oars, when it wishes to swim at the surface of the water.

The known species (*Clio helicina* of Phips and Gm.) *Argonauta arctica*, Fab., Faun., Groen., 387, is but little less abundant than the Clio Borealis in the icy sea, and is also considered as one of the principal aliments of the whale.

HYALEA, *Lam.* CAVOLINA, *Abild.*,

Have two very large wings, no tentacula, a mantle cleft at the sides, lodging the gills in the bottom of its fissures, and invested with a shell likewise cleft at the sides, whose central face is very gibbous, the dorsal plate larger than the other, and the transverse line which unites them behind provided with three short denticulations. In the living state the animal puts out, through the lateral clefts of its shell, some stripes, more or less long, which are productions of the mantle.

The most known species (*Anomia tridentata*, Forskahl; *Cavolina natans*, Abildgard; *Hyalea cornea*, Lam.), Cuv., Ann. du Mus. IV. pl. lix., et Péron, ib. XI. pl. iii., fig. 13, has a small yellowish shell, semi-transparent, and is found in the Mediterranean and the ocean.

CLEODORA, *Péron*,

For which Brown originally created the genus *Clio*, appear analogous to hyalea, in the simplicity of their wings and the absence of tentacula between them. It is probable that their gills are also concealed in their mantle. Nevertheless, their conical, or pyramidal shell, is not cleft upon the sides.

M. Rang distinguishes the CLEODORA, proper, with pyramidal shell.

CRESEIS with conical elongated shell.

CUVIERIES with cylindrical shell.

PSYCHE with globular shell.

EURYBIA with hemispheric shell.

It has been thought we may approximate to hyalea,

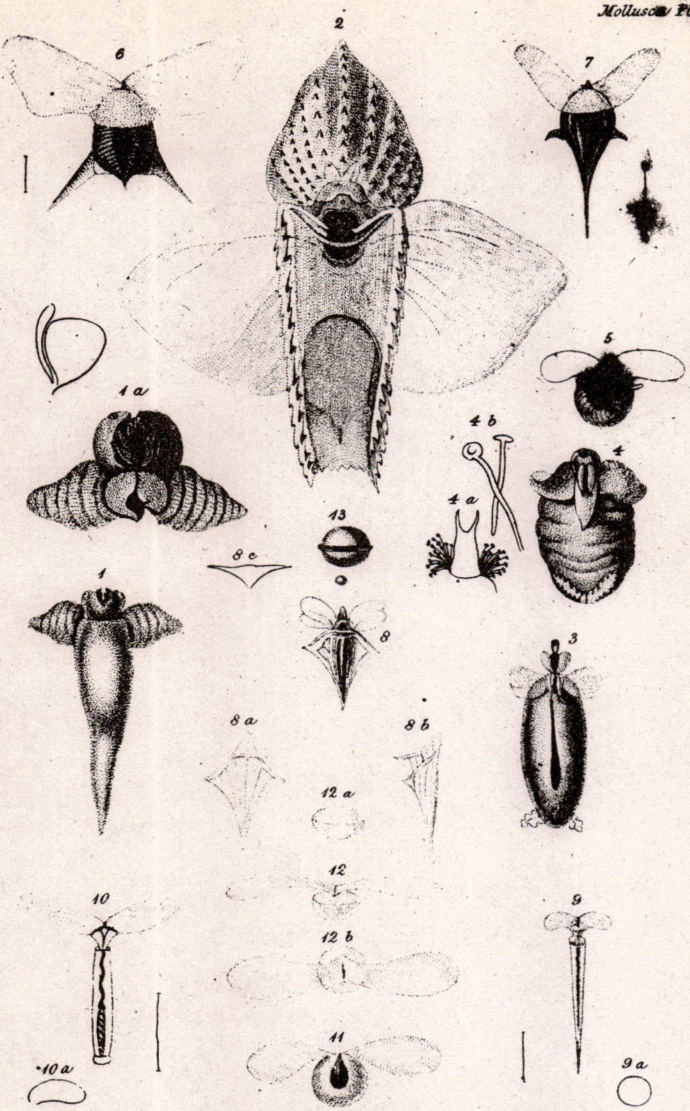

1. *Clio borealis.*
2. *Cymbulia Peronii.*
3. *Pneumodermon diaphanum.*
4. *Pneum. Peronii.*
5. *Limacina helicina.*
6. *Hyalea globulosa.*
7. *Hy. trespinosa.*
8. *Cleodora lanceolata.*
9. *Creseis virgula.*
10. *Cuvieria columnella.*
11. *Psyche globulosa.*
12. *Eurybia hemispherica.*
13. *Pyrgo lævis.*

PYRGO,

A very small fossil shell, discovered by M. Defrance, globular, very thin, divided by a transverse cleft, very narrow, except in front, where it widens a little.

THIRD CLASS OF THE MOLLUSCA.

THE GASTEROPODA

Constitute a very numerous class of mollusca, of which some notion may be formed from the slug and snail.

They generally crawl upon a fleshy disk placed under the belly, but which sometimes assumes the form of a furrow, or that of a vertical lamina. The back is furnished with a mantle, which extends more or less, assumes various figures, and produces a shell in the greater number of genera. Their head, placed in front, is shown more or less according to the degree in which it is engaged under the mantle. It has only small tentacula, which are above the mouth, and do not surround it; their number varies from two to six, and they are sometimes wanting; their use is only for touch, or at most for smell. The eyes are very small, sometimes adherent to the head, sometimes to the base, or the side, or the point of the tentacula; they are also sometimes wanting. The position, the structure, and the nature of the respiratory organs, vary, and give rise to their division into several families. But they have only an aortic heart, that is placed between the pulmonary vein and the aorta.

The position of the apertures through which the organs of

generation come forth, and that of the anus, vary. Nevertheless, they are almost always on the right side of the body.

Several are absolutely naked. Others have but a concealed shell; but the greater number have a shell which can receive and shelter them.

These shells are produced in the thickness of the mantle: some of them are symmetrical, with several pieces; some symmetrical, with a single piece; and some non-symmetrical, which, in the species where they are concave, and grow for a long time, necessarily produce an oblong spiral form.

Let us, in fact, figure to ourselves an oblique cone, in which other cones are successively placed, always broader in a certain direction than in others: it will be necessary that the whole shall be rolled towards the side which is the smallest. The part on which the cone is rolled is named the *columella*, and it is sometimes full, sometimes hollow. When it is hollow, its aperture is named *umbilicus*.

The whorls of the shell may remain pretty nearly in the same plane, or tend always towards the base of the columella.

In this last case, the preceding whorls rise one above the other, and form what is called a *spire*, which is so much the sharper as the whorls descend more rapidly and are less wide. These shells, with projecting spires, are named *turbinated*.

When, on the contrary, the whorls remain pretty nearly in the same plane, and do not envelope each other, the *spire* is *flat*, or even concave. These shells are called *discoïd*.

When the top of each whorl envelopes the preceding, the spire is *concealed*.

The part from which the animal appears to come forth is named the aperture.

When the whorls remain pretty nearly in the same plane when the animal creeps, it has its shell placed vertically, the columella crosswise over the hinder part of the back, and the head passes under the edge of the aperture opposite to the columella.

When the spire is projecting, it is directed obliquely from the right side in almost all the species. A small number only have their spire projecting at the left side when they walk, and are named reversed.

It is remarked that the heart is always on the side opposite to that on which the spire is directed. Thus it is usually on the left, but in the *reversed* it is on the right. The contrary is the case with the organs of generation.

The organs of respiration, which are always in the last whorl of the shell, receive the ambient element from under its edge, sometimes because the mantle is entirely detached from the body along that edge, sometimes because it is pierced there with a hole.

Sometimes the edge of the mantle is prolonged into a canal, that the animal may be able to seek the ambient element without putting forth its head and foot from the shell. Then the shell has also in its edge, near the end of the columella, opposite to that towards which the spire tends, a notch, or a canal, to lodge that of the mantle. Consequently, the canal is on the left in the ordinary species, on the right in the *reversed*.

For the rest, the animal, being very flexible, causes the direction of the shell to vary, and most frequently, when there is a notch or a canal, it directs the canal forward, which causes the spire to be behind, the columella towards the left, and the opposite edge towards the right. The contrary takes place in the *reversed*. This is the reason why it is said that their shell turns to the left.

The aperture of the shell, and consequently, also, the last whorl, are more or less large, in relation to the other whorls, according as the head or foot of the animal, which are constantly to pass out and in there, are more or less large in relation to the mass of viscera that remains fixed in the shell. This aperture is so much broader or narrower as those same parts are more or less thick. There are some shells whose aperture is narrow and long. This is when the foot is thin, and folds in two to enter.

Most of the aquatic gasteropoda with spiral shells have an *operculum* or a piece, sometimes horny, sometimes calcareous, attached on the hinder part of the foot, which closes the shell when the animal has entered and is folded up in it.

There are gasteropoda with the sexes separated, and others hermaphrodite, of which some can generate of themselves, while others have need of reciprocal coition. Their organs of digestion do not differ less than those of respiration.

This class is too numerous not to be divided into a certain number of orders, which we have derived from the position and form of the gills.

The Pulmonaria

Respire the atmospheric air in a cavity whose narrow orifice they open and close at will. They are hermaphrodites, but require reciprocal copulation. Some have no shell, others have, and often even completely turbinated; but they have no operculum.

The Nudibranchiata

Have no shell, and have naked gills of divers forms, on some portion of the back.

The Inferobranchiata

In other respects similar to the preceding, carry their gills under the reflected edges of their mantle.

The Tectibranchiata

Have gills on the back or side, covered by a lamina of the

mantle, which contains almost always a shell more or less developed. Sometimes they are merely enveloped by the folded-back edge of the foot.

These four orders are hermaphrodites, with reciprocal copulation.

The HETEROPODA

Have the gills on the back, where they form a transverse range of little plumes, and in some are protected, as well as a part of the viscera, by a symmetrical shell. What distinguishes them best is a foot compressed into a thin and vertical fin, at the edge of which is often seen a small cupper, the only vestige of the horizontal foot of the rest of the class.

The PECTINIBRANCHIATA

Have the sexes separate. Their respiratory organs consist almost always of gills, composed of small plates, united in the form of combs, and are concealed in a dorsal cavity, widely open above the head.

Pretty nearly all of them have turbinated shells, with the mouth sometimes entire, sometimes emarginated, sometimes provided with a siphon, and most frequently capable of being more or less completely closed, with an operculum attached to the foot of the animal behind.

The SCUTIBRANCHIATA

Have gills analogous to those of the pectinibranchiata; but the sexes are united, so that they fecundate themselves without mutual co-operation, like the class of the acephala. Their shells are very open, and in several in the form of a shield, not turbinated. They never have an operculum.

The CYCLOBRANCHIATA

Hermaphrodites, like the scutibranchiata, have a shell of one or several pieces, but never turbinated or operculated. Their gills are attached under the reflected edges of their mantle, as in the inferobranchiata.

THE FIRST ORDER OF GASTEROPODA.

THE PULMONARIA

Are distinguished from the other mollusca in respiring the elastic air through an open hole under the reflected edge of the mantle, which they dilate or contract at pleasure. Accordingly, they have no gills, but merely a net-work of pulmonary vessels, which spread over the parietes, and principally over the roof of their respiratory cavity.

Some are terrestrial, others live in the water, but are obliged from time to time to come to the surface to open the orifice of their pectoral cavity to respire.

All these animals are hermaphrodites.

The TERRESTRIAL PULMONARIA

Have almost all four tentacula; two or three only, of very small size, do not show the lower pair.

Those among them which have no apparent shell, formed in Linnæus the genus

LIMAX, *L.*,

Which we divide as follows:

LIMAX (proper),

Have the body elongated, and for a mantle a fleshy compact

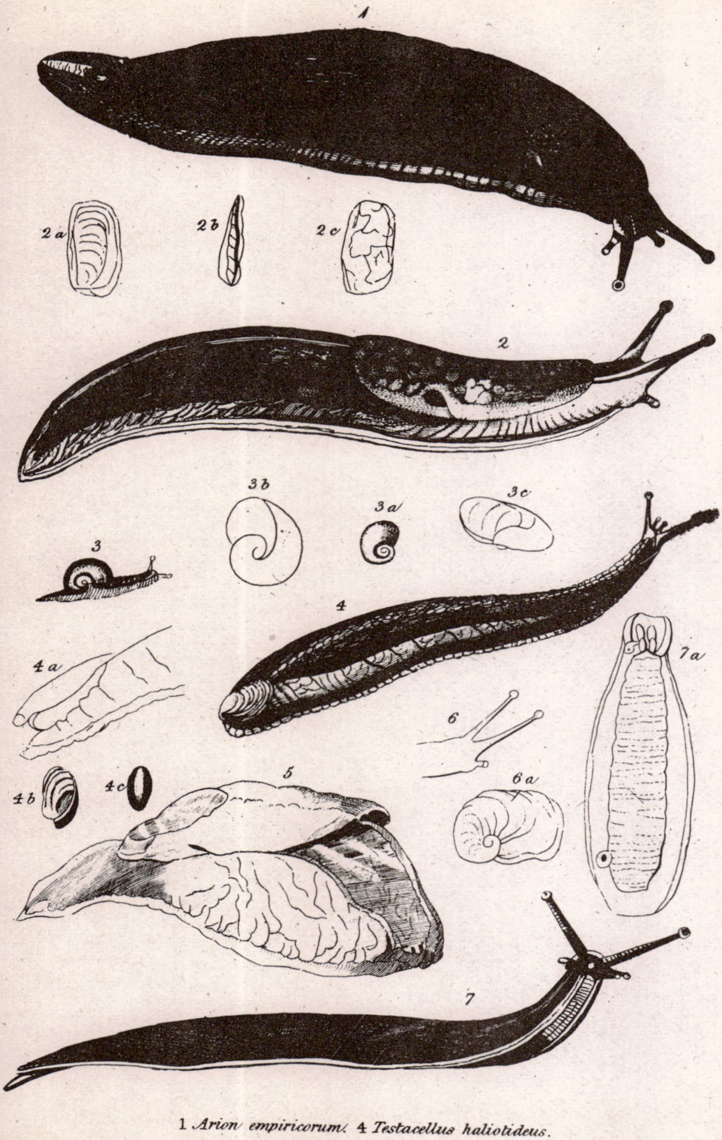

1 *Arion empiricorum.* 4 *Testacellus haliotideus.*
2 *Limas variegatus.* 5 *Parmacella Olivieri.*
3 *Vitrina pellucida.* 6 *Head &c. of Parmacella palliolum.*
7 *Vaginula Taunaysii.*

disk, which occupies only the forepart of the back, and only covers the pulmonary cavity. It contains, in many species, a small oblong and flat shell, or at least a calcareous concretion, which holds the place of it. The orifice of respiration is on the right side of this sort of sucker, and the anus is found at the edge of this orifice. The four tentacula issue out and re-enter by being turned inside out, like the finger of a glove, and the head itself can be partly withdrawn under the disk of the mantle. The organs of generation open under the superior right tentaculum. There is but one upper jaw to the mouth, in the form of a denticulated crescent, which enables them to gnaw herbs and fruits with great voracity, among which they make great havock. Their stomach is elongated, simple, and membranaceous.

M. de Férussac distinguishes the ARIONS, in which the orifice of respiration is towards the anterior part of the shield; there is nothing in the shield but calcareous grains. Such is

Limax rufus, L., Férussac, Mull. Terr. et Fluv., pl. i. and iii. which is to be met with at every step we take in wet weather, and which is sometimes almost entirely black.—*Ibid*, ii. 1, 2. The soup of this slug is made use of in maladies of the chest.

And the LIMAS, in which this orifice is towards the posterior part; their shell is often better defined. Such is

L. maximus, L.; *Lim. antiquorum*, Féruss. pl. iv. and viii. A. fi. 1.; *L. Sylvaticus*, Drap. Müll. ix. 10.

Often spotted, or striped with black. Found in cellars and dark forests.

L. agrestis, L., Féruss. pl. v. fi. 5—10.

Small, without spots. One of the most abundant and most mischievous.

VAGINULUS, *Féruss.*,

Have the mantle compact, without shell, and stretched over the whole length of the body; four tentacula, the inferior a little forked. The anus is altogether at the posterior extremity, between the point of the mantle and that of the foot; and the same orifice conducts to the pulmonary cavity, situated along the right side. The orifice of the male organ of generation is under the inferior right tentaculum, and that of the female organ under the middle of the right side. These organs, as well as those of digestion, are very similar to those of the snail.

These mollusca belong to both Indies, and are very similar to our slugs.

TESTACELLA, *Lam.*,

Have the orifice of respiration and the anus at the posterior extremity; their mantle is very small, and situated on this same extremity. It contains a small oval shell, with very wide aperture, and a very small spire, not equalling the tenth part of the length of the body. For the rest, these animals resemble slugs.

A species is found tolerably abundant in our southern departments (*Testacella haliotoidea*, Draparn.), Cuv. Ann. Mus. V. xxvi. 6—11, which lives under ground, and feeds chiefly on earth-worms. M. de Férussac has observed that its mantle is extraordinarily developed when it finds itself in too dry a place, and thus affords it a kind of shelter.

PARMACELLA, *Cuv.*,

Have a membranaceous mantle, with loose edges, placed on the middle of the back, and containing, in its posterior part, an oblong flat shell, which exhibits behind a slight commencement of spire. The orifice of respiration and the anus are under the right side of the middle of the mantle.

The first known species is from Mesopotamia, *Parmacella Olivieri*, Cuv. Ann. Mus. V. xxix. 12—15.

There is one belonging to Brazil (*P. palliolum*, Féruss. pl. vii. A.), and some others of the Indies.

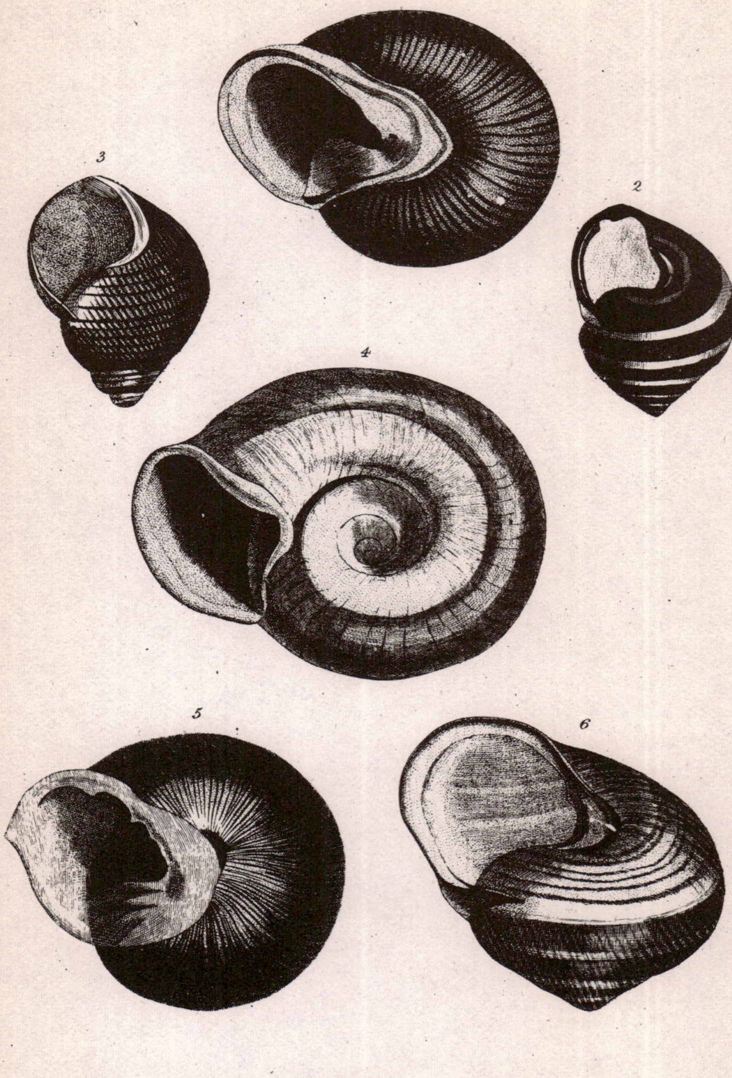

1 *Helix mora.*

2 *Helix viridis Desm.*

3 *Paludina subcostata.*

4 *Helix Cunninghami.*

5 *Helix Lamarckii.*

6 *Helix Fraseri.*

In the terrestrial pulmonaria with complete and apparent shell, the edges of the aperture are most frequently raised like a pad in the adult.

Linnæus refers to his genus

HELIX,

All the species in which the aperture of the shell, a little intruded upon by the projection of the last whorl but one, thus assumes a circumscribed and crescented form.

When this crescent of the aperture is as broad, or broader than it is high, they are

HELIX (properly so called), *Brug.* and *Lam.*

Some have the shell globular. In this number, every one knows the *common snail* (*Hel. Pomatia*, L.), common in gardens, vineyards, with reddish shell, marked with paler bands, a food very much in request in some places; and *H. nemoralis*. L., with a shell diversely and vividly coloured. In wet seasons it is very injurious to wall-fruit.

Add *Helix glauca*, *Citrina*, *Rapa*, *Castanea*, *Globulus*, *Lactea*, *Arbustorum*, *Fulva*, *Epistylium*, *Cincta*, *Ligata*, *Aspersa*, *Extensa*, *Nemorensis*, *Fruticum*, *Lucena*, *Vittata*, *Rosacea*, *Itala*, *Lusitanica*, *Aculeata*, *Turturum*, *Cretacea*, *Fuscescens*, *Terrestris*, *Nivea*, *Hortensis*, *Lucorum*, *Grisea*, *Hæmastoma*, *Pulla*, *Venusta*, *Picta*, Gm., &c.

There is no one who has not heard of the curious experiments which have been made on the reproduction of their amputated parts.—Consult Spallanzani, Schœffer, Bonnet, &c.

Others have the shell depressed, that is, with flatted spire. Such are *Hel. lapicida*, *H. cicatrosa*, *H. ægophthalmus*, *H. albella*, &c.

Among them some may be remarked that have ribs projecting interiorly, *Hel. sinuata*, *H. lucerna*, &c.

And particularly those in which the last whorl is suddenly

recurved in the adult, and assumes an irregular and plaited form. (*H. ringens*, Chemn.)

VITRINA, *Drap.* HELICO-LIMAX, *Férus.*,

Are snails with a very thin flatted shell, no umbilicus, large aperture, and margin not swelled. The body is too large to be withdrawn entirely into the shell. The mantle has a double border; the upper one, which is divided into several lobes, may extend considerably beyond the shell, and fold back on it to rub and polish it.

Those which are known in Europe live in humid places, and are very small. (*Helix pellucida*, &c.)

There are larger ones in the hot climates.

We should approximate to these some snails, which, without having the double border, do find some difficulty in withdrawing into their shell. (*Helix rufa et brevipes*, Féruss.)

When the crescent of the aperture is higher than it is wide, a disposition which always obtains when the spire is oblong or elongated, it constitutes the

BULIMUS TERRESTRIS, *Brug.*,

Which requires a still further sub-division: as follows.

BULIMUS (proper), *Lam.*

Margin of the aperture tumid in the adult, but without denticulations.

Hot climates produce large and beautiful species; some of these are remarkable for the size of their eggs, the shell of which is of a stony hardness; others for their awkward-looking shell.

Several moderate sized or small species are found in France, one of which (*Helix decollata*, Gm.) Chemn. cxxxvi. 1254—1257, has the singular habit of successively fracturing the whorls of the top of the spire. This is the example referred to as a proof that the muscles of the animal can be detached from the shell; for, at a particular epoch, of all the whorls of the spire originally possessed by this bulimus, not a single one remains. Add *Helix ovalis*, Gm.; *H. oblonga*, Chemn., &c.

PUPA, *Lam.*

Summit of the shell very obtuse; the last whorl in the adult narrower than the others, giving it the form of an ellipsoïd, and sometimes almost that of a cylinder. The surrounding margin of the aperture is tumid, and encroached on at the side next to the spire by the preceding whorl. The species are small, inhabiting humid places, among mosses, &c.

Sometimes there is no denticulation. (*Bulimus labrosus*, Oliv., &c.)

More commonly there is one in that portion of the aperture which is closed by the penultimate whorl. (*Turbo uva*, L.)

It is frequently observed, also, inside the external edge. (*Hel. vertigo*, Gm., &c.)

CHONDRUS, *Cuv.*

The aperture, as in the last-mentioned pupæ, indented on the side next to the spire, by the preceding whorl, and bordered with salient laminæ or teeth. But the form is more ovoïd, like that of a common bulimus.

Some of them have teeth on the margin of the aperture. (*Bulimus zebra*, Oliv., &c.)

Others are furnished with more deeply-seated laminæ. (*Bulimus avenaceus*, Brug., &c.)

Here terminates that series of terrestrial helices, the adult shells of which have a tumid margin round the aperture.

SUCCINEA, *Drap.*

Shell oval; the aperture higher than it is broad, as in bulimus, but larger in proportion; margin of the aperture not tumid, and the side of the columella almost concave. The shell will not receive the entire animal, and it might almost be considered as a large-shelled testacella. Its inferior tentacula are very small, and it lives on the plants and shrubs which line the borders of rivulets, a circumstance which has caused the genus to be considered as amphibious. *Succinea amphibia*, Drap. (*Helix putris*, L., &c.)

CLAUSILIA, *Drap.*

These mollusca formerly belonged to the genus *Turbo* of Linnæus, from which it has been found necessary to separate them, in order to approximate them to the terrestrial helices. The shell is long, slender, and pointed; the last whorl in the adult narrowed, compressed, slightly detached, and terminated by a complete aperture, with a tumid margin, frequently dentated, or furnished with laminæ. In the contraction of the last whorl we usually find a little plate, bent into an S, the use of which to the living animal is unknown.

The species are very small, living in masses at the foot of trees, &c. A great many of them are reversed. (*Turbo per Versus* Lin., &c. &c.)

ACHATINA, *Lam.*,

Necessarily separated from the bullæ of Linnæus, and placed here. The aperture of the oval or oblong shell is higher than it is broad, as in the bulimi, but wants the tumid margin; the extremity of the columella also is truncated, the first indication of the emarginations, which we shall find in so many marine gasteropoda. These achatinæ are large helices, which devour trees and shrubs in hot countries. (*Bulla zebra*, Lin., Chem., &c.)

Montfort distinguishes such as have in the last whorl a callus or peculiar thickening (*Liguus*, Mont.) This whorl is proportionably lower in them than in the others. (*Bulla virginea*, L., &c.)

1 *Bulimus Guadalupensis.* 5 *Achatina Mulleri.* 9 *Chonarus avenaceus.*
2 *Lymnæus stagnalis.* 6 *Clausilia inflata.* 10 *Planorbis Guadalupensis.*
3 *Pupa striatella.* 7 *Helix corocolla.* 11 *Scarabus impium.*
4 *Physa Novæ-Hollandiæ.* 8 *Auricula Midæ.* 12 *Succinea rubescens.*
13 *Conovulus fasciatus.*

And those in which the extremity of the columella is curved towards the inside of the aperture (*Polyphemus*, Mont.), the last whorl is higher. (*Bulimus glans*, Brug.)

The Aquatic Pulmonaria

Have but two tentacula, as already stated; they are continually compelled to rise to the surface for the purpose of breathing, so that they cannot inhabit very deep water. They are usually found in fresh-water or salt-ponds, or at least in the vicinity of the sea-coast, and of the mouths of rivers. Some of them have no shell; such as

Onchidium, *Buchanan.*

A broad fleshy mantle, in the form of a shield, overlapping the foot at all points, and even covering the head when it contracts; two long retractile tentacula; and on the mouth an emarginated veil, formed of two triangular and depressed lobes.

The anus and respiratory orifice are under the posterior edge of the mantle, where, a little more deeply, we also find the pulmonary cavity. Close to them, on the right side, opens the female organ of generation; that of the male, on the contrary, is under the right great tentaculum, the two openings being united by a furrow, which extends along the under part of the whole right margin of the mantle. These mollusca, destitute of jaws, have a muscular gizzard, followed by two membranous stomachs. Several of them inhabit the sea-shore, but in places where the ebb leaves them uncovered, so that they can readily breathe the air as is natural to them.

The name of Onchidium was given to this genus because the first species (*Onchidium typhæ*, Buchan.), was tuberculous. I now know one that is smooth (*O. lævigatum*, Cuv.), and four or five that are tuberculous.

The aquatic pulmonaria with complete shells were also

placed by Linnæus in his genera HELIX, BULLA, and VOLUTA, from which it has been found necessary to separate them.

In the first were comprized the two following genera, where we find the internal edge of the aperture crescent-shaped, as in helix.

PLANORBIS, *Brug.*

The planorbes had been already distinguished from the helices by Bruguières, and even previously by Guettard, on account of the slight increase of the whorls of their shell, the convolutions of which are nearly in one plane, and because the aperture is wider than it is high. It contains an animal with long, thin, filiform tentacula, at the inner base of which are the eyes, and from the margin of whose mantle exudes a quantity of a red fluid, which is not, however, its blood. Its stomach is muscular, and its food vegetable, like that of the Limnæi, of which, in all our stagnant waters, it is the faithful companion. (*Hel. vortex*, *H. cornea*, *H. spirorbis*, &c.)

LIMNÆUS, *Lam.*,

Separated from the bulimi of Bruguière by M. de Lamarck, has, like a bulimus, an oblong spire, and the aperture higher than it is wide ; but the margin, like that of succinea, is not reflected, and there is a longitudinal fold in the columella, which runs obliquely into the cavity. The shell is thin ; the animal has two compressed, broad, triangular tentacula, near the base of whose inner edge are the eyes. They feed on plants and seeds, and their stomach is a very muscular gizzard, preceded by a crop. Like all the pulmonariæ, they are hermaphrodites, and the female organ of generation being far from the other, they are compelled so to copulate, that the individual which acts as a male for one serves as a female for a third ; long strings of them may be observed in this position.

They inhabit stagnant waters in great numbers; they also abound with the planorbes, in certain layers of marl and calcareous strata, which they evidently prove were deposited in fresh water. (*Hel. stagnalis*, Lin., of which *H. fragilis* is a variety; *H. palustris, H. limosa*, &c.)

PHYSA, *Drap.*

The physæ, which were gratuitously placed among the bullæ, have a shell very similar to that of a limnæus, but devoid of the fold in the columella, as of the reflected edge, and very thin. When the animal swims or crawls, it covers its shell with the two notched lobes of its mantle, and has two long, slender, and pointed tentacula, on the greatly enlarged internal base of which are the eyes. They inhabit springs, &c.

One of them, *Bulla fontinalis*, L., which is convoluted towards the left, is found in France.

The neighbouring species, *Bull. hypnorum*, L.; and *Physa acuta;* and *Scaturiginum*, Drap.; require a fresh examination of their animals.

According to the observations of Van Hasselt, we should place here the

SCARABÆUS, *Montf.*,

Which has an oval shell, the aperture narrowed by projecting and stout denticulations, both on the side next the columella and that towards the external margin. This margin is enlarged, and as the animal renews it after each semi-whorl, the shell projects most on two opposite lines, and has a compressed appearance.

They feed on aquatic plants in the Indian Archipelago. (*Helix Scarabæus*, Lin.)

The two following genera were among the *volutæ*.

AURICULA, *Lam.*,

Differing from all the preceding aquatic pulmonaria in the

columella, which is marked with wide and oblique grooves. Their shell is oval or oblong, the aperture elevated as in bulimus, and the margin tumid. Several are large. We are not certain whether they inhabit marshes, like the limnæi, or simply their borders, like the succineæ.

Auricula myosotis, Drap. iii. 16, 17.; *Carychium myosotis,* Féruss., is the only species in France. The animal has but two tentacula, and the eyes are at their base. From the shores of the Mediterranean. Add *Voluta auris Midæ*, L.; *Voluta auris Judæ*, &c. &c.

CONOVULUS, *Lam.*, MELAMPES, *Mont.*

Projecting folds in the columella, as in the auriculæ, but the margin of the aperture is not tumid, and the internal lip is finely striated. The general form of the shell is that of a cone, of which the spire forms the base. They inhabit the rivers of the Antilles. (*Voluta minuta*, Lin. &c.)

THE SECOND ORDER OF GASTEROPODA.

THE NUDIBRANCHIATA

HAVE neither shell nor pulmonary cavity; their gills are exposed on some part of the back. They are all hermaphrodites and marine animals, frequently swimming in a reversed position, with the foot on the surface, concave, like a boat, and employing the margin of their mantle and their tentacula as oars.

Doris, *Cuv.*

The anus opens on the posterior part of the back, the gills being arranged in a circle round it, under the form of little arbusculæ, the whole resembling a sort of flower; the mouth is a small proboscis, situated under the anterior margin of the mantle, and furnished with two little conical tentaculæ. Two other claviform tentacula arise from the anterior and superior part of the mantle. The openings of the genital organs are approximated under its right margin; the stomach is membranous; a gland, interlaced with the liver, excretes a peculiar fluid through a hole near the anus. The species are numerous, and some of them large. They are found in every sea, where their eggs, like gelatinous bands, are diffused over stones, sea-weed, &c.

Species with an oval mantle projecting beyond the foot, *Doris verrucosa*, L.; *Doris argo*, L.; *Doris obvelata*, &c.

Prismatic species, where the mantle is almost as narrow as the foot, *Doris lacera*, Cuv.; *D. atromarginata*, Id., &c.

Onchidora, *Blainv.*,

Only differs from doris in the separation of the genital organs, the orifices of which communicate by a furrow running along the right side, as in onchidium. (*Onchidora Leachii*, Blainv.)

Plocameros, *Leuckard*,

Have all the characters of the onchidoræ, in addition to which the anterior margin of their mantle is ornamented with numerous branched tentacula. (*Plocameros ocellatus*, Leach.)

Polycera, *Cuv.*

The gills, as in doris, on the hind part of the body, but more simple, and followed by two membranaceous laminæ, which cover them in moments of danger. Anterior to the claviform tentacula, similar to those in doris, are four, and sometimes six others, simply pointed. (*Doris quadrilineata*, Müll., &c.)

TRITONIA, *Cuv.*

The body, superior tentacula, and genital organs, as in doris; but the anus, and orifice through which the peculiar liquid is secreted, are on the right, behind the organs of generation.

The gills, which resemble little trees, are arranged along the sides of the back; and the mouth, provided with broad membranous lips, is armed inside with two horny and trenchant lateral jaws, which may be compared to a pair of sheep-shears.

Trit. Hombergii, Cuv. Ann. du Mus. I. xxxi. 1, 2; a large species, of a copper colour, from the coast of France.

The same locality produces many others, which vary greatly in size and the form of their gills. Several of them are very small. (*Trit. elegans*, *Trit. rubra*, *Trit. glauca*, &c.)

THETHYS, *Lin.*

Two rows of gills, resembling branching tufts, along the back, and a very large, membranous, and fringed veil, upon the head, which shortens as it curves under the mouth. This latter is a membranous proboscis, without jaws. On the base of the veil are two compressed tentacula, from the margin of which projects a small conical point. The orifices of the genital organs, of the anus, and of the peculiar fluid, are situated as in the tritoniæ; the stomach is membranous, and the intestine very short.

T. fimbria, L., Cuv. Ann. du Mus. XII. xxiv.: grey, spotted with white. A beautiful species, from the Mediterranean.

SCYLLÆA, *Lin.*

Body compressed; the foot narrow, and marked with a

furrow, which enables it to clasp the stems of the fuci; no veil; the mouth resembling a little proboscis; orifices as in thethys; the compressed tentacula terminated by a cavity, from which issues a little uneven point; two pairs of membranous crests on the back, bearing on their internal surface pencils of filaments, which are the gills. The middle of the stomach is invested with a fleshy ring, internally armed with horny and trenchant laminæ like knives.

S. Pelagica, L., Cuv. Ann. du Mus. VI. lxi. 1. 3. 4: common on the floating fucus of almost every sea.

GLAUCUS, *Forster*.

Body elongated; orifices of the anus, and of the genital organs, as in the preceding; four very small conical tentacula; and on each side three gills, each of which is formed of long slips, arranged like the sticks of a fan, which also aid them in swimming. These are the beautiful little animals that inhabit the Mediterranean and the Atlantic, prettily coloured with blue and mother-of-pearl. They swim on their back with great swiftness. Their anatomical structure is very similar to that of the tritonia; but the species are not yet well ascertained. (*Doris radiata*, Gm., &c.)

LANIOGERUS, *Blainv*.

Two series, on each side, of small and finely-pectinated laminæ, which are the gills; the body, shorter and thicker than that of a glaucus; but there are four small similar tentacula. (*Laniogerus Elfortii*, Blainv. Malac. pl. xlvi. f. 4.)

EOLIDIA, *Cuv*.,

Resembles a small limax in form: four tantacula above and two on the sides of the mouth; the gills, composed of laminæ, arranged like scales, more or less crowded, on each side of the back. Found in every sea. (*Doris papillosa*, Zool. Dan., &c.)

Cavolina, *Brug.*

The tentacula of the eolidiæ, with retiform gills, arranged in transverse rows on the back. (*Doris peregrina*, Gm., &c.)

N. B. This genus must not be confounded with the *cavolina* of Abildgaard, which is the *hyalæa*.

Flabellina, *Cuv.*

The tentacula of the eolidiæ, with radiating retiform gills, supported by five or six pedicles on each side. They are closely related to the glauci; and, in fact, all the nudibranchiata, whose gills are situated on the sides of the back, are nearly allied. (*Doris affinis*, Gm., Cavol., Polyp., Mar. vii. 4.)

Tergipes, *Cuv.*

The form of the eolidia, but only two tentacula, with a range of cylindrical gills on both sides of the back, each of which is terminated by a little sucker or cup, and which are used by the animal as feet to walk upon its back. The species known are very small. (*Limax tergipes*, Forskal, &c.)

Busiris, *Risso*,

The body oblong, and back convex; two filiform tentacula, and behind them, on the nape, two plumiform gills. (*Busiris griseus*, Risso.)

Placobranchus, *Van Hasselt.*

Two tentacula, and as many labial lobes. The whole back, widened by its margin, is covered with numerous radiating striæ, which are the gills. In its ordinary condition the widened borders of the mantle are turned up, and cross each other, to form an envelope for the gills, which are thus enclosed, as in a cylindrical case.

They are small mollusca from the Indian Ocean. In the species known, (*Placobranchus Hasselti*, Cuv.) the branchial striæ are green, and the body a brown grey, sprinkled with little ocelli. Van Hasselt; Bullet. Univ. Oct. 1824, p. 340. Messrs. Quoy and Gaymard found it at the Friendly Islands.

THIRD ORDER OF GASTEROPODA.

INFEROBRANCHIATA.

The Inferobranchiata have nearly the same form and organization observed in doris and tritonia, but their gills, instead of being placed upon the back, resemble two long series of leaves, situated on the two sides of the body, under the projecting margin of the mantle.

Phyllidia, *Cuv.*

The mantle naked, usually coriaceous, and without any shell; the mouth a small proboscis, each side of which is furnished with a tentaculum; two others project above from two small cavities in the mantle. The anus is on the hind part of the mantle, and the genital orifices forward, under the right side; the heart near the middle of the back; the stomach simple and membranous, and the intestine short.

Several species inhabit the Indian Ocean. (*Phyllidia trilineata*, Seb., &c.)

DIPHYLLIDIA, *Cuv.*

The gills similar to those of the phyllidiæ, but the posterior part of the mantle more pointed. On each side of the semicircular head, a pointed tentaculum and a slight tubercle; the anus on the right side. (*Diphyllidia Brugmansii*, Cuv.)

FOURTH ORDER OF GASTEROPODA.

TECTIBRANCHIATA.

THE gills along the right side, or on the back, in the form of leaves, more or less divided, but not symmetrical: they are more or less covered by the mantle, in the thickness of which a small shell is generally contained. They approach the pectinibranchiata in the form of the organs of respiration, and like them inhabit the ocean; but they are all hermaphrodites, like the nudibranchiata and the pulmonaria.

PLEUROBRANCHUS, *Cuv.*

The body equally overlapped by the mantle and by the foot, as if it were between two shields. In some species, a little oval calcareous lamina is contained in the mantle, and a horny one in that of others; the mantle is emarginated above the head; the gills are attached along the right side in the furrow, between the mantle and the foot, forming a series of pyramids, divided into triangular leaflets; the mouth, a small proboscis, is surmounted by an emarginated lip, and by two tubular and cleft tentacula; the genital orifices are before, and the anus behind the gills. There are four stomachs, the second of which is fleshy, and sometimes armed with bony appendages, and the third furnished internally with salient longitudinal laminæ; the intestine is short.

Various species inhabit both the Mediterranean and the Atlantic, some of which are large, and marked with the most beautiful colours. (*Pleurobranchus Peronii*, Cuv., &c.)

PLEUROBRANCHÆA, *Meckel*. PLEUROBRANCHIDIUM, *Blainv.*

The gills and genital orifices situated as in pleurobranchus, but the anus is above the gills; the margin of the mantle and foot project but little, and on the fore-part of the former are four short distant tentacula, forming a square, that reminds the observer of the anterior disk of the akeræ. I can find but one stomach, which is merely a dilatation of the canal, with thin parietes. A multifid glandular organ opens behind the genital orifices. There is no vestige of a shell.

Pleurob. Meckelii (Leve, diss. de Pleur. 1813), is the only species known in the Mediterranean.

APLYSIA, *Linn.*

The margin of the foot turned up into flexible crests, surrounding the back, and even susceptible of being reflected over it; the head supported by a neck more or less long; two superior tentacula, excavated like the ears of a quadruped, with two flattened ones on the edge of the lower lip; the eyes beneath the former; the gills are on the back, and consist of highly complicated lamellæ, attached to a broad membranous pedicle, and covered by a small membranous mantle, in the thickness of which is a flat and horny shell; the anus opens behind the gills, and is frequently concealed under the lateral crests; the vulva is before, on the right, and the penis projects from under the right tentaculum; the seminal fluid is conducted from the penis to the vulva by a groove which extends from one to the other. An enormous membranous crop leads to a muscular gizzard, armed internally with cartilaginous and pyramidal corpuscles; which is followed by a third stomach, sown with sharp hooks, and by a fourth, in the form of a cœcum. The intestine is voluminous; and the animal feeds on fucus. A limpid humour, secreted by a particular gland, and which in certain species is said to be extremely acrid, is exuded through an orifice near the vulva; and from the edges of the mantle oozes an abundant liquid, of a deep purple colour, with which, when in danger, the animal tinges the water to a considerable extent. The eggs are deposited in long glairy strings, interlaced, and fine as packthread.

In the seas of Europe we have *Apl. fasciata*, Poiret, Rang., Apl. pl. vi. vii.: black, bordered with lateral red crests. One of the large species.

Apl. punctata, Cuv. Ann. du Mus. tome ii. p. 287, pl. 1. f. 2—4; Rang. Apl. pl. xviii. f. 2: lilac, sprinkled with greenish points.

Apl. depilatus, L., Bohatch. Anim. Mar. pl. i. and ii.; Rang. pl. xvi.: blackish, with large greyish, clouded spots.

Several other species are found in distant seas (*Apl. brasiliana*, &c.). It is well, however, to observe, that most of the aplysiæ, having been drawn from specimens preserved in spirits, the truth of the specific characters of some of them may be doubted.

DOLABELLA, *Lam.*

The dolabellæ only differ from the aplysiæ in the position of the branchiæ, and their surrounding envelope; they are at the posterior extremity of the body, which resembles a truncated cone. Their lateral crest presses closely on their branchial apparatus, merely leaving a narrow furrow. Their shell is calcareous. They are found in the Mediterranean and in the Indian Ocean. (*Dolabella Rumphii*, Cuv. Ann. du Mus., &c.)

NOTARCHUS, *Cuv.*

The lateral crests united and covering the back, a longitudinal emargination excepted, that leads to the branchiæ, which have no mantle to cover them, but are otherwise like those of the aplysiæ; the rest of their organization is always the same. (*Notarchus gelatinosus*, Cuv., &c.)

BURSATELLA, *Blainv.*

The lateral crests are united in front in such a manner as only to leave an oval aperture for the transmission of water to the gills, which are also deprived of a protecting mantle. (*Bursatella Leachii*, De Blainv.)

These two genera, however, probably form but one,

AKERA, *Müller.*

The branchiæ covered as in the preceding genera, but their tentacula are so shortened, widened, and separated, that they seem to be totally wanting, or rather to form a large fleshy and nearly rectangular shield, under which are the eyes. In other respects, the hermaphroditism of these animals, the position of their genital organs, the complication and armature of their stomach, and the purple liquid effused by several of their species, approximate them to the aplysiæ. The shell, of such as have any, is more or less convoluted, but with little obliquity, and without a projecting spire, emargination, or canal; the columella, making a convex projection, gives a crescented figure to the aperture, the part opposite to the spire being always the broadest and most rounded.

M. de Lamarck names those in which the shell is concealed in the thickness of the mantle **BULLÆA**. It has but very few whorls, and the animal is much too large to enter it.

Bullæa aperta, Lam.; *Bulla aperta* and *Lobaria quadriloba*, Gm.; *Phyline quadripartita*, Ascan. Müll. Zool. Dan. III. pl. ci.; Planc. Conch. Min. Not. pl. xi.; Cuv. Ann. du Mus. t. I. pl. xii. 6. The animal is whitish, and about an inch long; the fleshy shield formed by the vestiges of its tentacula, the lateral swellings of its foot, and the mantle occupied by the shell, seem to divide its upper surface into four lobes. Its thin, white, semi-diaphanous shell, is nearly all aperture, and its gizzard is armed with three very thick rhomboïdal pieces of bone. It is found in almost every sea, where it lives on oozy bottoms.

The *Sormet* of Adanson is a species closely allied to Bullæa; but I cannot establish a genus, or even a species, upon so imperfect a document.

M. de Lamarck leaves the name of BULLA to those species whose shell, merely covered with a slight epidermis, is large enough to shelter the animal. It is somewhat more convoluted than in Bullæa.

The genus *bulla* of Linnæus not only comprised the *akeræ*, but also the *auriculæ*, *agatinæ*, *physæ*, *ovulæ*, and *terebellæ*, animals between which there is much difference. Bruguières commenced the work of reformation, by separating the *agatinæ* and the *auriculæ*, which he united with the *lymnæi* to the genus *bulimus*. M. de Lamarck finished it by creating all the genera we have just named.

Bulla lignaria, L., Martini. I. xxi. 194—95; Cuv. Ann. du Mus. XVI. 1.; Pol. Test. Map. III. pl. xlvi. The oblong shell, with its concealed spire and ample aperture, very wide anteriorly, resembles a loosely rolled lamina, stretched in the direction of its whorls. The stomach of the animal is armed with two large semi-oval osseous pieces, and with a small compressed one. Gioëni having observed this stomach separate from the animal, took it for a shell, and made a genus of it, to which he gave his own name. He even went so far as to describe its pretended habits. Draparnaud was the first who perceived this mixture of error and fraud.

Bulla ampulla, L., Martini. I. xxii. 20. 204; Cuv. Ann. du Mus. XVI. i. The shell oval, thick, clouded with grey and brown; the stomach furnished with three black, very convex, rhomboidal pieces.

Bulla hydatis, L., Chemn. IX. cxviii. 1019; Cuv. Ann. du Mus. xvi. 1. Shell round, thin, and semi-diaphanous, the last whorl, and consequently the aperture, higher than the spire: three small scutelliform pieces in the gizzard.

Add *Bulla naucum*, *Bulla physis*, &c.

We reserve the name of AKERA, properly so called, (DORIDIUM, *Meck.*; LOBARIA, *Blainv.*) for those species which have no shell whatever, or only a vestige of one behind, although their mantle has its external form.

A small species, *Bulla carnosa*, Cuv. Ann. du Mus. xvi. 1; Meck. Anat. Compar. II. vii. 1. 3; Blainv. Malac. pl. xlv. f. 3; is found in the Mediterranean. The stomach is no more armed than the mantle; its fleshy œsophagus is extremely thick.

We also find in the same sea a tuberculous species, *Doridium Meckelii*, Delle Chiaie Memor. pl. x. f. 1—5.

GASTROPTERON, *Meck.*,

Appears to be an akera, the margin of whose foot is developed into broad wings, serving for swimming, which it performs on its back. It has no shell nor stony armature to the stomach. A slight fold of the skin is the only vestige of a branchial operculum that is visible.

The known species is also from the Mediterranean, *Gastropteron Meckelii*, Kosse. Diss. de pteropodum ordine. Halæ. 1813, f. 11—13; and Blainv. Malacol. pl. xv. f. 5; or *Clio amati*, Delle Chiaie. Memor. pl. ii. f. 1—8; a small animal, an inch long and two broad, when the wings are extended.

GASTROPLAX, *Blain.* UMBRELLA, *Lam.*

Until the anatomy of this singular genus be more closely studied, we must place it in the order tectibranchiata, and even near pleurobranchus. The animal is large and circular, the foot projects considerably beyond the mantle, and its upper surface is studded with tubercles; the viscera are in a round, superior, and central part; the mantle is visible only by its slightly projecting and trenchant edges along the whole of the front and of the right side; the lamellated pyramidal gills, like those of pleurobranchus, are under this slight margin, and behind them is a tubular anus. Under this same margin, and forwards, are two tentacula, longitudinally cleft, still as in pleurobranchus; and at their internal base are the eyes: between them is a sort of proboscis, which perhaps is the organ of generation. There is a large concave space in the anterior margin of the foot, the edges of which can be drawn up like a purse, and at the bottom of which is a tubercle pierced by an orifice, which perhaps is the mouth, and surmounted by a fringed membrane. The inferior surface of the foot is smooth, and serves the animal to crawl on, as in the other gasteropoda.

The shell is strong, flat, and irregularly rounded, thickest in the middle, with trenchant edges, and marked with slightly concentric striæ. It was at first thought to be attached to the foot, but more recent observation has proved that it is on the mantle, and in the usual place.

In the specimen from the British Museum, described by M. de Blainville, under the name of GASTROPLAX, the shell is in fact attached to the under part of the foot, and it is difficult to determine by what means; the mantle is, however, so thin, that it seems as if it must have been protected by the shell.

M. Reynaud has just brought an individual which had lost its shell, but where, it appears, traces of the membranes which attached it to the mantle may be perceived, notwithstanding which, no remains of muscles are visible. A similar shell is also found in the Mediterranean : its animal, however, has not yet been observed.

FIFTH ORDER OF GASTEROPODA.

HETEROPODA, Lam.

These are distinguished from all the others by their foot, which, instead of forming a horizontal disk, is compressed into a vertical muscular lamina, which they use as a fin, and on the edge of which, in several species, is a dilatation forming a hollow cone, that represents the disk of the other orders. Their gills, formed of lobes like feathers, are situated on the hind part of the back, directed forward; and immediately in their rear are the heart and a small liver, with part of the viscera, and the internal organs of generation. Their body, a gelatinous and transparent substance, lined with a muscular layer, is elongated, and usually terminated by a compressed tail. There is a muscular mass belonging to the mouth, and a tongue furnished with little hooks; the œsophagus is very long, their stomach thin; two prominent tubes on the right side of the visceral bundle afford a passage to the fœces, semen, and eggs. They usually swim on their backs, with the foot upwards. This mode of swimming having induced Peron to think that the natatory lamina was on the back, and the heart and gills under the belly, has given rise to many errors respecting the place of these animals. A simple inspection of their nervous system led me to suppose, in my memoirs on the mollusca, that they were analogous to the gasteropoda. A more exact anatomical investigation, made since then, and that given by M. Poli in his third volume, fully confirm my opinion. The fact is, that there is but little difference between the *heteropoda* and the *tectibranchiata;* notwithstanding which M. Laurillard believes their sexes to be separated. They have the faculty of distending their body, by filling it with water, in a way not well understood. Forskal comprized them all in his genus,

PTEROTRACHEA, *Forsk.*,

But it has been found necessary to sub-divide them.

CARINARIA, *Lam.*

The nucleus, formed of the heart, liver, and organs of generation, covered by a slender symmetrical shell, the point of which is bent backwards, and frequently raised by a crest, under the anterior edge of which float the feathers of the gills. On the head are two tentacula, and behind their base the eyes.

There is one species in the Mediterranean, *Carinaria cymbium*, Lam., Péron, Ann. du Mus. XV. iii. 15; Poli. III. xliv.; Ann. des Sc. Nat. tome XVI. pl. i.

Another, the *Carinaria fragilis*, Bory St. Vincent, (Voy. aux isles d'Afr. I. vi. 4.), is found in the Indian Ocean.

Add *Carinaria depressa*, Rang. Ann. des Sc. Nat. Feb. 1829, p. 136.

The *Argonauta vitrea* of authors, Favanne vii. c. 2; Martini, I. xiii. 163., must be the shell of a large carinaria, but the animal is not yet known.

Atlanta, *Lesueur*.

These mollusca, according to the recent observations of M. Rang, should belong to this order. Their shell, instead of being widened like that of the carinariæ, has a narrow cavity, spirally convoluted on one plane; its contour is raised by a thin crest.

They are extremely small shells, from the Indian Ocean, in one of which Lamanon thought he had discovered the original Cornu ammonis. *Atlanta Peronii* and *Atlanta Keraudrenii*, Lesueur, Journ. de Phys. lxxxv. November, 1817, and Rang. de la Soc. d'Hist. Nat. tome. III. p. 373, and pl. ix.

N. B. We must not confound the Atlantæ of Lesueur with the *Atlas* described by him in the same place, and which, so confined is his description, I do not know where to class.

Firola, *Peron*.

The body, tail, foot, gills, and visceral mass, as in the carinariæ, but no shell has ever been observed; the snout is elongated into a recurved proboscis, and the eyes are not preceded by tentacula. From the end of the tail is frequently observed to proceed a long articulated filament, which Forskal took for a tænia, and the nature of which is not yet very clearly understood.

One species, the *Pterotrachea coronata*, Forskal, Peron, Ann. du Mus. XV. ii. 8, is very common in the Mediterranean; and M. Lesueur describes several from the same sea, which he considers as different, Journ. Acad. Nat. Soc. Philad. vol. i. p. 3, but which require further comparison. (*Firola mutica*, *F. gibbosa*, &c.)

M. Lesueur distinguishes the Firoloidæ, where the body, instead of terminating in a compressed tail, is abruptly truncated behind the visceral bundle.—*Ibid.* p. 37. (*Firoloida*, *Demarestia*, &c.)

To these two, now well known genera, I presume we must add, when better known,

TIMORIENNA, *Quoy et Gaim.*, Zool. de Freyc. pl. lxxx. f. 1, which appear to be firolæ deprived of their foot and bundle of viscera.

And the

MONOPHORA, id. ib. f. iv. and v., which has nearly the form of a carinaria, but is without a foot, distinct bundle of viscera, and shell.

We must not confound them with the monophoræ of M. Bory St. Vincent (Voy. aux Isles d'Afr.), which are pyrosomæ.

It is not certain that we should place here

PHYLLIROE, *Péron*, Ann. du Mus. XV. pl. ii. f. 1., where the transparent and strongly compressed body has a snout before, surmounted by two long tentacula without eyes, a truncated tail behind, and which allows its heart, nervous system, stomach, and the genital organs of both sexes, to be seen through the integuments. The genital orifices and the anus are on the right side, and sometimes a tolerably long penis is visible. I can find no other organ of respiration than its thin and vascular skin.

THE SIXTH ORDER OF GASTEROPODA.

THE PECTINIBRANCHIATA

FORM, beyond all comparison, the most numerous division, inasmuch as they include the whole of the spiral univalves, and several that are simply conical. Their gills, composed of numerous lamellæ or strips, laid parallel with each other, like the teeth of a comb, are attached on one, two, or three lines, according to the genus, to the ceiling of the pulmonary cavity, which occupies the last whorl of the shell, and which has a large opening between the edge of the mantle and the body.

In two genera only, *Cyclostoma* and *Helicina*, do we find, instead of gills, a vascular net-work covering the ceiling of a cavity that is otherwise similar; they are the only ones that respire the natural air, all the others respire water.

All the pectinibranchiata have two tentacula and two eyes, sometimes placed on particular pedicles, and a mouth, resembling a more or less elongated proboscis. The sexes are separate. The penis of the male, attached to the right side of the neck, cannot usually be retracted within the body, but is reflected into the cavity of the gills; it is sometimes very stout. The paludina is the only one which can retract it through an orifice perforated in its right tentaculum. The rectum and oviduct of the female also run along the right side of this cavity, and between them and the gills is a peculiar organ, composed of cells, from which exudes an extremely viscid fluid, serving for the formation of a common envelope which contains the eggs, and is deposited with them. The figure of this envelope is often very complex and singular.

Their tongue is armed with little hooks, and by slow and repeated rubbing acts upon the hardest bodies.

The greatest difference in these animals consists in the presence or absence of the little canal formed by a prolongation of the edge of the pulmonary cavity, on the left side, and which passes through a similar canal or emargination in the shell, to enable the animal to breathe without leaving its shelter. There is also this distinction between the genera, that some of them have no operculum. The species differ from each other by the filaments, fringes, and other ornaments of the head, foot, or mantle.

These mollusca are arranged in several families, according to the forms of their shells, which appear to be constantly in relation to that of the animal.

The first family of the gasteropoda pectinibranchiata, or the

TROCHOIDA,

Is known by the shell, the aperture of which is entire, without an emargination or canal for a siphon of the mantle, as the animal has none, and is furnished with an operculum or some organ in place of it.

TROCHUS, *Linn.*,

Have shells, in which the external margin of the angular aperture approaches more or less to a perfect quadrangular figure, and is in an oblique plane with respect to the axis of the shell, because the part of the margin next to the spire projects more than the rest. Most of these animals have three filaments on each edge of the mantle, or at least some appendages to the sides of the feet.

Of those which have no umbilicus, there are some in which the columella, having the form of a concave arch, is continuous with the external margin without any projection. It is the angle and projection of this margin which distinguish them from turbo. These are the TECTARIA, *Montf.; Troch. inermis*, Chemnitz, &c.

Several are flattened with a trenchant edge, which has caused them to be compared to the rowel of a spur. (CALCAR, *Montf.; Turbo calcar*, L., &c.)

Some, again, are slightly depressed, orbicular, and shining, with a half-rounded aperture, the columella convex and callous. (ROTELLA, *Lam.; Troch. vestiarius*, L.)

The columella of others is distinguished near the base by a

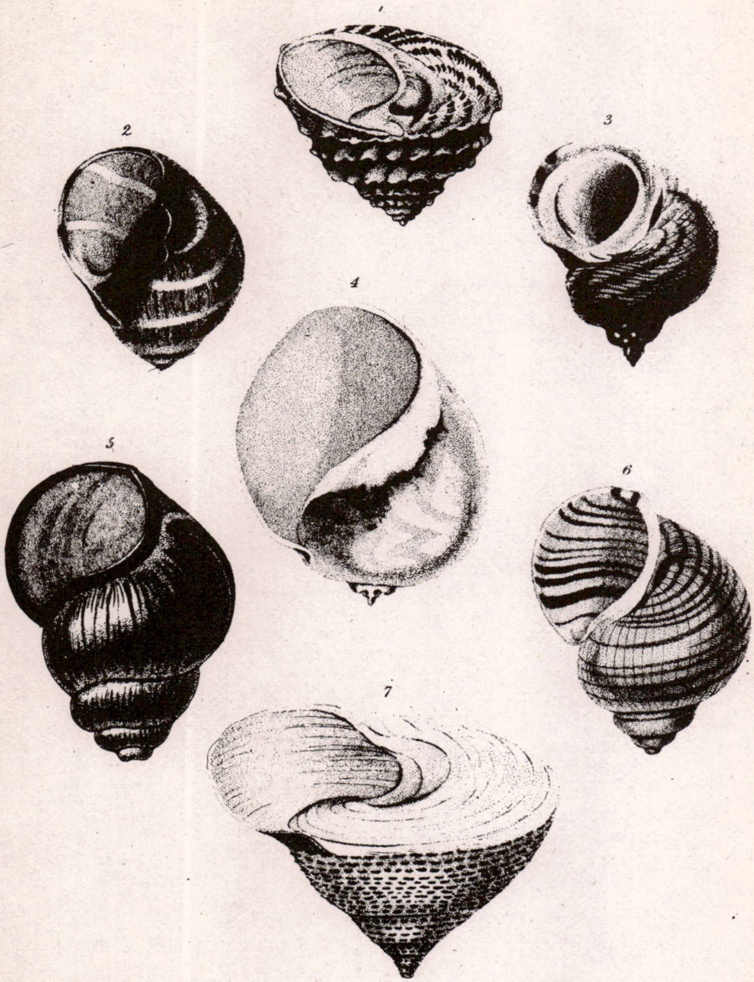

1 *Trochus bicarinatus.* 4 *Natica fluctuata.*
2 *Natica bifasciata.* 5 *Paludina Chinensis.*
3 *Litteraria pulchra.* 6 *Paludina pulchra.*
7 *Trochus Cunninghami.*

little prominence or vestige of a tooth, similar to that of the monodontes, from which these trochi only differ in the angle of their aperture, and the projection of their margin. The aperture is usually about as high as it is wide. (CANTHARIS, *Montf.; Troch. iris*, Chemn., &c.)

In some of them, on the contrary, the aperture is much wider than it is high, and their convex base approximates them to the calyptracea. (INFUNDIBULUM, *Montf.; Trochus concavus*, Chemn.)

In others, where the aperture is also much wider than it is high, the columella forms a spiral canal. (*Trochus foveolatus*, Chemn.)

Those which have a turreted shell approach cerithium. (TELESCOPIUM, *Montf.; Trochus telescopium*, Chemn.)

Among the umbilicated trochi there are some in which there is no projection in the columella; most of them are flattened, and have the external angle trenchant. Of this number is

Tr. agglutinans, L., Chemn. V. clxxii. 1688—9, remarkable for the habit of glueing to its shell, and even incorporating with it, as fast as it increases in size, various foreign bodies, such as little pebbles, fragments of other shells, &c. It frequently covers its umbilicus with a testaceous plate. Add *Troch. indicus*, Chemn., &c.

The margin of others, however, is rounded; such as

Tr. cinerarius, L., Chemn. V. clxxi. 1686: a small species, and the most common on the coast of France; greenish, obliquely streaked with violet.

Other umbilicated trochi have a prominence near the bottom of the columella. (*Tr. virgatus*, Chemn., &c.)

And finally, there are some in which it is longitudinally crenate. (*Tr. maculatus*, Chemn., &c.)

SOLARIUM, *Lam.*,

Is distinguished from all other trochi by a very broad conical spire, at the base of which is an extremely wide umbilicus, in which may be seen the external edges of all whorls, marked by a crenated cord. (*Tr. perspectivus*, L.)

EVOMPHALUS, *Sowerby.*

Fossil shells, resembling a *solarium*, but wanting the dentations on the internal whorls of the umbilicus. (*Evomphalus pentangulatus*, Sowerb.) The genus

TURBO, *Lin.*,

Comprehends all the species with a completely and regularly turbinated shell, and a perfectly round aperture. Close observation has caused them to be greatly subdivided.

TURBO (proper), *Lam.*

The shell is round or oval, and thick ; the aperture is completed on the side next to the spire by the penultimate whorl. The animal has two long tentacula, and the eyes placed on pedicles at their external base ; the sides of the foot are provided with membranous wings, sometimes simple, at others fringed, and occasionally furnished with one or two filaments. To some of these belong those petrous and thick opercula which we see in cabinets, and which were formerly employed in medicine under the name of *unguis odoratus*.

Some of them, MELEAGER, *Montf.*, are umbilicated. (*Turbo pica*, L., &c.)

Others are not. (TURBO, *Montf.; Turbo petholatus*, List., &c.)

DELPHINULA, *Lam.*

The shell thick, as in turbo, but convoluted nearly in the

same plane; the aperture completely formed by the last whorl, and the margin not tumid: the animal similar to that of a turbo.

The most common species, *Turbo delphinus*, Lin., List. 608. 45, takes its name from the ramous and convoluted spines, which have caused it to be compared to a dried fish. Add *Turbo nodulosus*, Chemn., &c.

Pleurotoma, *Defrance.*

Fossil shells, with a round aperture, on the external margin of which is a narrow incision, ascending considerably. It is probable that it corresponded, like that of the siliquariæ, to some cleft in the branchial part of the mantle.

M. Deshayes already enumerates upwards of twenty fossil species. The Scissurellæ of *M. d'Orbigny* are living species of the same.

Turritella, *Lam.*

The same round aperture as in turbo properly so called, and completed also by the penultimate whorl; but the shell is thin, and is so far from being convoluted in one plane, that its spire is prolonged into a *turreted* obelisk. The eyes of the animal are placed on the external base of its tentacula; the foot is small. (*Turbo imbricatus,* Martini, &c.)

They are found in great numbers among fossils. The Proto, *Defr.*, should be approximated to them.

Scalaria, *Lam.*

The spire, as in turritella, elongated into a point, and the aperture, as in delphinula, completely formed by the last whorl; it is, moreover, surrounded by a ridge, which the animal renews from space to space as its shell increases in size, so that it forms as it were steps. The tentacula and penis of the animal are long and slender.

One species, celebrated for the high price it commands, the *Turbo Scalaris*, L., Chemn., IV. clii. 1426, &c.—vulgo, *Scalata*,—is distinguished by the whorls only coming in contact at the point where the ribs unite them, the intervals being open.

A second species, the *Turbo clathrus*, Linn., List. 588, 50, 51, is not marked by this peculiarity: it is more slender, and is commonly found in the Mediterranean.

Some terrestrial or fresh-water subgenera, in which the aperture is entire, round, or nearly so, and operculated, may be placed here. Of this number is the

Cyclostoma, *Lam.*

The cyclostoma should be distinguished from all the others because they are terrestrial, as, instead of gills, the animal has merely a vascular net-work spread over the parietes of its pectoral cavity. In every other respect, however, it resembles the other animals of this family: the respiratory aperture opens in the same way above the head by a great solution of continuity; the sexes are separated; the penis of the male is large, fleshy, and reflected into the pectoral cavity; the two tentacula are terminated by blunt tubercles, and two other tubercles placed on their external base support the eyes.

The shell is a spiral oval, with complete whorls, transversely and finely striated; and its aperture in the adult is surrounded with a small ridge. It is closed by a small round operculum. Found in woods, under moss, stones, &c.

The most common is the *Turbo elegans*, List. 27, 25, about six lines in length, and of a greyish colour; found under all the mosses. Add *Turbo lincina*, List., &c.

Valvata, *Müll.*

The valvata inhabit fresh water; their shell is convoluted in almost one plane, like that of a planorbis, but the aperture

1

2

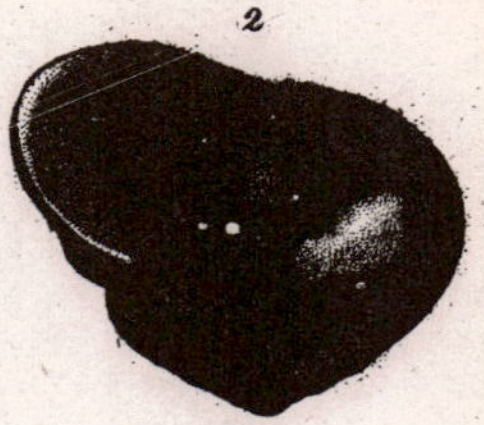

3

4

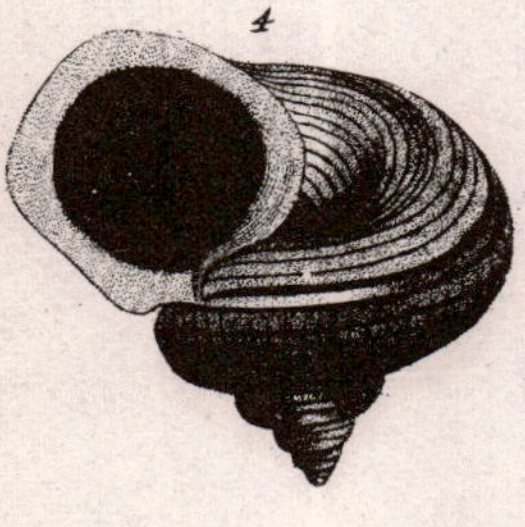

5

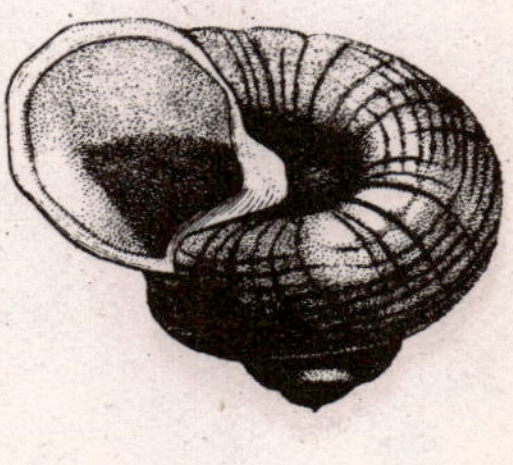

1 *Cyclostoma articulata.*

2 *Helix argellacea.*

3 *Cyclostoma pulchra.*

4 *Cyc. Madagascariensis.*

5 *Cyc. auriculare.*

is round, and furnished with an operculum; the animal, which has slender tentacula, with the eyes at their anterior base, respires by means of gills. In a species found in France,

Valv. cristata, Müll., Drap. I. 32, 33; Gruet-Huysen. Act. Nat. Cur. X. pl. xxxviii., the gills, formed like a feather, project from under the mantle, and float externally, vibrating with the breathing of the animal; on the right side of the body is a filament, which resembles a third tentaculum; the foot is divided anteriorly into two hooked lobes; the penis of the male is slender, and reflected into the branchial cavity; the shell, which is hardly three lines broad, is greyish, flat, and umbilicated. Found in stagnant water. Add *Valvata planorbis*, Drap., &c.

It is here that we must place the completely aquatic shells, or those respiring by gills, which belonged to the old genus HELIX: *i. e.* those in which the penultimate whorl forms, as in the helices, lymnææ, &c., a depression, which gives the aperture more or less the form of a crescent.

The first three genera are still closely allied to turbo.

PALUDINA, *Lam.*

This genus has been lately separated from the cyclostoma, because there is no ridge round the aperture of the shell; because there is a small angle to that aperture, as well as to the operculum; and finally, because the animal, being provided with gills, inhabits the water like all other genera of this family. It has a very short snout, and two pointed tentacula; eyes at the external base of the latter, but on no particular pedicle; and a small membranous wing on each side of the fore-foot of the body. The anterior edge of the foot is double, and the wing of the right side forms a little canal, which introduces water into the respiratory cavity, the incipient indication of the siphon in the following family.

The common species, *Helix vivipara*, L., Drap., whose smooth and greenish shell is marked with two or three longitudinal bands, and which abounds in stagnant waters in France, produces living young ones: in the spring of the year they may be found in the oviduct of the female, in every stage of development. Spallanzani assures us, that if the young ones be taken at the moment of their birth, and be reared separately, they will reproduce without fecundation, like those of the aphis. The males, however, are nearly as common as the females: they have a large penis, which protrudes and retracts, as in helix, but through a hole pierced in the right tentaculum, a circumstance which renders that tentaculum apparently larger than the other, and which furnishes us with a mode of recognizing the male.

Add *Cyclost. achatinum*, Drap. i. 18, &c.

The ocean produces some shells which differ only from the paludinæ in being thick. They form the

LITTORINA, *Féruss.*,

Of which the common species, *Le vigneau* (*Turbo littoreus*, L.), Chemn. V. clxxxv. 1852, abounds on the coast of France, where it is eaten. The shell is round, brown, and longitudinally streaked with blackish.

MONODON, *Lam.*,

Only differs from littorina in having a blunt and slightly salient tooth at the base of the columella, which sometimes also has a fine notch; the external edge of the aperture is crenulated in several species. The animal is more highly ornamented, and is generally furnished with three or four filaments on each side, as long as its tentacula; the eyes are planted on particular pedicles, at the external base of the tentacula; the operculum is round and horny.

A small species, the *Trochus tesselatus*, L., Adanson, Seneg. XII. 1; List. 642. 33, 34, with a brown shell, spotted with whitish, is very abundant on the coast of France. Add *Trochus labeo*, Adans., &c. &c.

Phasianella, *Lam.*

An oblong or pointed shell, similar to that of several bulimi and lymnææ; the aperture also higher than it is wide, and furnished with a strong operculum; base of the columella sensibly flattened, but no umbilicus.

They inhabit the Indian Ocean, and are much sought for by collectors, on account of the beauty of their colours. The animal is provided with two long tentacula, with eyes placed on two tubercles at their external base, and with double lips that are emarginated and fringed, as well as the wings, each of which has three filaments. (*Buccinum tritonis*, Chemn., &c.)

Ampullaria, *Lam.*,

Have the shell round and bellied; a short spire, as in most of the helices; the aperture higher than it is wide, and provided with an operculum; the columella umbilicated.

They inhabit the fresh or brackish waters of hot countries. The animal has long tentacula, and eyes placed on pedicles at their base. In the roof of the respiratory cavity, by the side of a branchial comb, according to the observations of Messrs. Quoy and Gaymard, is a large pouch, without an issue, that is filled with air, and may be considered as a natatory bladder. (*Helix ampullacea*, L., List. 130, &c.)

The Lanistæ, *Montf.*,

Are ampullariæ, with a large, spiral, convoluted umbilicus. *Ampulla carinata*, Olivier, Voy. en Turquie.

Helicina, *Lam.*

Judging by the shell, the helicinæ are ampullariæ, in

which the margin of the aperture is reflected. *Hel. striata*, Blainville, Malac. XXXV. iv.

When this reflected margin is trenchant, they are the AMPULLINÆ, *Blainv.;* and when it is an obtuse ridge, the OLYGIRÆ, *Say.*

There is one species which is remarkable for a border, and a stony traverse, on the internal face of its operculum—the *Hel. neritella*, List.

The organs of respiration in those animals are arranged as in the cyclostomæ, and like the latter they can live out of water.

MELANIA, *Lam.*

A thicker shell; the aperture, higher than it is wide, enlarges opposite to the spire; the columella without folds or umbilicus; length of the spire very various.

The melaniæ inhabit rivers, but are not found in France; the animal has long tentacula, the eyes being on their external side, and at about the third of their length. *Mélanie thiare* (*Melania amarula*, Lam.), Chemn. Tab. 134, f. 1218 and 1219, from the Isle of France and Madagascar. Add *Mel. truncata*, Lam., &c.

RISSOA, *Fremin.* ACMEA, *Hartm.*

Differs from melamia, because the two edges of the aperture unite above. (*Riss. Freminvilii*, &c.)

MELANOPSIS, *Féruss.*

Where the form, nearly that of a melania, differs from it in a callus on the columella, and in a vestige of an emargination near the bottom of the aperture, which seems to indicate a relation with the terebræ of Bruguières. (*Mel. buccinoidea*, Féruss.) In the

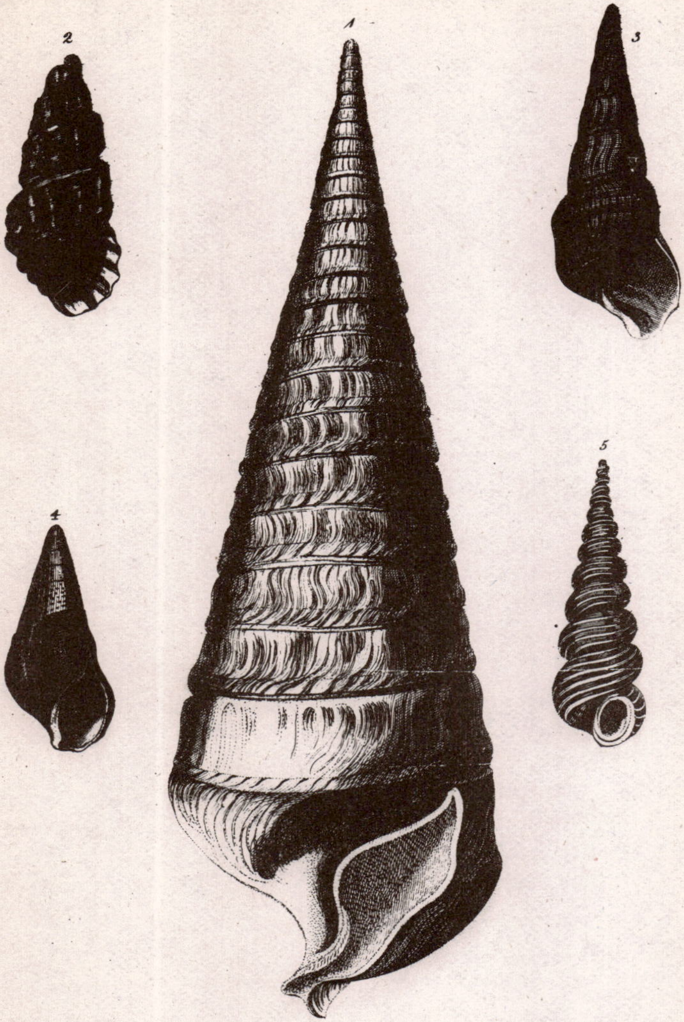

1 *Cerithium truncatum.* 3 *Melania Carolinæ.*

2 *Melania Henriettæ.* 4 *Mel. lineolata.*

5 *Turritella suturnalis.*

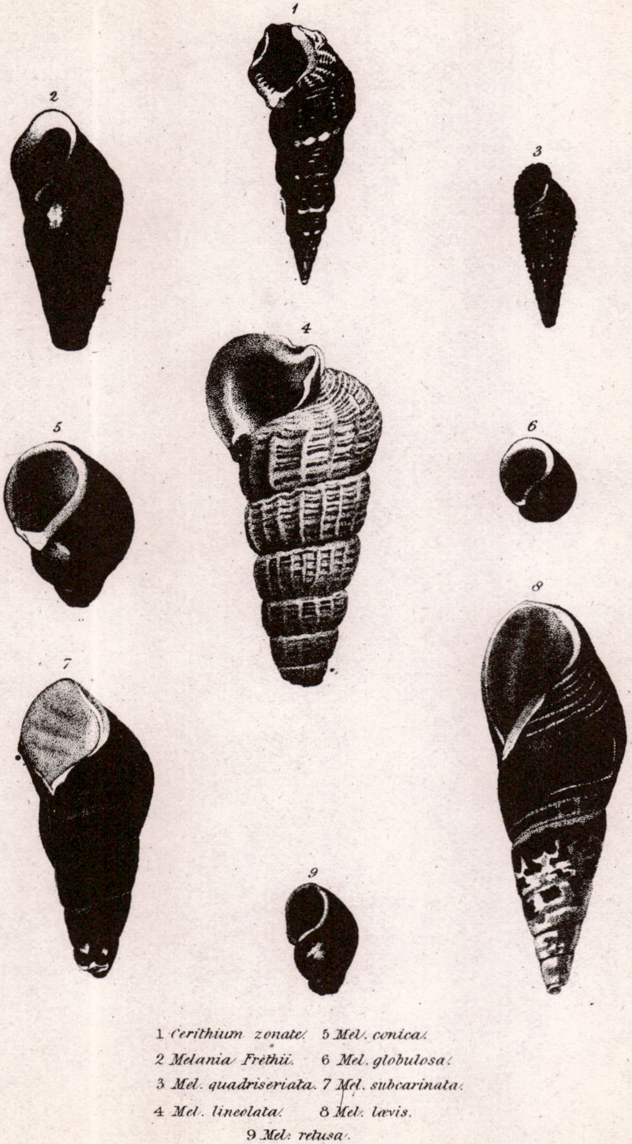

1 *Cerithium zonate.* 5 *Mel. conica.*
2 *Melania Frethii.* 6 *Mel. globulosa.*
3 *Mel. quadriseriata.* 7 *Mel. subcarinata.*
4 *Mel. lineolata.* 8 *Mel. lævis.*
9 *Mel. retusa.*

PIRENA, *Lam.*,

We find not only this little sinus below, but likewise a second on the opposite side. (*Pir. terebralis*, Lam.)

These two subgenera, as well as the melaniæ, inhabit the rivers of Southern Europe and of all hot countries.

There are two genera detached from the volutæ, which, but that they are operculated, and have two tentacula, would resemble the auriculæ. We think they may be placed here.

ACTÆON, *Montf.* TORNATELLA, *Lam.*,

Where the shell is elliptical, the spire but slightly salient, the aperture lengthened into a crescent and widened below, and the base of the columella marked by one or two large folds or oblique callosities. They must be carefully distinguished from the *actæons* of Oken, which appear to be allied to the aplysiæ. (*Voluta tornatilis* and *bifasciata*, Lin., &c.) And the

PYRAMIDELLA, *Lam.*,

Where the spire is turreted, the aperture crescent-like and wide, and the base of the columella obliquely contorted and marked with sharp spiral plicæ. (*Trochus dolabratus.*)

JANTHINA, *Lam.*

The form of the animal separates the Janthinæ from all the preceding genera; their shell, however, is similar to that of the terrestrial limaces, the columellar margin being also indented, but slightly angular at the external edge, and the columella, somewhat extended beyond the half oval, which, without this prolongation, would be formed by that edge.

The animal has no operculum, but the under surface of its foot is furnished with a vesicular organ resembling a bubble of foam, but composed of a solid substance, which prevents it from crawling, but allows it to float on the surface of the water. The head, a cylindrical proboscis, terminated by a vertically cleft mouth, and armed with little hooks, has a bifurcated tentaculum on each side.

The common species, *Helix Janthina*, Lin., has a pretty violet shell, and is very abundant in the Mediterranean. When the animal is touched, it sheds a thick fluid, of a deep violet colour, that dyes the surrounding water.

NERITA, *Lin.*

The columella of the neritæ being in a straight line, renders the aperture semi-circular or semi-elliptical. This aperture is generally large in comparison with the shell, but is always furnished with an operculum that completely closes it. The spire is almost effaced, and the shell semi-globular.

NATICA, *Lam.*,

Are neritæ with an umbilicated shell. The animal of the species known has a large foot, simple tentacula, with the eyes at their base, and a horny operculum.

NERITA, *Lam.* PELORONTA, *Oken.*,

The umbilicus is wanting, the shell thick, columella dentated, and operculum stony; the eyes of the animal on pedicles, by the side of the tentacula, and a moderate foot.

VELATA, *Montf.*,

Where the side of the columella is covered with a calcareous, thick, and convex layer, is distinguished from the above, but perhaps without any good reason. (*Nerita perversa,* Gmel.) A large fossil species. Also the

NERITINA, *Lam.*,

Where the shell has no umbilicus, and is thin, with a horny operculum; the animal is like a true nerita, and most generally the columella is not dentated. It inhabits fresh water.

A small species, very prettily coloured, abounds in the rivers of France; it is the *Nerita fluviatilis*, L., Chemn. IX. cxxiv. 188. Add *Nerita turrita*, Chemn.

The columella in others, however, is finely crenulated (*Nerita pulligera*, Chemn.), and of this number there are some in which the spire is armed with long spines. (CLITHON. *Mont.; Nerita corona*, Chemn.)

According to our recent investigations, it is to the trochoida that we must approximate a family, which we shall call

CAPULOIDA,

And which comprehends five genera, four of which have been dismembered from the patellæ: they all have a widely opened, scarcely turbinated shell, with neither operculum, emargination, nor siphon. The animal resembles the other pectinibranchiata, and has the sexes separate. There is but one branchial comb, transversely arranged on the roof of the cavity, and its filaments are frequently very long.

CAPULUS, *Mont.* PILEOPIS, *Lam.*

A conical shell, with a recurved and spiral summit, which has long caused it to be placed with the patellæ; the gills are in one range, under the anterior margin of the branchial cavity; the proboscis is long, and there is a closely-plaited membranous veil under the neck; the eyes are at the external base of the conical tentacula. (*Patella Hungarica*, List. 544. 32.)

HIPPONYX, *Defr.*,

Would appear from the shell to be a fossil capulus, very remarkable, however, for a bed formed of calcareous matter, on which it rests, and which probably exuded from the foot of the animal. (*Patella cornucopiæ*, Lam.)

Crepidula, *Lam.*

The shell oval, with an obtuse horizontal point, directed obliquely backwards and laterally, the aperture forming the base of the shell, which is half closed beneath, and behind with a horizontal plate ; the abdominal sac which contains the viscera is on this plate, the foot beneath, and the head and gills forwards : the latter consist of a range of long filaments, attached under the anterior margin of the branchial cavity ; the eyes are at the external base of two conical tentacula. (*Patella fornicata*, List. 545. 33. 35, &c.)

Pileolus, *Sowerby*,

Appears a crepidula, in which the transverse plate occupies half the aperture ; the shell, however, is more like that of a patella. (*Pileolus plicatus*, Sowerb.) They are only found fossil.

Septaria, *Fér.* Navicella, *Lam.*

Cimber, *Mont.*, 82.

The shell resembles a crepidula, except that the summit is symmetrical, and laid on the posterior margin, and that the horizontal plate is less salient. The animal is also provided with an additional irregularly shaped testaceous plate, attached horizontally to the superior surface of the muscular disk of its foot, and covered by the abdominal sac, which it partially supports. It is probably analogous to an operculum, but does not exercise its functions, being, in a measure, situated internally. The animal has long tentacula, at whose external base are pedicles which support the eyes. They inhabit the rivers of hot countries. (*Patella neritoidea*, List. 545. 36, &c.)

Calyptræa, *Lam.*,

Have a conical shell, in the hollow of which is a little lamina that projects inwards, forms, as it were, the commencement of a columella, and interposes itself between a fold of the abdominal sac; the gills are composed of a range of numerous filaments, long and slender, like hairs.

In some of them this lamina adheres to the bottom of the cone, being itself bent into a portion of a cone or of a tube, and descending vertically. (*Patella equestris*, L., List. 546. 38, &c.)

In others it is almost horizontal, and adheres to the sides of the cone, which is marked above by a spiral line, that establishes some relation between this shell and that of a trochus. (*Patella contorta*, &c.)

SIPHONARIA, *Sowerb.*

The shell of the siphonariæ, which have been recently separated from the patellæ, at the first glance seems very similar to a flattened patella with radiating furrows; but its margin projects rather more on the right side, and is excavated beneath by a slight furrow, terminating at this projection, to which also corresponds a lateral hole in the mantle, for the introduction of water into the branchial cavity placed on the back; at all other points it is completely closed. The respiratory organ consists of a few small lamellæ, arranged in one transverse line on the roof of this cavity; the tentacula seem to be wanting, the head being merely furnished with a narrow veil. (*Patella sipho*, Sowerb., &c.)

There are some species in which even this slight appearance of the canal in the shell is effaced, resembling altogether that of a patella, except in its summit, which is behind. (*Siphonaria tristensis*, Sowerb.)

SIGARETUS, *Adans.*

The shell is flattened, its aperture ample and round, and the spire very moderate; its whorls rapidly enlarging, and seen within, but concealed, during the life of the animal, in the thickness of a fungous shield, which projects considerably beyond it, as well as the foot, and which is the true mantle. In front of this mantle are an emargination and a semi-canal, which serve to conduct water into the branchial cavity, and which form a transition to the following family, but of which there are no impressions on the shell; the tentacula are conical, with the eyes at their external base; the penis of the male is very large

Some species are found on the coast of France.

CORIOCELLA, *Blainv.*,

Consists of sigareti, the shell of which is horny and almost membranous, like that of the aplysiæ. (The *Coriocelle noire*, Blainv.) This animal is not without a shell, as the author of this genus imagined, but it is thin and flexible.

CRYPTOSTOMA, *Blainv.*

The shell, resembling that of a sigaretus, borne with the head and abdomen, which it covers, on a foot four times larger, cut square behind, and forming before a fleshy oblong bundle that constitutes nearly one-half of its mass. The animal has a flat head, two tentacula, a broad branchial comb on the roof of its dorsal cavity, and a penis under the right tentaculum; but I can find no emargination in the mantle. Besides the species in the British Museum, *Cr. Leachii*, Blainv., we have one, *Cr. Carolinum*. Cuv., sent from Carolina by M. L'Herminier.

The third family of the *gasteropoda pectinibranchiata*, or the

BUCCINOIDA,

Has a spiral shell, in the aperture of which, near the extremity of the columella, is an emargination, or a canal, for transmitting the siphon or tube, which is itself but an elongated fold of the mantle. The greater or less length of the canal, when there is one, the size of the aperture, and the form of the columella, furnish the grounds of its division into genera, which may be variously grouped.

Conus, *L.*,

So called from the conical shape of the shell ; the spire, either perfectly flat, or but slightly salient, forms the base of the cone ; its point is at the opposite extremity ; the aperture is narrow, rectilinear, or nearly so, extending from one end to the other, without enlargement or fold, either on its edge or on the columella. The thinness of the animal is proportioned to the narrowness of the aperture through which it issues. Its tentacula and proboscis have great capacity of elongation. The eyes are placed on the outer side of the former, and near the point ; the operculum, situated obliquely on the hind part of the foot, is too narrow and short to close the whole of the aperture.

In placing here the genera with a straight aperture, we must not be understood as meaning to approximate them to the preceding family, but only to present them first, as possessing the most striking characters of all those which are furnished with a siphon.

The shells of this genus being usually ornamented with the most beautiful colours, are very common in cabinets. The seas of Europe produce very few.

They are distinguished by the flatness or slight projection of the spire ; by the whorls being tuberculated or not ; by its being more salient, and even pointed, and furnished, or not, with tubercles.

There are some in which the spire is sufficiently salient to give them a cylindrical appearance, in which case it may be either smooth or tuberculated. Species with a crowned spire: *Con. cedonulli*, L., &c. Species with a simple spire: *Con. litteratus*, L., &c. The appellation of *crowned spire* is applied to that which is studded with tubercles.

Cypræa, *L.*

The spire projecting but little, and the aperture narrowed, and extending from one extremity to the other; but the shell, which is protuberant in the middle, and almost equally narrowed at both ends, forms an oval, and the aperture in the adult animal is transversely wrinkled on each side. The mantle is sufficiently ample to fold over and envelope the shell, which, at a certain age, it covers with a layer of another colour, so that this difference, added to the form acquired by the aperture, may easily cause the adult to be taken for another species. The animal has moderate tentacula, with the eyes at their external base, and a thin foot without an operculum.

The colours of these shells also are extremely beautiful: they are very common in cabinets, though with very few exceptions they all inhabit the seas of tropical countries. They are termed *Porcelainous*.

Anula, *Brug.*

The shell is oval, and the aperture narrowed and long, as in cypræa, but without folds on the side next to the columella; the spire is concealed, and the two ends of the aperture equally emarginated, or equally prolonged in a canal. Linnæus confounded them with the bullæ, from which Bruguières has very properly separated them. The animal has a broad foot, an extended mantle which partly folds over the shell, a moderate and obtuse snout, and two long tentacula, on which, at about a third of their length, are the eyes.

Montfort particularly designates, by the name OVULÆ, those in which the external margin is transversely wrinkled. (*Bulla ovum*, L.)

Those in which the two extremities of the aperture are prolonged into a canal, and in which the external margin is not wrinkled, he denominates VOLVÆ. (*Bulla volva*, L.)

When this external margin is not wrinkled, nor the extremities of the aperture prolonged, he styles them CALPURNÆ. (*Bulla verrucosa*, L.)

TEREBELLUM, *Lam.*

An oblong shell, with a narrowed aperture, without folds or wrinkles, and increasing regularly in width to the end opposite to the spire, which is more or less salient, according to the species. The animal is not known. (*Terebellum subulatum*, Lam., &c.)

VOLUTA, *Lin.*,

Varies as to the form of the shell and that of the aperture, but is recognized by the emargination, without a canal, which terminates it, and by the salient and oblique folds of the columella. From this genus Bruguières first separated

OLIVA, *Brug.*,

So named from the oblong or elliptical shape of the shell, the aperture of which is narrowed, long, and emarginated opposite to the spire, which is short; the folds of the columella are numerous, and resemble striæ; the whorls are sulciform. These shells are quite as beautiful as the cypræa. (*Oliv. subulata*, Lam., &c.)

The animal has a large foot, the anterior part of which (before the head), is separated by an incision on each side; its tentacula are slender, and bear the eyes on their sides, about the middle of their length; the proboscis, siphon, and penis, are tolerably long, but it has no operculum. MM. Quoy and Gaymard have observed an appendage on its posterior portion, which enters the furrow of the whorls.

The remainder of the genus voluta was afterwards divided into five by M. de Lamarck, exclusive of the tornatellæ and pyramidellæ already mentioned.

VOLVARIA, *Lam.*,

Closely resembles the oliva in its oblong or cylindrical form; but the aperture is narrow, and its anterior edge ascends to the top of the spire, which is excessively short. There is one fold, or several, at the foot of the columella. The lustre and whiteness of this shell are such that on the coasts it is used for making necklaces. (*Volv. monilis*, Lin.) A small fossil species is found in the vicinity of Paris. (*Volvaria bulloides.*)

VOLUTA (proper), *Lam.*

The aperture is ample, and the columella marked with large folds, the one farthest from the spire being the largest. The degree of projection in the spire varies greatly.

In some of them, CYMBIUM, *Montf.*; CYMBA, *Sowerb.*, the last whorl is bellied. The animal has a large thick and fleshy foot, without operculum, and a veil on the head, from the sides of which issue the tentacula; their eyes are on this same veil, outside of the tentaculum; the proboscis is tolerably long, and there is an appendage on each side of the base of the siphon. They attain a large size, and many of them are extremely beautiful. (*Volv. Æthiopica*, List. 797. 4., &c.)

In others, VOLUTA, *Montf.*, the last whorl is conical, becoming narrower at the extremity opposite the spire. (*Voluta musica*, List., &c.) The foot of the animal is not so large as that of the preceding ones; their shells are frequently remarkable for the beauty of their colours, or the figures traced on them.

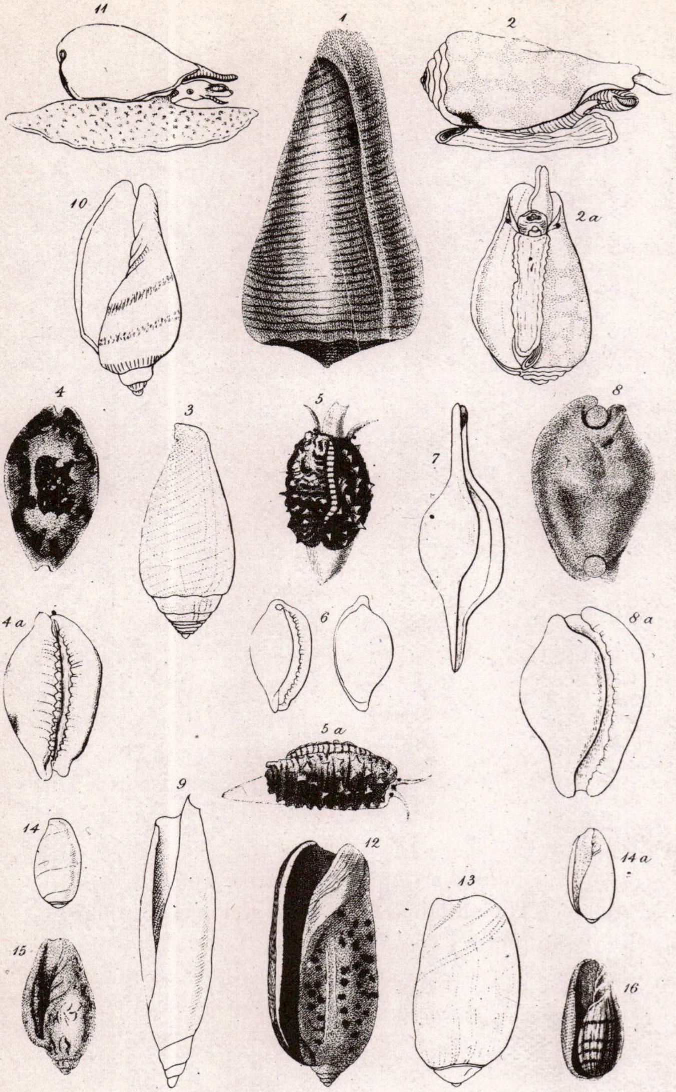

1 *Conus caledonicus.*
2 *Animal of Conus bandanus.*
3 *Conus tendineus.*
4 *Cypræa stolida.*
5 *Animal of Cyp. pediculus.*
6 *Ovula triticea.*
7 *Ovula volva.*
8 *Calpurnus verrucosus.*
9 *Terebellum subulatum.*
10 *Voluta nivosa.*
11 *Animal of Vol. æthiopica.*
12 *Oliva ispidula.*
13 *Ol. auricularia*
14 *Volvaria pallida.*
15 *Marginella nubeculata.*
16 *Marg. bullata.*

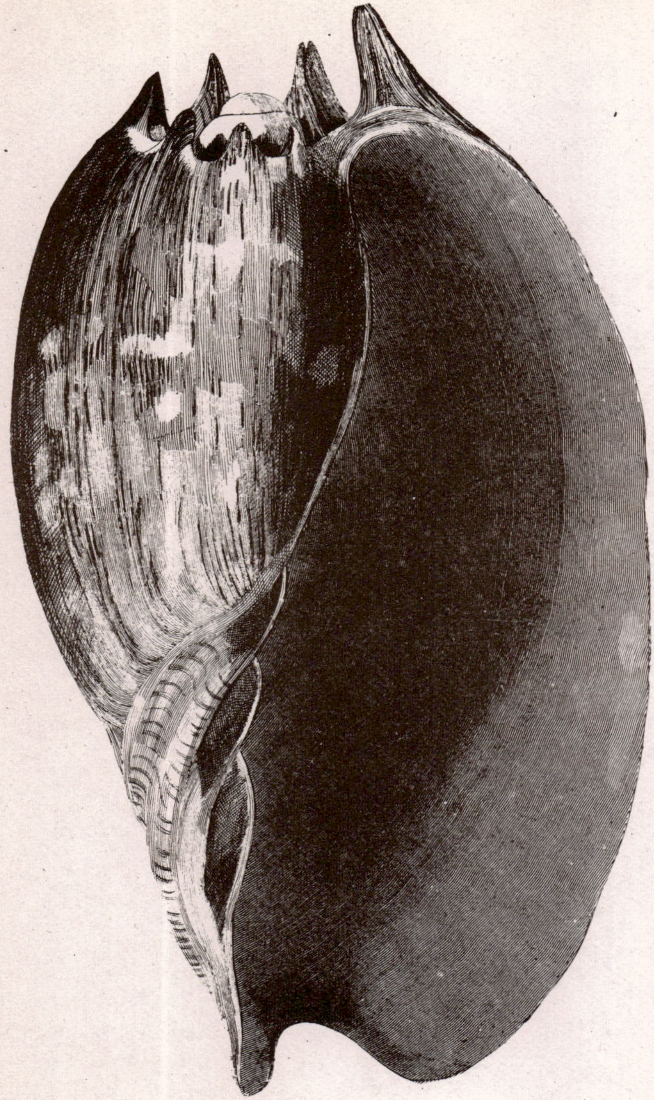

Voluta Georginæ.

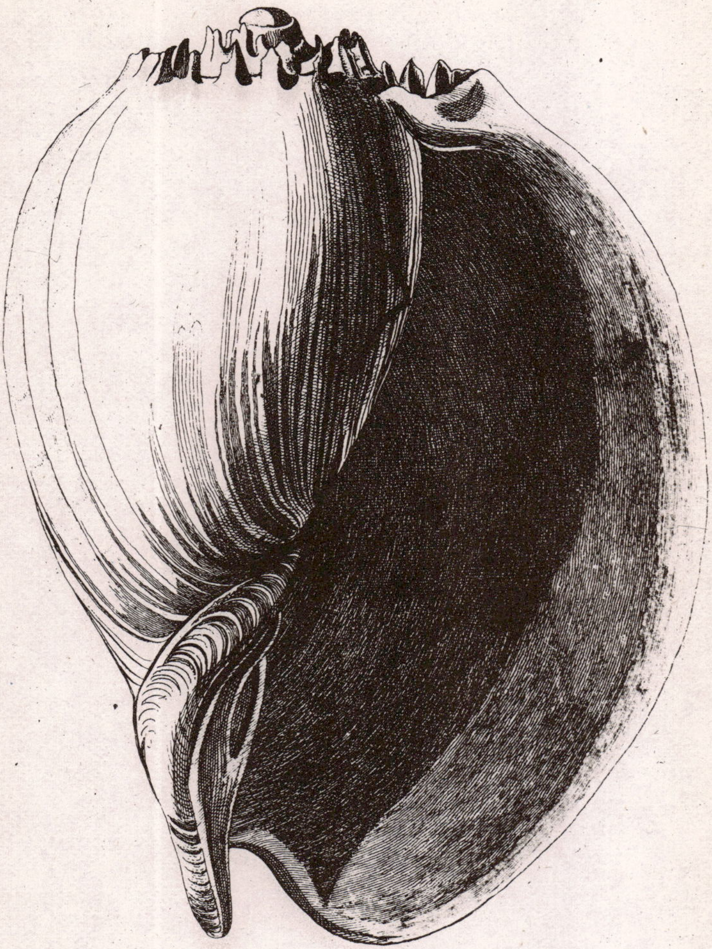

Voluta Broderippii.

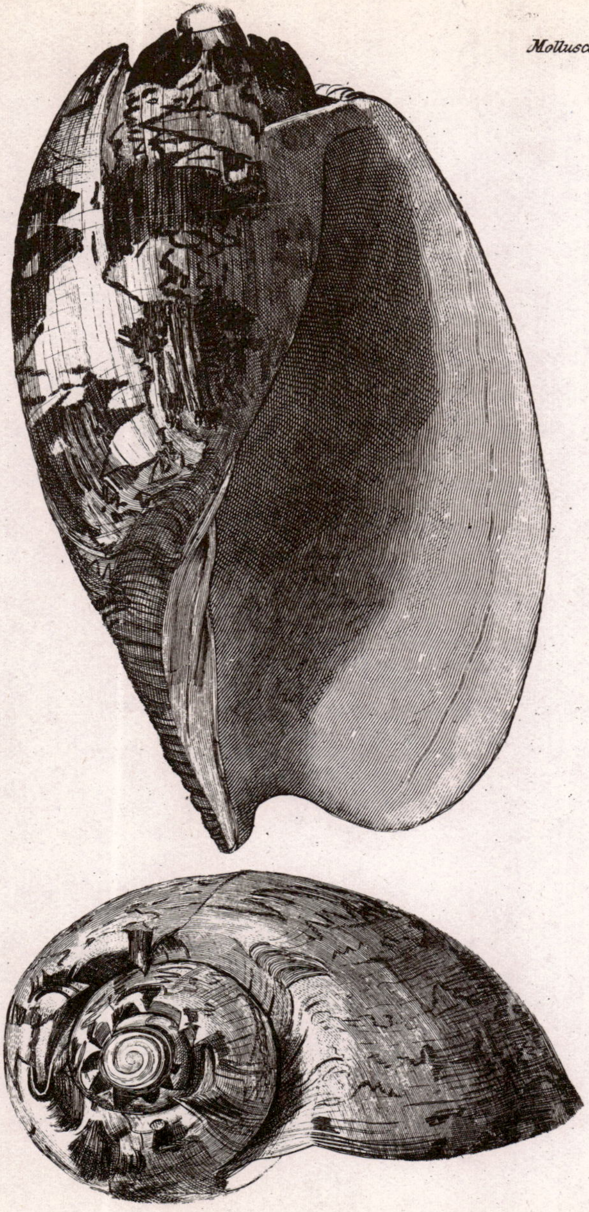

Voluta Miltonis.

Marginella, *Lam.*

Form of the shell similar to that of a true voluta, but the external margin of the aperture is tumid; the emargination is but slightly marked. The foot of the animal, according to Adanson, is very large, and has no operculum. By turning up the lobes of its mantle, it partly covers the shell. The eyes are on the external side of the base of its tentacula. (*Voluta glabella*, Adans., &c.)

M. de Lamarck also distinguishes the Colombella, in which the folds are numerous, and the rim of the external margin is enlarged at its centre. It appears that the operculum is wanting. (*Voluta mercatoria*, List., &c.)

Mitra, *Lam.*

The aperture oblong, with a few large folds on the columella, the one nearest the spire being the largest; the spire usually pointed and elongated. Several species are brilliantly spotted with red on a white ground. The foot of the animal is small; the tentacula are of a moderate length, with the eyes on the side, near their inferior third; the siphon is also of a moderate length, but it frequently protrudes a proboscis longer than its shell.

Cancellaria, *Lam.*

The last whorl swollen out, aperture ample and round, the internal margin forming a plate on the columella; the spire is salient and pointed, and the surface of the shell marked with crossing furrows. (*Voluta cancellata*, L.)

Buccinum, *Lin.*,

Comprises all the shells furnished with an emargination, or a short canal inflected to the left, and in which the columella is destitute of folds.

Bruguières has divided them into four genera, of *Buccinum*, *Purpura*, *Cassis*, and *Terebra*, part of which have been again subdivided by Messrs. de Lamarck and Montfort.

Buccinum, *Brug.*,

Includes the emarginated shells without any canal, whose general form, as well as that of the aperture, is oval. The animals, all such as are known, are deprived of the veil on the head, but are furnished with a proboscis, two separated tentacula, on the external side of which are the eyes, and a horny operculum. Their siphon extends out of the shell.

The name of *Buccinum* is especially applied by M. de Lamarck to those in which the columella is convex and naked, and the margin without folds or rim. Their foot is moderate, their proboscis long and thick, and their penis frequently excessively large. (*Buccinum undulatum*, L., List. 662. 14, &c.)

Nassa, *Lam.*

The side of the columella is covered by a more or less broad and thick plate, and the emargination is deep, but without a canal. The animal resembles a true buccinum, and there are gradual transitions among the shells from one subgenus to the other. (*Buccinum arcularia*, List. 970. 24, 25, &c.)

Eburna, *Lam.*

Those which to a smooth shell, without a folded margin, add a widely and deeply-umbilicated columella. The general form of the shell is closely allied to that of the Olivæ. Their animal is unknown. (*Buccinum glabratum*, List. 974. 29. &c.)

Ancillaria, *Lam.*

The same smooth shell, and at the lower part of the columella a marked lip. There is no umbilicus, neither is the spire furrowed. The animal of several species resembles that of the Olivæ, the foot being still more developed. (*Ancillaria cinnamomea*, Lam., &c.)

DOLIUM, *Lam.*

Those in which projecting ribs that follow the direction of the whorls render the margin undulated ; the inferior whorl is ample and swollen. Montfort subdivides them into

DOLIUM, properly so called, when the lower part of the columella is twisted. (*Buc. olearium*, List., &c.) And

PERDIX, when it is trenchant. (*Bucc. Perdix*, List.)

Their animal has a very large foot widened before, a proboscis larger than its shell, and slender tentacula, on the external side of which, and near the base, are the eyes ; the head has no veil, nor has the foot an operculum.

HARPA, *Lam.*

The harpæ are easily recognized by the projecting transverse ribs on the whorls, the last of which forms a lip on the margin. The shell is beautiful, and the animal has a very large foot, pointed behind, and widened in its anterior portion, which is distinguished by two deep emarginations. The eyes are on the sides of the tentacula, and near their base. It has neither veil nor operculum. *Buccinum harpa*, L., and the other species long confounded with it, &c. Messrs. Reynaud, Quoy, and Gaymard have observed, that, under certain circumstances, the posterior part of the foot is spontaneously detached.

PURPURA, *Brug.*,

Is known by its flattened columella, which is trenchant near the end opposite to the spire, and which, with the external margin, forms a canal there, sunk in the shell, but not salient. The purpuræ were scattered among the buccina and the murices of Linnæus. The animal resembles that of a true buccinum. (*Buccinum persicum*, List., &c.)

The genus LICORNE, *Montf.*, MONOCEROS, *Lam.*, consists of shells similar to the purpuræ, but in which the external edge of the emargination is furnished with a salient spire. (*Buccinum monodon*, Gm., &c.)

Others, also resembling purpuræ, in which the columella, or at least the margin, is provided, in the adult, with teeth, which narrow the aperture, form the SISTRA, *Montf.*, or the RICINULA, *Lam.* (*Murex ricinus*, L., &c.)

CONCHOLEPAS, *Lam.*

The general characters of the purpuræ; but the aperture is so enormous, and the spire so small, that the shell has almost the appearance of a capulus, or one of the valves of an arca. A small salient tooth is visible on each side of the emargination. The animal resembles that of a true buccinum, with the exception of its foot, which is enormous in width and thickness, and that it is attached to the shell by a muscle, shaped like a horse-shoe, as in the capuli. It has a thin, narrow, and horny operculum.

But a single species is known, the *Buccinum concholepas*, Brug., Argenv. pl. ii. f. F. D.; and Sowerb. Gen. of Shells, No. VI. From the coast of Peru.

CASSIS, *Brug.*

The shell oval; aperture oblong or narrow; the columella covered with a plate, as in Nassa, and that plate transversely folded, as well as the external margin; the emargination terminating in a short canal that is reflected and pushed back, as it were, to the left; varices are frequently observed upon it. The animal resembles that of a true buccinum, but its horny operculum is denticulated, in order to pass between the folds of the external margin.

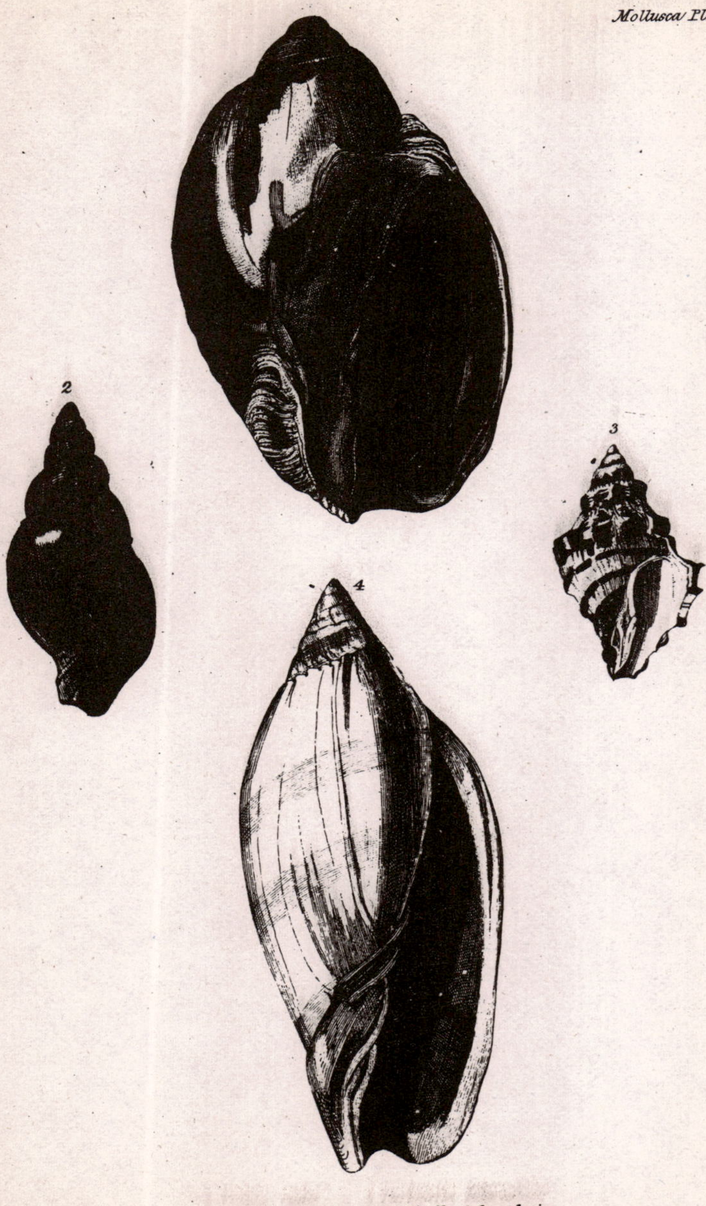

1. *Voluta rudis.* 3. *Turbinella tubercularis.*

2. *Nassa Northiæ.* 4. *Voluta pallida.*

In some, the lip of the margin is denticulated externally near the emargination. (*Buccinum vibex*, Martini, &c.)

In others it is entire: the *Buccinum* of the second division of Gmelin, except the *B. echinophorum*, *strigosum*, No. 26, and *tyrrhenum*, which are cassidariæ. It must also be recollected that among the true cassides, Gmelin appears to have made several repetitions.

MORIO, *Montf.* CASSIDARIA, *Lam.*

Was separated from cassis by Montfort. The canal curves less suddenly, and the whole shell leads directly to certain murices. The animal resembles that of a buccinum, but its foot is more developed. (*Buccinum caudatum*, L., &c.)

TEREBRA, *Brug.*

The aperture, emargination, and columella of a true buccinum, but the general form is turriculated; that is to say, the spire is lengthened into a point. The whole of the last subdivision of the *Buccina*, Gm.: such as *Buc. maculatum*, L. In the

CERITHIUM, *Brug.*

Very properly separated from the MUREX of *Linnæus*, we observe a shell with a turriculated spire. The aperture is oval and the canal short, but well marked, and reflected to the left or backwards. The animal has a veil on its head, and is furnished with two separated tentacula, on the side of which are the eyes, and with a round horny operculum.

Many are found fossil, *Murex vertagus*, List., &c., with the numerous fossil species described by M. de Lamarck, Ann. du Mus.

It is also near the cerithia that we must place several fossil shells which form the genus *nerinea* of M. Defrance. This genus is distinguished by strongly marked folds on each whorl, and on the columella, the centre of which, besides, is hollow throughout. Nine species are already ascertained.

M. Brongniart separates from the cerithia, the

POTAMIDA, *Brong.*,

Which, with the same form of shell, has a very short and scarcely emarginated canal, no furrow on the upper part of the right margin, and the external lip dilated. The potamida inhabit rivers, or at least their mantles, and fossil specimens are found in strata which contain other fresh-water or land species only.—See Brongn, Ann. du Mus. xv. 367. In this subgenus should be placed the *Cerithium atrum*, Brug., and some other cerithia of the same writer, and among the fossils the *Potamida Lamarckii*, Brongn.

MUREX, *Lin.*,

Comprises all those shells in which there is a salient and straight canal. The animal of each subgenus is furnished with a proboscis, long approximated tentacula, on the external side of which are the eyes, and with a horny operculum: the veil on the head is wanting; and, the length of the siphon excepted, it otherwise resembles that of the buccina.

To this genus Linnæus has added several *purpuræ*, in which the canal is not salient, and all the cerithia, in which it is recurved.

Bruguières divides them into two genera, which have been since subdivided by Messrs. Lamarck and Montfort.

MUREX, *Brug.*,

Includes all those which have a salient and straight canal, with varices across the whorls. These varices are projecting rims, with which the animal edges its mantle, at each interruption in the growth of its shell.

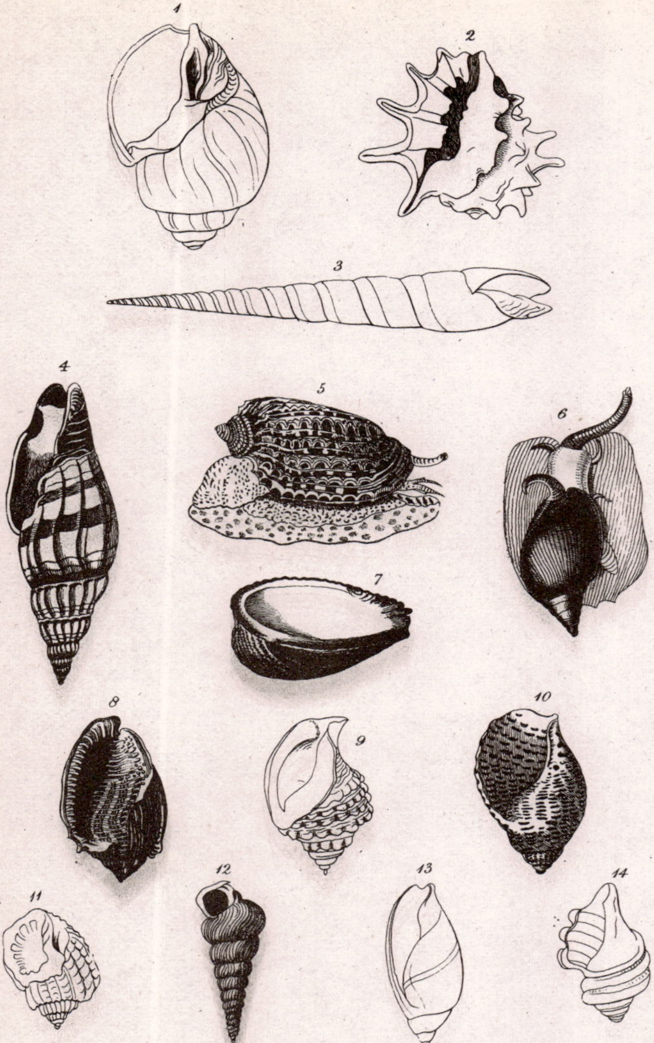

1 *Eburna spirata.*

2 *Ricinula arachnoides.*

3 *Terebra muscaria.*

4 *Mitra vulpecula.*

5 *Harpa ventricosa*

6 *Nafsa reticulata.*

7 *Concholepas Peruvianus.*

8 *Cafsis decufsata.*

9 *Cafsidaria echinophora.*

10 *Dolium perdix.*

11 *Cancellaria asperula.*

12 *Potamis fragilis.*

13 *Ancillaria cinnamomæa.*

14 *Purpura trochlea.*

Lamarck appropriates this name to those in which the varices are not contiguous on two opposite lines.

If their canal be long and slender, and the varices armed with spines, they become the MUREX, properly so called, of *Montfort*. (*Murex tribulus*, List., &c.)

When, with this long canal, the varices are mere knobs, they form the BRONTIS, *Montf.* (*Murex haustellum*, List., &c.)

Some of them, which, with a moderate canal, have projecting tubes, which penetrate into the shell between spiny varices, constitute the TYPHIS, *Montf.* (*Murex tubifer*, Roissy. Brug. Journ. d'Hist. Nat. I. xi. 3.)

When, instead of spires, the varices are furnished with folded lamellæ, slashed, or divided into branches, they are the CHICORACEA, *Montf.* Their canal is long or moderate, and their foliaceous productions vary infinitely in figure and complication. (*Murex ramosus*, List., and all its varieties. Mur. Scorpio, Mart., &c.)

When, with a moderate or short canal, the varices are merely knotty, and the base is provided with an umbilicus, they form the AQUILLA, *Montf.* Several species inhabit the coast of France. (*Murex cretaceus*, L., &c.)

If the umbilicus be wanting, they are his LOTORIUM. (*Mur. lotorium*, L., &c.)

Finally, when the canal is short, the spire elevated, and the varices simple, they are his TRITONIUM. Their mantle is usually folded transversely on both margins. Very large ones inhabit the seas of Europe. (*Mur. tritonis.* L.)

The varices are sometimes numerous, compressed, and almost membranous, constituting the TROPHONA, *Montf.* (*Mur. magellanicus*, Martini.)

At other times they are compressed, very salient, and but few in number. (*Mur. tripterus*, Born. x. 18, 19, &c.)

M. de Lamarck separates from all the murices of Bruguières, the

RANELLA, *Lam.*,

Characterised by opposing varices, so that the shell is burdened with them on both sides; their canal is short, and their surface studded with mere tubercles; margins of the aperture folded. (*Mur. Bufo*, Mont., &c.)

The APOLLES, *Montf.*, are merely umbilicated ranellæ. (*Murex gyrinus*, List. 939. 34.)

FUSUS, *Brug.*,

Comprises all shells with a salient and straight canal, which are destitute of varices.

When the spire projects, the columella is without folds, and the margin is entire, they are the FUSUS, properly so called, Lam., which Montfort again subdivides; when they have no umbilicus, they are his FUSUS. (*Mur. cochlidium*, Seb., &c.)

The shortest and most swollen gradually approach the form of the buccina. (*M. islandicus*, Martini, &c.)

When provided with an umbilicus, they are his LATHIRA. (*Mur. Vespertilio*, Martini.)

The STRUTHIOLARIÆ are distinguished from the true fusi by a border which surrounds their aperture, and which covers the columella. The margin of the adult is inflated, which connects them with the murex. (*Mur. stramineus*, Gm., &c.)

When the spire is salient, the columella without folds, and there is a small indentation or well-marked notch in the margin near the spire, they are the PLEUROTOMA, *Lam.* (*Mur. babilonius*, L., &c.)

The CLAVATULÆ, in which the emargination is wide, and reaches to the spire, are also properly distinguished.

When the spire is but slightly marked, flattened, or rounded,

80A

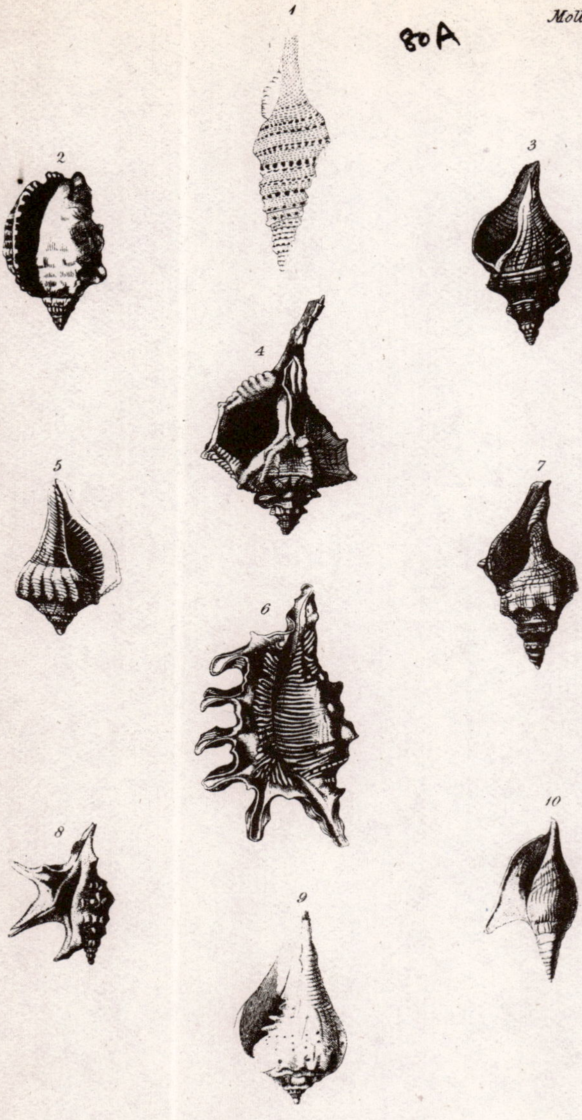

1 *Pleurotoma babylonia.*

2 *Strombus pupilio.*

3 *Fusus morio.*

4 *Murex brandaris.*

5 *Pyrula perbersa.*

6 *Pterocera scorpio.*

7 *Fasciolaria trapezium.*

8 *Rostellaria pespelicani.*

9 *Turbinella pyrum.*

10 *Hippocrenes macroptera.*

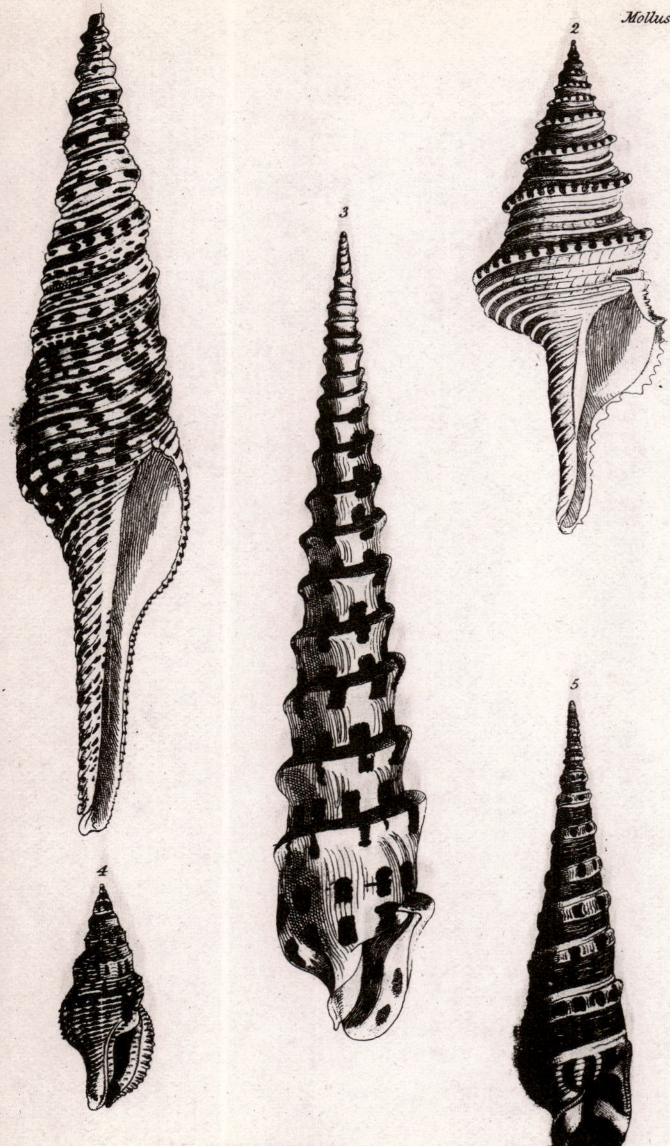

1 Pleurotoma grandis. 3 Terebra subulata.
2 Pleu. carinata. 4 Triton iostoma.
5 Terebra Africana.

London, Published by Whittaker & C°. Ave Maria Lane. 1833.

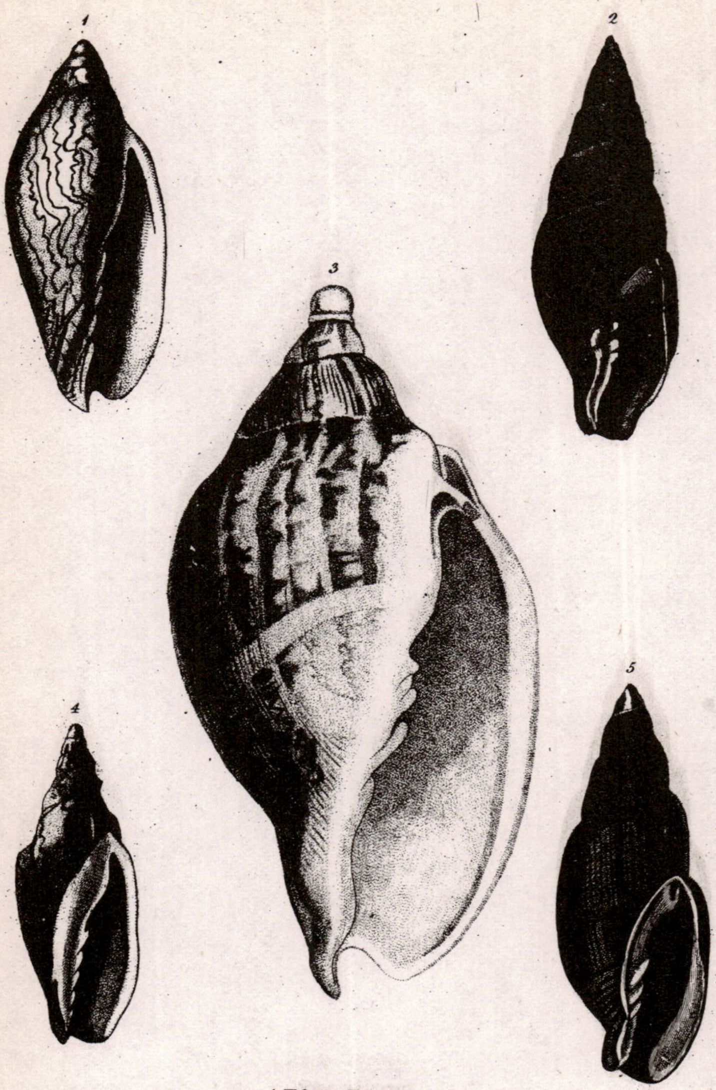

1 *Voluta Turneri.*

2 *Mitra Chinensis.*

3 *Vol. papillosa.*

4 *Vol. gracilis.*

5 *Mitra orientalis.*

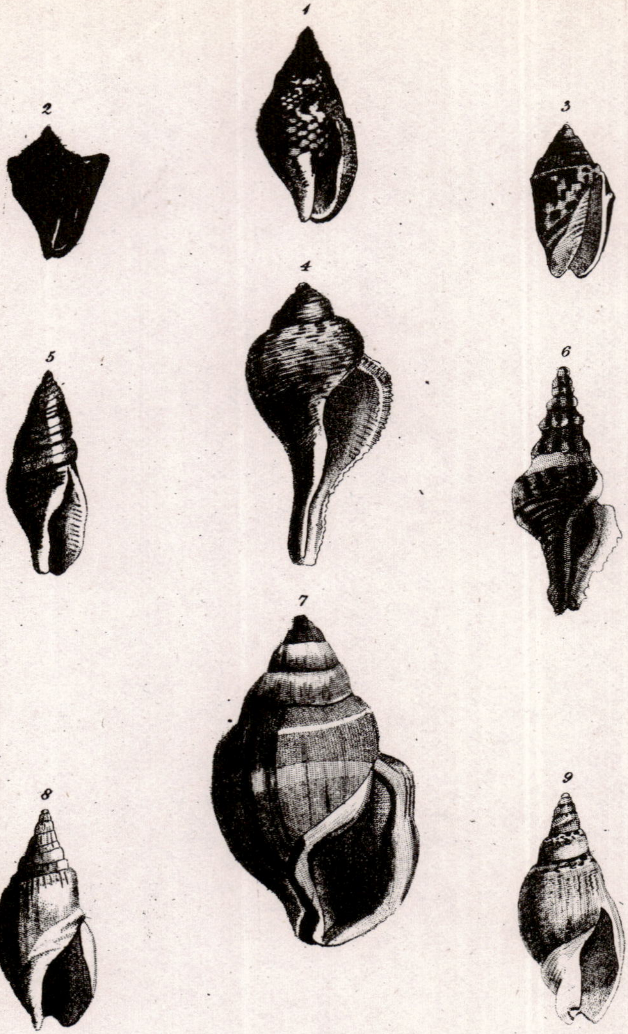

1 *Columbella Tyleræ.*
2 *Col. harpaformis.*
3 *Oliva tessalata var.*
4 *Pyrula striata.*
5 *Triton (Pusio) vexillum.*
6 *Clavatula Griffithii.*
7 *Bulimus auris-vulpina.*
8 *Bullia semiplicata.*
9 *Ancillaria.*

and the columella is without folds, they are the PYRULA, *Lam.*

Some are umbilicated: *Mur. rapa*, Martini, &c.

And others not: *Bulla ficus*, L., &c.

From these pyrulæ Montfort again separates the species with a flattened spire, internally striated near the lip, by the name of FULGUR. They are a sort of pyrulæ with a folded columella, the folds being sometimes almost insensible. (*Murex perversus*, L., &c.)

Among these divisions of the fusi of Bruguières, the FASCIOLARIÆ, *Lam.*, are distinguished by some oblique and well-marked folds on the columella, near the origin of the siphon. (*Mur. tulipa*, L., &c.)

TURBINELLA, *Lam.*,

Also consists of shells with a straight canal, but without varices, distinguished by the large transverse folds on their columella, which extend the whole length of the aperture, and which closely approximate them to the conical volutæ; they only differ from the latter in the elongation of their aperture into a sort of canal: the line that separates them is not easily traced. (*Mur. Scolymus*, Martini, &c.)

STROMBUS, *Lin.*,

Includes those shells with a canal that is either straight or reflected towards the right, of which the external margin of the aperture dilates with age, but still preserves a sinus towards the canal, under which passes the head of the animal, when it extends itself.

In most of them the sinus is at some distance from the canal. They are subdivided by M. de Lamarck into two subgenera.

STROMBUS (proper), *Lam.*,

In which the margin expands into a wing of more or less extent, but not digitated. The foot is proportionally small, and the eyes are supported on lateral pedicles of the tentacula, thicker than the tentacula themselves; the operculum is horny, long, and narrow, and placed on a thin tail. Nearly all the strombi comprised in the second and third division of Gmelin, observing, that owing to the various degrees of development acquired by the external margin, there are several repetitions.

PTEROCERA, *Lam.*

The margin in the adult is divided into long and slender digitations, varying in number, according to the species. The animal is the same as that of the true strombus. (*Strombus lambis*, Rondel. 79, &c.)

In other strombi the sinus of the external margin is contiguous to the canal, forming the ROSTELLARIA, *Lam.* There is usually a second canal ascending the spire, formed by the external margin, and by a continuation of the columella.

In some of them the margin is still digitated. Their animal resembles that of a murex, but has only a very small operculum. (*Strombus pes pelicani*, L., &c.)

In others we merely observe a detached margin. Their canal is long and straight. (*Strombus fusus*, L.)

In some, again, that margin is entire; they are the HIPPOCRENES, *Montf.* (*Strombus amplus*, L.)

SEVENTH ORDER OF THE GASTEROPODA.

TUBULIBRANCHIATA.

THE tubulibranchiata should be detached from the pectinibranchiata, with which they are very closely allied, because the shell, which resembles a more or less irregularly shaped tube, only spiral at the commencement, attaches itself to various bodies; they consequently are deprived of copulatory organs, and fecundate themselves. In the

VERMETUS, *Adans.*,

We remark a tubular shell, whose whorls, at an early age, still form a kind of spire, but afterwards continue in a tube more or less irregularly contorted, or bent like the tubes of a serpula. This shell usually attaches itself by interlacing with others of the same species, or is partly enveloped by lithophytes. The animal having no power of locomotion, is deprived of a foot properly so called; but the part which in ordinary gasteropoda forms the tail, is here turned under it, and extends to beyond the head, where its extremity becomes enlarged, and finished with a thin operculum. When the animal withdraws into its shell, it is this mass which closes the entrance. It is sometimes seen with various appendages, and in certain species the operculum is spiny. The head of the animal is obtuse, and has two moderate tentacula, on the external sides of which, at the base, are the eyes. The mouth is a vertical orifice, beneath which is a filament on each side, that has all the appearance of a tentaculum, but belonging in reality to the foot.

The branchiæ form but a single range along the left side of the roof of the branchial cavity. The right side is occupied by the rectum and the spermatic canal, which also transmits the ova. There is no penis, the animal fecundating itself.

The species are numerous, but not very distinct. Linnæus left them among the serpulæ.

The VERMILIÆ, also left by M. de Lamarck near the serpulæ, are similar to the vermeti.

MAGILUS, *Montf.*

The magili have a longitudinally carinated tube, which is at first regularly spiral, and then extends itself in a line more or less straight. Although the animal is unknown it is highly probable that it should be placed near the vermeti. The

SILIQUARIA, *Brug.*,

Resembles Vermetus in the head, the position of the operculum, and in the tubular and irregular shell; but there is a fissure on the whole length of this shell which follows its contour, and which corresponds to a similar cleft in that part of the mantle which covers the branchial cavity. Along the whole side of this cleft is a branchial comb, composed of numerous loose and tabular-like lamellæ. Linnæus left them with the serpulæ, and till very lately they were considered as belonging to the class of the annelides.

EIGHTH ORDER OF GASTEROPODA.

SCUTIBRANCHIATA.

THE scutibranchiata comprise a certain number of gasteropoda, similar to the pectinibranchiata in the form and position of the gills, as well as in the general form of the body, but in which the sexes are united, in such a way, however, as to allow them to fecundate themselves. Their shells are very open, without an operculum, and for the most part without a trace of turbination, so that they cover these animals, and particularly the gills, in the manner of a shield. The heart is traversed by the rectum, and receives the blood from two auricles, as is the case in the greater number of bivalves.

HALYOTIS, *Lin.*,

Is the only genus of this order in which the shell is turbinated; it is distinguished from that kind of shell by the excessive amplitude of the aperture, and the flatness and smallness of the spire, which is seen from within. This form has caused it to be compared to the ear of a quadruped.

HALYOTIS (proper), *Lam.*,

Or the true Halyotis. The shell is perforated along the side of the columella by a series of holes; when the last hole is not terminated, it gives to that part the look of an emargination. The animal is one of the most highly ornamented of all the gasteropoda. A double membrane, cut into leaves, and furnished with a double range of filaments, extends, at least in the most common species, round the foot, and on to the mouth. Outside its long tentacula are two cylindrical pedicles, which support the eyes. The mantle is deeply cleft on the right side, and the water, which passes through the holes of the shell, penetrates through this cleft into the branchial cavity. Along its edges we observe three or four filaments, which the animal can protrude through these holes. The mouth is a short proboscis. All the *halyotides* of Gmelin, except *imperforata* and *perversa*. This genus has its counterpart among the fossils.

The PADOLLÆ, *Montfort*, have a circular shell, in which the holes are nearly obliterated; and there is a deep furrow that follows the middle of the whorls, and is marked externally by a salient ridge.

STROMATIA, *Lam.*

The shell more hollow, the spire more salient, and the holes wanting; otherwise, like that of the halyotides, which it thus connects with certain species of turbo. The animal is much less ornamented than that of halyotis.

In the following genera, which are separated from the patellæ, the shell is perfectly symmetrical, as well as the position of the heart and branchiæ. In

FISSURELLA, *Lam.*,

We perceive a broad fleshy disk under the belly, as in the patellæ, a conical shell placed on the middle of the back, but not always completely covering it, and perforated at its summit by a small orifice, which affords at once an issue for the fæces and a passage to the water, required for respiration; this orifice penetrates into the cavity of the branchiæ, situated on the fore part of the back, and in the bottom of which terminates the anus; a cavity otherwise widely opened above the head. A branchial comb is symmetrically arranged on each side; the eyes are on the external base of the conical tentacula, and the sides of the foot are furnished with a range of filaments.

EMARGINULA, *Lam.*

The structure of the emarginulæ is similar to that of a fissurella, except that instead of the hole in the summit, there is a small cleft or emargination in the anterior margin of their mantle and shell, which also penetrates to the branchial cavity; the margin of the mantle envelopes and covers a great part of that of the shell; the eyes are placed on a tubercle of the external base of the conical tentacula, and the margin of the foot is furnished with a range of filaments.

PARMOPHORUS, *Lam.*

A great portion of the shell curved by the reflected margin of the mantle, as in the emarginulæ; the shell itself oblong, slightly conical, and without hole or emargination; the branchiæ and other organs, as in the preceding genera.

NINTH ORDER OF GASTEROPODA.

CYCLOBRANCHIATA.

THE branchiæ of the cyclobranchiata resemble small lamellæ, or little pyramids forming a cordon, more or less complete, under the borders of the mantle, very nearly as in the inferobranchiata, from which they are distinguished by the nature of their hermaphroditism; for, like the preceding genus, they have no copulating organ, but fecundate themselves. Their heart does not embrace the rectum, but varies as to situation. But two genera of this order are known, in both of which the shell never approaches in the least to the turbinated form.

PATELLA, *Lin.*

The entire body covered with a shell, formed of a single piece, in the form of a broad-based cone; a cordon of little branchial lamellæ under the margin of the mantle; the anus and genital orifices somewhat to the right and above the head, which is furnished with a thick and short snout, and two pointed tentacula, on the external base of which are the eyes; the mouth is fleshy, and containing a spiny tongue, which inclines backwards, and is reflected deeply in the interior of the body. The stomach is membranous, and the intestine long, thin, and greatly flexed; the heart is forwards, above the neck, and a little to the left.

Some species abound on the coast of France.

CHITON, *Lin.*

A range of testaceous and symmetrical scales along the back of the mantle, but not occupying its whole breadth; edges of the mantle coriaceous, and furnished either with a naked skin or little scales, which give it the appearance of shagreen, or with spines, hairs, or setaceous fasciculi. Under these edges, on each side, is a range of lamellar, pyramidal branchiæ; and before, a membranous veil on the mouth supplies the want of tentacula. The anus is under the posterior extremity. The heart is situated behind, on the rectum; the stomach is membranous, and the intestine very long and greatly contorted. The ovary is situated over the other viscera, and appears to open on the sides by two oviducts.

A few small species are found on the coast of France; very large ones abound in the seas of hot climates.

FOURTH CLASS OF THE MOLLUSCA.

ACEPHALA.

THEY have no apparent head, but a mere mouth concealed in the bottom, or between the folds of their mantle: the latter is almost always doubled, and encloses the body, as a book is clasped by its cover; but it frequently happens that in consequence of the two lobes uniting before, it forms a tube. Sometimes it is closed at one side, and then it represents a sac. This mantle is generally provided with a calcareous bivalve, and sometimes multivalve shell, and in two genera only is it reduced to a cartilaginous or even membranous nature. The brain is over the mouth, and there are also one or two other ganglia; the gills usually consist of large leaves, covered with vascular net-work, under or between which the water passes; they are more simple, however, in the genera without a shell. From these gills the blood proceeds to the heart, generally simple, which distributes it throughout the system, returning to the pulmonary artery, without the aid of another ventricle.

The mouth is always toothless, and can only receive the molecules brought to it by the water; it leads to a first stomach, to which is sometimes added a second; the length of the intestines is very various; the bile is thrown into the stomach by several pores, and the stomach itself is surrounded by the mass of the liver.

All these animals fecundate themselves, and in several testacea the young ones, which are innumerable, pass some time in the thickness of the gills, previously to being brought to light. All the acephala are aquatic.

THE FIRST ORDER OF ACEPHALA.

TESTACEA.

TESTACEOUS acephala, or acephala with *four branchial leaflets*, are, beyond all comparison, the most numerous: all the bivalves, and some genera of the multivalves, belong to this order. Their body, which contains the liver and viscera, is placed between the two laminæ of the mantle; forwards, and still between these laminæ, are the four branchial leaflets, transversely and regularly striated by the vessels; the mouth is at one extremity, the anus at the other, and the heart towards the back; the foot, when it exists, is inserted between the four gills; on the sides of the mouth are four other triangular leaflets, which are the extremities of the two lips, and serve as tentacula; the foot is a mere fleshy mass, the motions of which are effected by a mechanism analogous to that of the tongue of the mammalia; its muscles are attached to the bottom of the valves of the shell: other muscles which form sometimes one and sometimes two masses, cross transversely from one valve to the other, to keep them closed, but when the animal relaxes these muscles, an elastic ligament, placed behind the hinge, opens the valves by its contraction.

A considerable number of bivalves are provided with what is termed a *byssus*, or a bundle of threads, more or less loosely connected, which issue from the base of the foot, and by which the animal adheres to various bodies. It uses its foot to direct the threads, and to agglutinate their extremities: it even reproduces them when cut; but the nature of the production is not thoroughly ascertained. Reaumur considered these threads as a secretion, spun and drawn from the furrow of the foot. Poli thinks that they are mere prolongations of tendinous fibres.

The shell essentially consists of two pieces, called valves, to which, in certain genera, are added others, connected by a hinge that is sometimes simple, and sometimes composed of a greater or smaller number of teeth and plates, which are received into corresponding cavities.

There is usually a projecting part near the hinge, called the summit, or *nates*.

Most of these shells fit closely when the animal approximates them, but there are several which exhibit a gaping portion either before or at the extremities.

The first family of the testaceous acephala, or

The Ostracea,

Have the mantle open, without tubes, or any particular aperture.

The foot is either wanting in these mollusca, or is small; they are for the most part fixed by the shell or byssus to rocks and other submerged bodies. Those which are free seldom move, except by acting on the water, by suddenly closing their valves.

In the first subdivision, there is nothing but a muscular mass, reaching from one valve to the other, as seen by the single impression left upon the shell.

It is thought proper to class with them certain fossil shells, the valves of which do not even appear to have been held together by a ligament, but which lay together like a vase and its cover, and were connected by muscles only. They form the genus

ACARDA, *Brug.* OSTRACITA, *La Peyr.*,

Of which M. de Lamarck makes a family, that he names RUDISTA. The shells are thick, and of a solid or porous tissue. They are now divided into the

RADIOLITES, *Lam.*,

In which the valves are striated from the centre to the circumference: the one is flat, the other thick, nearly conical, and fixed.

SPHÆRULITES, *Lameth*,

Where the valves are roughened by irregularly raised plates.

It is also thought that we may add the

CALCEOLARIA,

One valve of which is conical but free, and the other flat, and even somewhat concave, so that they remind us of a shoe; and even the

HIPPURITES,

Where one valve is conical or cylindrical, with two obtuse longitudinal ridges on the inside. The base even appears to be divided into several chambers, by transverse partitions; the other valve fits like a cover.

BATOLITHES, *Montf.*, 334,

Are cylindrical and straight hippurites. They are frequently found greatly elongated. There is much uncertainty, however, with respect to all these bodies.

As to the well known living testaceous acephala, Linnæus had united in the genus

Ostrea, *Lin.*,

All those which have but a small ligament at the hinge, inserted into a little depression on each side, and without teeth or projecting plates.

Ostrea, *Brug.*

The true oysters have the ligament as above described, and irregular inequivalve, and lamellated shells. They adhere to rocks, piles, and even to each other, by their most convex valve.

The animal (Peloris, *Poli*) is one of the most simple of all the bivalves, possessing nothing remarkable but a double fringe round the mantle, the lobes of which are only united above the head, near the hinge; but there is no vestige of a foot.

Every one is acquainted with the *common oyster* (*Ostrea edulis*, L.), which is gathered upon the rocks, and reared in depots, to be ready when wanting. Its fecundity is as astonishing as its flavour is agreeable.

Among the neighbouring species we should notice

The little Mediterranean oyster. (*Ostrea cristata*, Poli. II. xx.)

Among the foreign species we should remark

The Parasite oyster (*Ostrea parasitica*, L.), Chemn. VIII. lxxiv. 681.

Round and flat, which fixes itself on the roots of the mangroves and other trees, which the salt water can reach.

The leaf oyster (*Ostrea folium*, L.) Ibid. lxxi. 662—666. Oval, with the margins folded in zig-zags, which attaches itself by the indentations of its convex valve to the branches of the Gorgoniæ and the other Lithophytes.

M. de Lamarck separates by the name of

GRYPHÆA, *Lam.*,

Certain oysters, mostly fossil, of the ancient calcareous and schistose strata, in which the summit of the most convex valve greatly projects and curves more or less into a hook or portion of a spiral; the other valve is frequently concave. The greater number of these shells appear to have been free; some of them, however, seem to have adhered to other bodies by their hook.

G. tricarinata. The only living species known.

PECTEN, *Brug.*,

Very properly separated from the oysters by Bruguières, although they have the same kind of hinge. They are easily distinguished by their inequivalve semi-circular shell, almost always regularly marked with ribs, which radiate from the summit of each valve to the edge, and furnished with two angular productions called *ears*, which widen the sides of the hinge. The animal (ARGUS, *Poli*) has but a small oval foot, placed on a cylindrical pedicle, before a sac-like abdomen that hangs between the gills. Some species, known by a deep emargination under their anterior *ear*, are furnished with a byssus; the others cannot adhere, and even swim with rapidity by suddenly closing their valves. The mantle is surrounded with two ranges of filaments, several of the external ones being terminated by a little greenish globule; the mouth has numerous branched tentacula, in place of the four usual labial leaflets. The shell is frequently tinged with the most lively colours.

The great species of the French coast (*Ostrea maxima*, Linn.), has convex valves, one whitish and the other reddish, each having fourteen ribs, broad, and longitudinally striated. The animal is eaten. We may also remark the *sole* of the Indian Ocean (*Ostrea solea*), Chemn. VII. lxi. 595, with extremely thin and almost equal valves, one brown, the other white; and internal ribs, fine as hairs, approximated two by two.

LIMA, *Brug.*

The limæ differ from the pectens, in the superior length of their shell, in a direction perpendicular to the hinge, the ears of which are shorter, and the sides less unequal, thus forming an oblique oval. The ribs of most of them are raised with scales. The valves cannot join during the life of the animal, whose mantle is furnished with numberless filaments of different lengths, without tubercles, and more internally with a large border, which closes the opening of the shell, and even forms a veil in front; the foot is small, and the byssus trifling. The limæ swim with rapidity, by means of their valves.

One species, the *Ostrea lima*, L., Chemn. VII. lxviii. 651, of a fine white, inhabits the Mediterranean. It is eaten.

PEDUM, *Brug.*

The oblong and oblique shell, with small ears of the limæ; but the valves are unequal, and only the most convex has a deep emargination for the byssus. The animal is similar to that of a lima, but its mantle is only furnished with a single range of small slender tentacula. Its byssus is larger.

But a single species is known; it inhabits the Indian Ocean. (*Ostrea spondyloidea*, Gm.)

Certain fossils may be placed here, which have the hinge, ligament, and central muscle of the ostrea, pectines, and limæ, but are distinguished by some of the details of the shell.

HINNITA, *Defr.*

The hinnitæ appear to be ostreæ, or limæ with small ears, and adhering, irregular, and very thick shells, the convex valve in particular. A depression is observed at the hinge for the ligament.

PLAGIOSTOMA, *Sowerb.*

The oblique shell of a lima, flattened on one side; very small ears; the valves more convex, striated, without scales; the opening for the byssus smaller. Found in formations anterior to chalk.

PACHYTES, *Defr.*

Nearly the same form as the pectines: shell regular, with small ears; a flattened transverse space between their summits, which in one of the valves is marked by a triangular notch, in which passed the ligament. Found in chalk. In the

DIANCHORA, *Sowerb.*,

The valves are oblique and irregular; one of them adherent, and with a perforated summit, the other free, and with ears.

PODOPSIS, *Lam.*

Regularly striated valves, without opercula; the summit of one of them more salient, truncated, and adherent, frequently very thick, and forming a sort of pedestal to the shell.

Although multivalve, we should approximate the

ANOMIA, *Brug.*,

To the ostreæ. The anomiæ have two thin, unequal, irregular valves, the flattest of which is deeply notched on the side of the ligament, which is similar to that of the ostreæ. The greater part of the central muscle traverses this opening, to be inserted into a third plate, that is sometimes stony and sometimes horny, by which the animal adheres to foreign bodies, and the remainder of it (the muscle) serves to join one valve to the other. The animal (ECHION, *Poli*) has a small vestige of a foot, similar to that of a pecten, which slips between the emargination and the plate that closes it, and perhaps serves to direct water to the mouth, which is close to it.

These shells are found attached to various bodies, like the ostreæ. They occur in every sea.

PLACUNA, *Brug.*,

A small genus, allied to the anomiæ, in which the valves are thin, unequal, and frequently irregular, as in the latter, but both entire. Two projecting ribs, like a chevron, are seen in the inside of one of them, near the hinge.

The animal is not known, but it most resembles that of the ostreæ, or that of the anomiæ.

SPONDYLUS, *Lin.*,

A rough and foliaceous shell, as in the ostreæ, and frequently spiny, but the hinge is more complex; besides the cavity for the ligament, analogous to that of the ostreæ, there are two teeth to each valve, that enter into fossæ in the opposite one. The two middle teeth belong to the most convex valve, which is usually the left one, and which has a projecting heel flattened as if sawed through behind the hinge. The animal, like that of a pecten, has the borders of its mantle furnished with two rows of tentacula, some of the external ones being terminated by coloured tubercles; before the abdomen is a vestige of a foot formed like a broad radiated disk, on a short pedicle, and endowed with the faculty of contraction and expansion. From its centre hangs a filament terminated by an oval mass, the use of which is unknown.

The spondyli are eaten like oysters; their shells are frequently tinged with the most brilliant colours; they adhere to all sorts of bodies.

Plicatula, *Lam.*

The plicatula, separated by Lamarck from the spondyli, have nearly the same kind of hinge, but no heel, and flat, almost equal, irregular, plicated, and scaly valves, as in many of the ostreæ.

Malleus, *Lam.*

A simple depression for the ligament, as in the ostreæ, where the mallei were left by Linnæus on account of their having the same irregular and inequivalve shell; but they are distinguished by a notch on the side of this ligament for the passage of a byssus.

The most known species, *ostrea malleus*, Lin., which ranks among the number of high priced and rare shells, has the two ends of the hinge extended, and forming something like the head of a hammer, of which the valves, elongated in a transverse direction, represent the handle. It inhabits the Archipelago of India.

There are some others, possibly young ones of the same species, in which the hinge is not prolonged: we must be careful not to confound them with the vulsellæ.

Vulsella, *Lam.*,

A little plate projecting inwards at each side of the hinge, from one of which to the other extends the ligament; otherwise similar to that of the ostreæ. By the side of this plate is a notch for the byssus, as in the mallei. The shell is elongated in a direction perpendicular to the hinge.

The most known species inhabit the Indian Ocean.

Perna, *Brug.*,

Several parallel cavities across the hinge, opposed to each

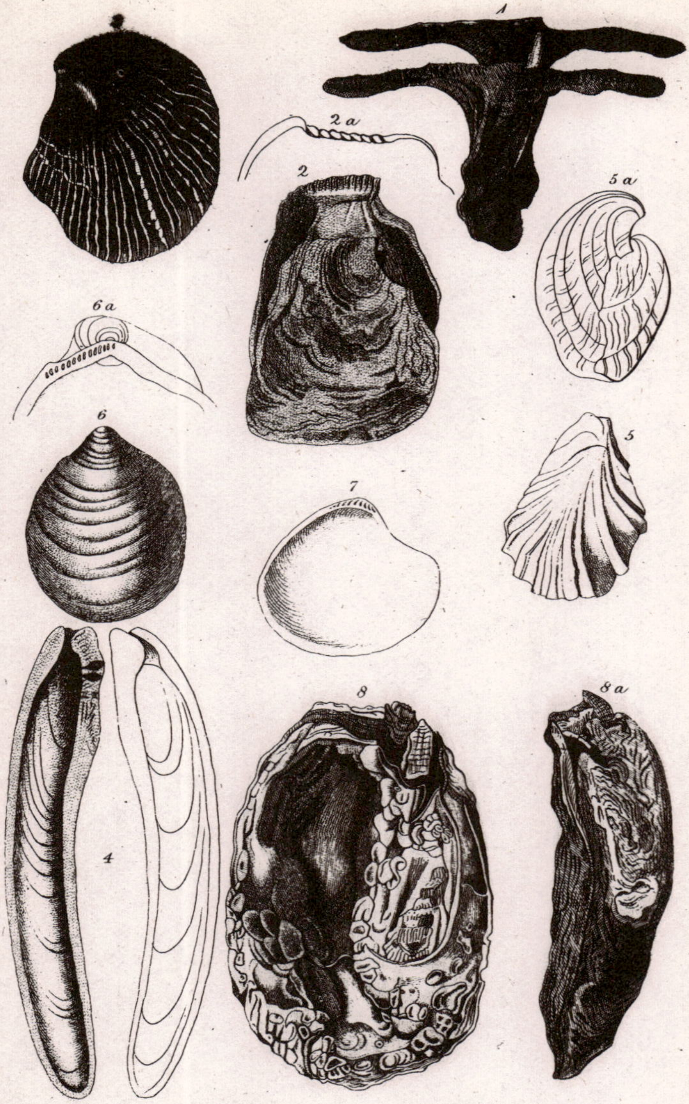

1 *Malleus vulgaris.* 5 *Inoceramus sulcatus.*
2 *Perna ephippium.* 6 *Catillus Cuvierii.*
3 *Crenatula avicularis.* 7 *Pulvinites Adansonii.*
4 *Gervilia soleneïdes.* 8 *Etheria elliptica.*

other in the two valves, and lodging as many elastic ligaments; the irregular and foliaceous shell marked on the anterior side beneath the hinge by a notch traversed by the byssus. The pernæ were also left by Linnæus among the ostreæ.

CRENATULA, *Lam.*

The crenatula, lately separated from the pernæ, instead of having transverse cavities on a broad hinge, are furnished with oval ones on the very margin, where they occupy but little of its breadth. The byssus seems to be wanting; and they are frequently found among sponges.

It is thought that we may approximate to the pernæ, certain fossil shells, in which the hinge is also furnished with cavities, more or less numerous, that correspond to each other, and thus appear to have furnished points of attachment to ligaments; thus, those of the

GERVILLIA, *Defr.*,

Have a shell closely resembling that of the vulsellæ, but with a kind of double hinge, externally with opposed cavities, receiving as many ligaments, and internally furnished with very oblique teeth in each valve. Their impressions are found along with ammonites in compact limestone.

INOCERAMUS, *Sowerb.*,

Is remarkable for the elevation and inequality of the valves, the summit of which curves in a hook towards the hinge; their texture is lamellated.

CATILLUS, *Brong.*,

Independently of the depressions for the ligament, the catilli have a conical furrow sunk in a rim, which is bent at a right angle, to form one of the margins of the shell. The valves are about equal, and of a fibrous texture. They appear to have had a byssus.

PULVINITES, *Defr.*,

A regularly triangular shell, in which the few depressions diverge from the summit on the inside. The impression is found in chalk.

In the second subdivision of the ostracea, as well as in almost all the bivalves which follow, besides the single transverse muscular mass of the preceding genera, there is a bundle which is placed before the mouth, and extends from one valve to the other. It is apparently in this subdivision that we must place the

ETHERIA, *Lam.*,

Large inequivalve shells, as irregular as those of the ostreæ, and more so; no teeth to the hinge; the ligament partly external and partly internal. They differ from the ostreæ chiefly in having two muscular impressions. The animal is not seen to produce a byssus.

They have lately been discovered in the upper Nile.

AVICULA, *Brug.*,

An equivalve shell with a rectilinear hinge, frequently extended into wings, at its extremities, furnished with a narrow and elongated ligament, and sometimes with small denticulations near the mouth of the animal. In the anterior side, a little beneath the angle of the side of the mouth, is a small notch for the byssus. The anterior transverse muscle is still excessively small.

The species with less salient ears form the PINTADINÆ, *Lam.*, or MARGARITÆ, *Leach.*

The most celebrated (*mytilus margaritiferus*, Lin.) has nearly a semicircular shell, greenish without, and ornamented

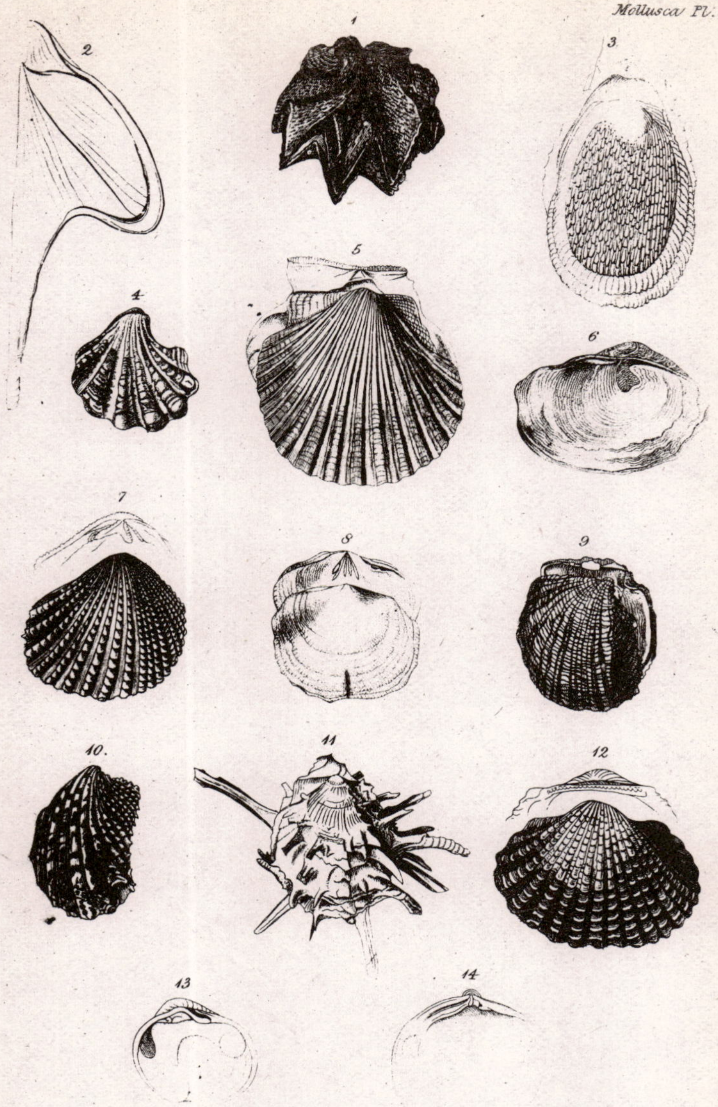

1 *Ostrea crista-galli.*
2 *Avicula heteroptera.*
3 *Lima glacialis.*
4 *Plicatula cristata.*
5 *Pecten gibbosus.*
6 *Anomia ephippium.*
7 *Trigonia pectinata.*
3 *Placuna placenta.*
9 *Pintadina margaritera.*
10 *Cardita caliculata.*
11 *Spondylus Americanus.*
12 *Arca granosa.*
13 *Petricola lucinalis.*
14 *Loripes lactea.*

with the most beautiful mother-of-pearl within. The latter is employed in the arts, and it is from the extravasation of this substance that are produced the oriental or fine pearls, taken by the divers at Ceylon, at Cape Comorin, and in the Persian Gulph.

The name oı AVICULA is appropriated to such as have more pointed ears and a more oblique shell. The vestige of a tooth, of which traces are visible in the pintadinæ, is observable on the hinge, before the ligament.

One species (*mytilus hirundo*, Lin.) that inhabits the Mediterranean is remarkable for the pointed ears which extend its hinge on each side. Its byssus is coarse and stout, resembling a little tree.

PINNA, *Lin.*

The pinnæ have two equal valves, forming a segment of a circle, or resembling a half-opened fan, which are closely united by a ligament along one of their sides. The animal, the CHIMÆRA, *Poli*, is elongated, like its shell; the lips, gills, and other parts are in the same proportion. Its mantle is closed along the side of the ligament; its foot resembles a little conical tongue excavated by a furrow; it is furnished with a small transverse muscle, situated at the acute angle formed by the valves, near which is the mouth, and with a very large one in their broader portion. By the side of the anus, which is behind this large muscle, is a conical appendage peculiar to the genus, susceptible of expansion and elongation, the use of which is unknown.

The byssus of several species of pinna is as fine and brilliant as silk, and is employed in fabricating the most precious stuffs; such is the

P. Nobilis, Lin., which is moreover recognized by the valves being roughened with recurved and semi-tabular plates.

It remains half buried in the sand, and anchored by its byssus. In the

ARCA, *Lin.*,

The valves are equal and transverse, that is to say, the hinge occupies the longest side. It is furnished with a large number of small teeth, which interlock with each other, and, as in the subsequent genera, with two bundles of transverse and nearly equal muscles inserted into the extremities of the valves, which seem to close them. In

ARCA, (properly so called) *Lam.*,

The hinge is rectilinear, and the shell most elongated in a direction parallel to it. The summits are generally convex, and curve over the hinge, but are separated from each other. The valves do not close perfectly in the centre, because there is a horny plate or tendinous band before the abdomen of the animal that serves for a foot, and by which it adheres to submerged bodies. They are found in rocky bottoms near the shore, and are usually covered with a hairy epidermis. They are not much esteemed for the table.

Some species are found in the Mediterranean, and a great many fossil, in strata anterior to the chalk, particularly in Italy.

Certain arcæ, in which the teeth of the two ends of the hinge assume a longitudinal direction, are distinguished by Lamarck under the name of CUCULLÆA, *(arca cucullata)*, Chemn.

We ought also, it is probable, to separate the species with well marked ribs, and completely closing and interlocking edges, for we may presume that their animal is not fixed, but rather resembles that of a pectunculus.

We have a still better warrant for removing the *arca tor-*

tuosa, Chemn., in its fantastic figure and unequally oblique valves.

PECTUNCULUS, *Lam.*,

The hinge forming a curved line, and the shell lenticular; its valves always close completely, and their summits are approximated. The animal AXIMEA, *Poli*, is furnished with a large compressed foot with a double inferior margin, which enables it to crawl. They live in ooze. Some species are found upon the coast of France.

NUCULA, *Lam.*

The nuculæ are arcæ in which the teeth are arranged in a broken line. Their form is elongated and narrowed near the posterior extremity. Their animal is unknown, but is probably not far removed from those of the preceding shells, (*arca pellucida*,) Chemn.

This has long been the place assigned to the

TRIGONIA, *Brug.*,

So remarkable for the hinge, which is furnished with two plates like a chevron, crenulated on both faces, each of which penetrates into two cavities, or rather between four plates of the opposite side, similarly crenulated on the internal surface. The internal impressions on the shell had already warranted the supposition that the animal was not provided with long tubes. Messrs. Quoy and Gaymard have lately discovered living specimens of this genus, and, in fact, its mantle, like that of the arcæ, is open, and without any separate orifice even for the anus. The foot is large, its anterior portion trenchant, and like a hook.

The living trigoniæ resemble the cardiæ in the form of their shell and the ribs which furrow it: its anterior is composed of mother-of-pearl. (*Trigonie nacrée*, Lam.)

The fossil trigonia are different. Their shell is flattened on one side, oblique, longest in a direction perpendicular to the hinge, and traversed in a contrary direction by a series of tubercles. (*Trig. Scabra,* Encyc. Meth.)

In the second family of the testaceous acephala, the

MYTILACEA,

The mantle is open before, but has a distinct aperture for the fæces.

All these bivalves have a foot, used in crawling, or at least, serving to draw out, direct, and place the byssus. They are commonly known under the generic name of mussels.

MYTILUS, *Lin.*

The true mytili, or sea-mussels, have a closed shell, with equal, convex, and triangular valves. One of the sides of the acute angle forms the hinge, and is furnished with a long narrow ligament. The head of the animal is in the acute angle; the other side of the shell, which is the largest, is the anterior one, and allows the passage of the byssus; it terminates in a rounded angle, and the third side ascends towards the hinge, to which it is joined by an obtuse angle. Near this latter is the anus, opposite to which the mantle forms an opening or small particular tube. The animal (CALLITRICHE, *Poli,*) has the edges of its mantle provided with branched tentacula, near the rounded angle, as it is there that the water enters required for respiration. Before, and near the acute angle, is a small transverse muscle, and a large one behind, near the obtuse angle. Its foot resembles a tongue.

In the true mytili the summit is close to the acute angle. Some of them are striated and others smooth.

Myt. Edulis, L. This common mussel is frequently seen suspended in extended clusters, along the whole coast of France, to rocks, piles, &c. It forms a considerable item of

1

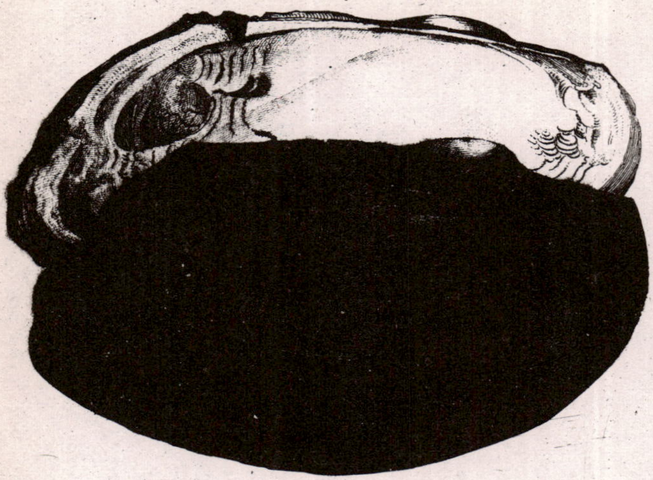

2

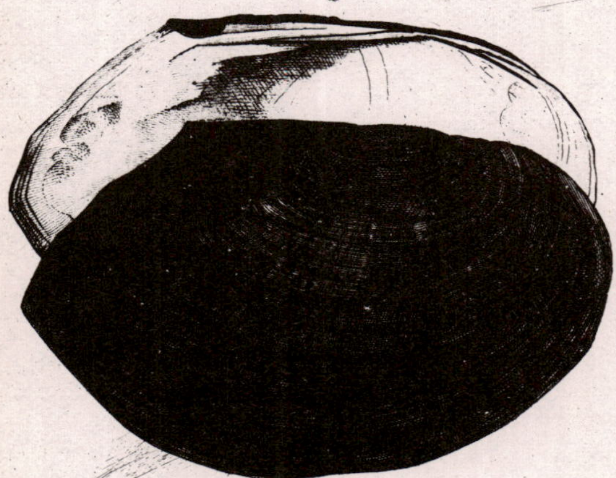

1 *Anodontea Susannæ.*

2 *Unio tenuis.*

food, but is dangerous if eaten to excess. Add *mytilus barbatus*, L., &c.

Some of them are found fossil, and have been formed into a sub-genus by M. Brongniart, under the name of MYTILOIDA. In the

MODIOLUS, *Lam.*,

Separated from the *mytili* by Lamarck, the summit is lower and nearer the third of the hinge. This summit is also more salient and rounded, approximating the modioli more closely to the ordinary form of the bivalves. We may also separate from the mytili,

LITHODOMUS, *Cuv.*,

In which the shell is oblong and almost equally rounded at the two ends, the summit being close to the anterior extremity. The species of this sub-genus at first simply attach themselves to stones, like the common mytili; subsequently, however, they perforate and excavate them in order to form cells, into which they enter, and which they never quit afterwards. Once entered, their byssus ceases to grow.

One of them, the *mytilus lithophagus*, Lin., is very common in the Mediterranean, where, from its peppery taste, it is esteemed as food.

A second (*modiola caudigera*, Encyc. pl. xxcci. f. 8.) has a very hard small appendage at the posterior extremity of each valve, which perhaps enables it to excavate its habitation.

ANODONTES, *Brug.*,

The anterior angle rounded like the posterior, and that next to the anus obtuse and almost rectilinear. The shell thin, and moderately convex, has no appearance whatever of a tooth at the hinge, being merely furnished with a ligament which extends along its whole length. The animal (LIMNÆA, *Poli*,) has no byssus; its foot, which is very large, compressed, and almost quadrangular, enables it to crawl upon the sand or ooze. The posterior extremity of its mantle is provided with numerous small tentacula. The anodontes inhabit fresh water.

Several species are found in France, one of which, (*mytilus cygneus*, Lin.) very large, is common in ponds, &c. with oozy bottoms. Its light and thin shells are used for milk-skimmers, but its flesh is not eaten in consequence of its insipidity. Add *M. Anatinus*, Chemn., &c.

An oblong species, in which the hinge is granulated throughout its entire length, is distinguished by M. de Lamarck under the name of IRIDINA (*Irid. exotica*, Encyc. Method.); the hind part of its mantle is somewhat closed.

And Dr. Leach distinguishes by the term DIPSADA another, which has its angles more decided, and in which there is a vestige of a tooth upon the hinge.

UNIO, *Brug.*,

These mollusca resemble the anodontes, both in their animal and shell, with the exception of their hinge, which is more complex. There is a short cavity in the anterior part of their right valve which receives a short plate or tooth from the left one, and on its posterior part is a long plate, which inserts itself between two others on the opposite side. They also inhabit fresh water, preferring running streams.

Sometimes the anterior tooth is more or less stout and unequal, as in

Mya margaritifera, L., a large thick species, the mother-of-pearl of which is so beautiful, that its concretions are used in dress as pearls. Found in France, as is the

Unio littoralis, Lam., a smaller and square species.

1

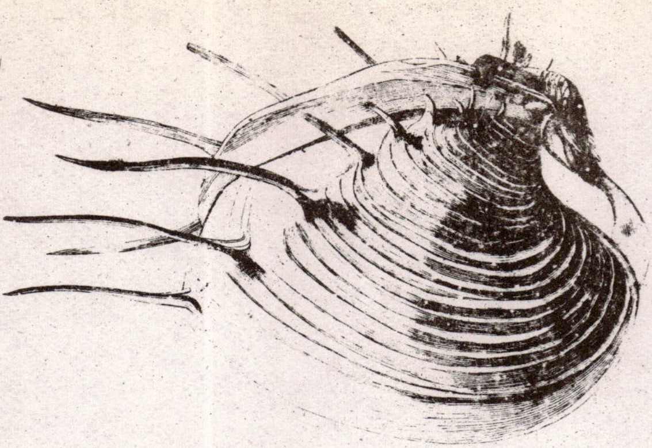

2

3

1 *Cytherea Dronea. var.*

2 *Tellina Guildfordiæ.*

3 *Anadon Georginæ.*

1 *Unio Childreni*. 2 *Cyrena similis*.
3 *Unio Smithii*.

At other times the anterior tooth is lamelliform, as in the *Mya pictorum*, L., an oblong and thin species known to every one.

Lamarck distinguishes the

HYRIA, *Lam.*,

In which the angles are so decided that the shell is nearly triangular. (*H. rugosa*, Encyc. Meth.)

CASTALIA, *Lam.*,

When the slightly heart-formed shell is striated in radii; the teeth and plates of the hinge are transversely sulcated, which gives them some affinity with the trigoniæ. (*Castalia ambigua*, Lam.)

There are certain marine mollusca which have a similar animal, and about the same kind of hinge, that should be placed near the unios; the summits of the shell, however, are more convex, and it is marked by projecting ribs, extending from the summits to the edge. They form the

CARDITA, *Brug.*,

Which are more or less oblong, or heart-shaped; the inferior margin in some gaping. (*Chama antiquata*, Chemn., &c.)

CYPRICARDIA, *Lam.*

Carditæ, in which the tooth under the summit is divided into two or three. Their form is oblong, and their sides unequal. (*Chama oblonga*, Gm., &c.)

M. de Blainville also separates the

CORALLIOPHAGA, *Blainv.*,

Where the shell is thin, and the lateral plate considerably effaced, which may cause their approximation to venus.

One of them is known, which excavates coral masses to form its habitation. (*Chama coralliophaga*, Gm., &c.)

VENERICARDIA, *Lam.*,

Only differ from the carditæ in the circumstance that the posterior plate of their hinge is shorter and more transverse, which caused their approximation to venus; their form is almost round. Judging from the impression of its muscles on them, their animal must resemble that of the carditæ and unios. (*Venus imbricata*, Chemn., &c.; and the fossil species, Lam.)

Both of them approach the cardia in their general form and the direction of their ribs. I suspect that this is also the place for the

CRASSATELLA, *Lam.* PAPHIA, *Roiss.*,

Which has sometimes been approximated to mactra, and at others to venus; the hinge has two slightly marked lateral teeth, and two very strong middle ones, behind which, extending to both sides, is a triangular cavity for an internal ligament. The valves become very thick by age, and the impression made by the margin of the mantle leads to the belief that, as in the preceding, there are no protractile tubes. (*Venus ponderosa*, Chemn. VII. lxix. A. D.)

The third family of the testaceous acephalæ, or

CHAMACEA,

Have the mantle closed, and perforated only by three holes, through one of which passes the foot; the second furnishes an entrance and exit to the water requisite for respiration, and the third for the excretion of fæces: these two latter are not prolonged into tubes, as in the subsequent family. It only comprises the genus

CHAMA, *Lin.*,

Where the hinge is very analogous to that of a unio; that is

1

2

3

1 *Unio Leaii.*

2 *Unio Douglasiæ.*

3 *Unio Grayii.*

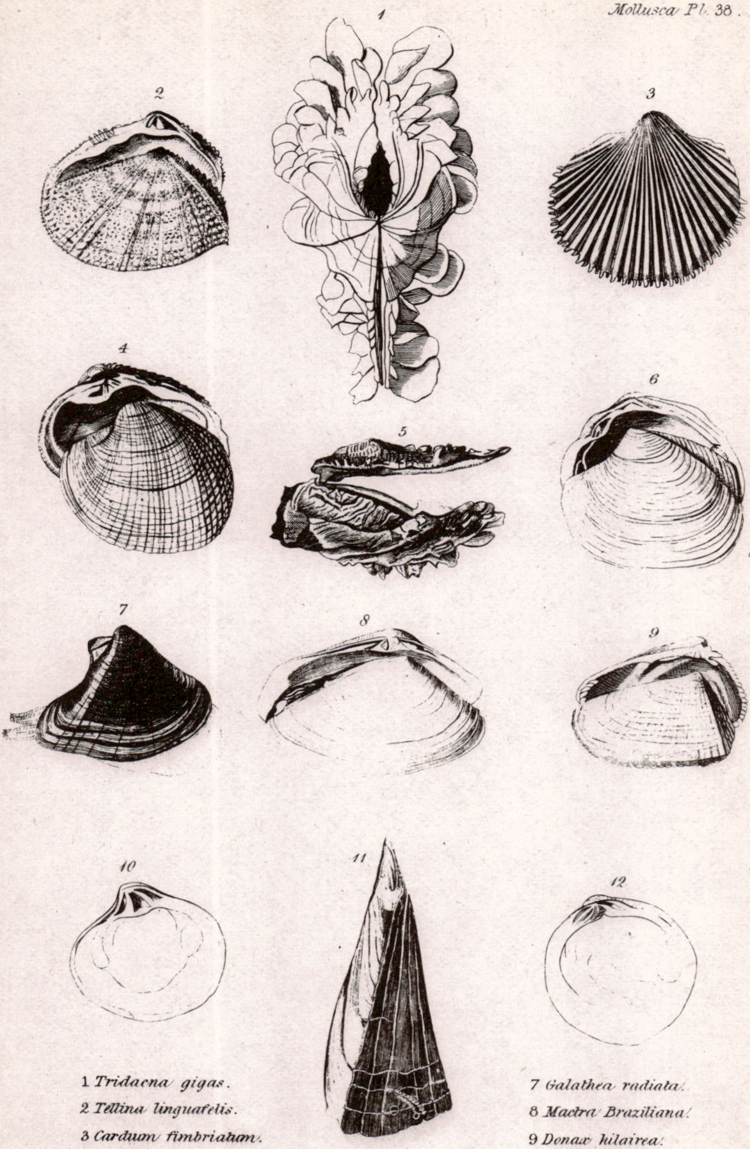

1 *Tridacna gigas.*
2 *Tellina linguafelis.*
3 *Cardium fimbriatum.*
4 *Venus.*
5 *Chama croceata.*
6 *Lucina Jamaicensis.*
7 *Galathea radiata.*
8 *Mactra Braziliana.*
9 *Donax hilairea.*
10 *Venus Dannonensis.*
11 *Pinna angustana.*
12 *Venus exoleta.*

to say, the left valve near the summit is provided with a tooth, and further back with a salient plate, which are received into corresponding fossæ of the right valve. This genus has necessarily been divided into the

TRIDACNA, *Brug.*

The shell greatly elongated transversely, and equivalve; the superior angle, which answers to the head and summit, very obtuse.

The animal is very singular, inasmuch as it is not, like many others of the kind, placed in the shell, but its parts are directed, or, as it were, pressed out towards the front. The anterior side of the mantle is widely opened for the passage of the byssus; a little belcw the anterior angle is another opening, which transmits water to the gills; and in the middle of the inferior side is a third and smaller one, which corresponds to the anus, so that the posterior angle transmits nothing, and is merely occupied by a cavity of the mantle, open only at the third orifice, of which we have just spoken.

There is but a single transverse muscle, corresponding to the middle of the margin of the valves. In

TRIDACNA, *Lam.*,

Or the tridacna properly so called, the front of the shell, as well as of the mantle, has a wide opening with notched edges for the transmission of the byssus, which latter is evidently tendinous, and is uninterruptedly continuous with the muscular fibres.

Such is the celebrated and enormous shell of India, the *Chama gigas*, L., Chemn. VII. xlix. which is decorated with broad ribs, relieved by projecting semi-circular scales. Specimens have been taken that weighed upwards of three hundred pounds. The tendinous byssus that attaches them to the rocks is so thick and stout that the axe is required to sever it. The flesh, though tough, is edible. In

Hippopus

The shell is closed and flattened before, as if truncated. (*Chama Lazarus*, Chemn.) In the

Chama, *Brug.*,

Or the true chamæ, the shell is irregular, inequivalve, usually lamellar and rough, adhering to rocks, corals, &c., like that of an oyster. Its summits are frequently very salient, unequal, and curled up; the internal cavity frequently has the same form, without any external indication of the fact. The animal (Psilopus, *Poli*) has a small foot, bent almost like that of a man; its tubes, if it have any, are short and disjointed, and the aperture in the mantle which transmits the foot is not much larger. Some species are found in the Mediterranean.

There are also several that are fossil.

Diceras, *Lam.*

Between diceras and the chamæ there is no essential difference, only the cardinal tooth of the former is very thick, and the spiral lines of the valves are sufficiently prominent to remind us of two horns. Fossil shells from the Jurassic strata. (*Dic. arietina*, Lam.) In the

Isocardia, *Lam.*,

The shell free, regular, and convex, with spirally-curled summits, divided anteriorly. The animal (Glossus, *Poli*) only differs from that of an ordinary chama in having a larger and more oval foot, and because the anterior opening of its mantle begins to resume its ordinary proportions.

A large smooth red species, (the *Chama cor*. L., Chemn. VII. xlviii. 483), inhabits the Mediterranean.

In the fourth family of the testaceous acephala, the

CARDIACEA,

The mantle is open before, and there are besides two separate apertures, one for respiration, and the other for the fæces, which are prolonged in tubes, sometimes distinct, and at others united in one single mass. There is always a transverse muscle at each extremity, and a foot generally used for crawling. It may be considered as a general rule, that those which are furnished with long tubes live in ooze or in sand. This mode of organization may be recognized on the shell, by the more or less depressed contour described by the insertion of the edges of the mantle, previously to its uniting with the impression of the posterior transverse muscle.

CARDIUM, *Lin.*

The cardia, like many other bivalves, have an equivalve convex shell, with salient summits curved towards the hinge, which, when viewing it sideways, give it the figure of a heart: hence its name of cardium, heart, &c. Ribs, more or less elevated, are regularly distributed from the summits to the edges of the valves; but what chiefly distinguishes the cardia is the hinge, where we see, on one side and the other, towards the centre, two small teeth, and at some distance, anteriorly and posteriorly, another tooth, or projecting lamina. The animal (CERASTES, *Poli*) has generally an ample aperture in the mantle, a very large foot bent in the middle, and with its point directed forwards; and two short or but moderately long tubes.

Numerous species of cardia are found upon the coast of France, some of which are eaten, such as the *C. edule*, L.,

Chemn. VI. xix. 194. Fawn-coloured or whitish, with twenty-six transversely-folded ribs.

Under the name of HEMICARDIUM we might separate those species in which the valves are compressed from before, backwards, and strongly carinated in the middle; for it seems almost certain that a modification of the animal must be a necessary consequence of this singular configuration. (*Cardium cardissa*, VI. xiv. 143—146, &c.)

DONAX, *Lin.*

The donaces have very nearly the same kind of hinge as the cardia, but their shell is of a very different form, being a triangle, of which the obtuse angle is at the summit of the valves, and the base at their edge, and of which the shortest side is that of the ligament, or posterior side, a rare circumstance in this grade among bivalves. They are generally small, and prettily striated from the summit to the edges. Their animal (PERONÆA, *Poli*) is furnished with long tubes, which are received into a sinus of the mantle. Some of them are found upon the coast of France. (*Donax rugosa*, Chemn. VI. &c.)

CYCLAS, *Brug.*

Separated from venus by Bruguières. Like the cardia and donaces, has two teeth in the middle of the hinge, and before and behind two salient and sometimes crenulated plates; but the shell, as in several species of venus, is more or less rounded, equilateral, and transversely striated. The animal has moderate tubes; the external tint is usually grey or greenish. The cyclades inhabit fresh water.

One species, the *Tellina cornea*, L., Chemn. VI. xiii. 133, is very common on the coast of France. M. de Lamarck separates the

CYRENA, *Lam.*,

Where the shell is thick, slightly triangular and oblique,

covered with an epidermis, and otherwise distinguished from the cyclades by having three cardinal teeth. The cyrenæ also inhabit rivers, but there are none in France. (*Tell. fluminea*, Chemn.)

CYPRINA, *Lam.*,

Also separated from the cyclades by Lamarck. The shell is thick, oval, with recurved summits, and three stout teeth; further back is a plate, and under the teeth a large cavity, which receives a part of the ligament. (*Venus islandica*, Chemn., &c.)

GALATHÆA, *Brug.*

The shell triangular; three teeth on the summit of one valve, and two on the other, like a chevron. The lateral plates approximated. (*Egeria*, Roiss.; or *Galathœa*, Brug., &c.)

But one is known, found in the fresh-water of India.

It is here also must be placed another genus, separated from venus—the

CORBIS, *Cuv.* FIMBRIA, *Megerl.*

Marine testaceous acephala, transversely oblong, which have also the stout middle teeth and well-marked lateral plates; their external surface is furnished with transverse ribs so regularly crossed by rays that it may be compared to wicker-work.

The impression of their mantle exhibiting no flexure, their tubes must be short. (*Venus Fimbriata*, Chemn.)

Some of them are fossil. In the

TELLINA, *Lin.*,

There are in the middle one tooth on the left and two teeth on the right, frequently forked, and at the same distance before and behind, on the right valve, a plate, which does not penetrate into a cavity of the opposite one. There is a slight fold near the posterior extremity of the two valves, which renders them unequal in that part, where they are somewhat open.

The animal of the Tellina (PERONÆA, *Poli*), like that of the donaces, has two long tubes for respiration and for the anus, which withdraw into the shell, and are concealed in a duplicature of the mantle.

Their shells are generally transversely striated and decorated with beautiful colours.

Some of them are oval and thick.

Others are oblong and strongly compressed.

Some, again, are lenticular, where, instead of a fold, there is frequently nothing but a slight deviation of the transverse striæ.

We might separate certain oblong species which have no lateral teeth, and others which, with the hinge of the tellinæ, have not the fold of the posterior extremity. They are the TELLINIDES, *Lam.*

It is necessary to distinguish from the tellinæ the

LORIPES, *Poli.*,

Which have the shell lenticular, the central teeth almost effaced, and a simple furrow for the ligament behind the nates. The animal is furnished with a short double tube, and its foot is prolonged into a kind of cylindrical cord. Besides the usual impressions, we may observe on the inside of the shell a line running obliquely from the print of the anterior muscle, which is very long towards the nates. There is no flexure in the print of the mantle for the retractor muscle of the tube. (*Tellina lactea.*)

LUCINA, *Brug.*

Separated lateral teeth, as in the cardia, cyclades, &c.,

penetrating between the plates of the other valve; in the middle are two teeth frequently but slightly apparent. The shell is orbicular, and without any impression of the retractor muscle of the tube; that of the anterior constrictor, however, is very long. Possessing similar traits of character with the loripedes, their animals must be analogous. (*Venus Pennsylvanica*, Chemn. VII. xxxvii. 394—396, &c.)

The living species are much less numerous than those that are fossil; the latter are very common in the environs of Paris. (*Lucina saxorum*, Lam., &c.)

We should approximate to the lucinæ the UNGULINÆ, which also have an orbicular shell and two cardinal teeth; the lateral ones, however, are wanting, and the anterior muscular impression is not so long. (*Ungulina transversa*, Kam., Sowerb., Gen. of Shells, No. X.) The genus

VENUS, *Lin.*,

Comprizes many testacea whose general character consists in the teeth and plates of the hinge being approximated under the summit in a single group; they are usually more flattened and elongated in a direction parallel to the hinge than the cardices. The ribs, when there are any, are almost always parallel to the edges, being directly the reverse of their arrangement in the cardia.

The ligament frequently leaves an elliptical impression behind the summits, which has received the appellation of vulva; and before these same summits there is almost always an oval impression termed the *anus* or lunula.

The animal is always furnished with two more or less protractile tubes, sometimes united, and with a compressed foot, which enables it to crawl.

M. Lamarck appropriates the name of VENUS to those which have three small diverging teeth under the summit.

This character is particularly well marked in the oblong and slightly convex species. (*Venus litterata*, Chemn.)

Some of them (the ASTARTÆ, *Sowerb.*, or CRASSINÆ, *Lam.*) have only two diverging teeth on the hinge, and approach the crassatellæ in their thickness and some other characters. (*Venus Scotica*, &c.)

Among the heart-shaped species, that is, those which are shorter, and have more convex nates, with more closely approximated teeth, we should remark those where the plates or transverse striæ terminate behind in crests or tuberosities, and those that have longitudinal ribs and crests elevated behind. (*Venus dysera*, Chemn.; *Venus puerpera*.)

We subsequently and gradually come to the CYTHEREÆ, *Lam.*, which have a fourth tooth on the right valve, projecting under the *lunula*, and received into a corresponding cavity in the right one.

Some of them have an elliptical and elongated form: *Venus gigantea*, &c.

Others are convex: *Ven. meretrix*, &c.

Among these we must place a celebrated species (*Venus Dione*, L.), from whose form originated the application of the name *Venus* to the genus. Its transverse plates terminate behind in salient and pointed spines.

There are some species of an orbicular form, and with slightly hooked summits, in which the impression of the retractor of the tubes forms a large and almost rectilinear triangle. (*Venus exoleta.*)

When their animals are better known, we shall most probably have to separate from cytherea—

1. Those species of a compressed lenticular form, in which the nates are united into a single point; the fold of the contour of the mantle is wanting, and shows that their tubes are not protractile: (*Venus scripta.*)

2. Those of a convexly orbicular form, in which the fold is not only wanting, but where, as in the lucinæ, the impression of the anterior muscle is very long. (*Venus tigerina*, Chem.)

3. The thick species with radiated ribs, in which the fold is also wanting, and which connect the genus Venus with that of the Venericardia. (*Venus pectinata.*) In the

Capsa, *Brug.*,

Already separated from the former, there are two teeth on the hinge at one side, one only, but bifid, at the other; the lunula is wanting, the shell convex, and the fold indicative of the retractor of the foot, considerable. (*Venus deflorata.*)

Petricola, *Lam.*,

Also separated from the same genus: the petricolæ on each side have two or three very distinct teeth on the hinge, one of which is forked; the shell is more or less heart-shaped, but as they inhabit the interior of stones, it sometimes becomes very irregular. Judging from the marginal impressions of the mantle, their tubes must be very large. (*Venus lapicida*, Chemn., &c.)

Corbula, *Brug.*

Similar in form to the triangular cythereæ or cordate: but a single stout tooth in the middle of each valve, corresponding to the side of its antagonist; the ligament is internal, the tubes must be short, and the valves but rarely equal.

The fossil species are much more numerous than the living ones. (*Corbula gallica*, &c.)

Some of them live in the interior of stones. (*Venus monstrosa*, Chemn.)

Mactra, *Lin.*

The mactræ are distinguished from the other testacea of

this family by their ligament being internal and lodged throughout in a triangular depression, as in the oysters; they all have a compressed foot fit for crawling. In the

MACTRA, *Lam.*,

Or the Mactræ properly so called, the ligament is accompanied on the left valve, before and behind, by a projecting plate, which is received by two others on the right one; close to the ligament, near the lunule, is, on each side, a little plate chevron-formed. The tubes are united and short.

After abstracting the *lavignons* and *lutrariæ*, the genus *mactra* of Gmelin may remain; the species, however, are far from being well distinguished. Add *Mya Australis*, Chemn.

Some of these mactræ are found on the coast of France.

In the LAVIGNONS the lateral plates are almost effaced; but a single small tooth is observable near the internal ligament; there is also a second and internal ligament. The posterior side of the shell is the shortest; the valves are somewhat open, and the tubes are separate and very long, as in the tellinæ.

There is one found on our coast (*Mya hispanica*, Chemn. VI. iii. 21), which lives in the ooze, at the depth of several inches.

The fifth family of testaceous acephala, or that of

INCLUSA,

Have the mantle open at the anterior extremity, or near the middle only, for the passage of the foot, and extended from the other end into a double tube, which projects from the shell, whose extremities are always gaping. Nearly all of them live buried in sand, stones, ooze, or wood. Those of the genus

MYA, *Lin.*,

Have but two valves to their oblong shell, the hinge of which varies; the double tube forms a fleshy cylinder, and the foot is compressed. The different forms of the hinge have furnished Messrs. Daudin, Lamarck, &c. with the following subdivisions, in the first three of which the ligament is internal.

LUTRARIA, *Lam.*

The lutrariæ, like the mactræ, have a ligament inserted into a large triangular cavity of each valve, and before this cavity a small chevron-formed tooth, but the lateral plates are wanting; the valves gape widely, particularly at the posterior extremity, through which passes the thick, double, fleshy, respiratory and anal tube, a disposition which attaches them to this family; the foot, which issues at the opposite end, is small and compressed.

Some of them are found in sand, at the mouths of rivers in France. (*Mactra lutraria*, List., &c.) In

MYA, *Lam.*,

Or Mya proper, one valve is furnished with a plate which projects into the other, furnished with a cavity; the ligament stretches from this cavity to that plate.

Some species are found in the sand along the coast of France. (*Mya truncata*, Lin.)

ANATINA, *Lam.*

The anatinæ of Lamarck should be approximated to the preceding myæ: each of their valves has a small projecting plate inside, with the ligament extending from one to the other.

One oblong and excessively thin species is known, the valves of which are supported by an internal ridge, *Solen anatinus*, Chemn.; and another of a squarer form, without the ridge, *Corbule*, Encyc. 230—6.

Solemya, *Lam.*

The ligament is seen on the outside of the shell, part of it remaining attached to a horizontal internal spoon-like impression on each valve. There is no other cardinal tooth, and a thick epidermis projects beyond the edges of the shell.

One species (the *Tellina togata*, Poli II. xv. 20), is found in the Mediterranean.

Glycymeris, *Lam.* Cyrtodaria, *Daud.*

Neither teeth, plates, nor cavities on the hinge, but a simple callous enlargement, behind which is an external ligament. The animal resembles that of the myæ.

The most common species (*Mya Siliqua*, L., Chemn. XI. 193. fi. 194.) is from the Arctic Ocean.

Panopea, *Mesnard, Lagr.*

A stout tooth anterior to the callous enlargement of the preceding sub-genus, and immediately under the summit, which crosses with a similar one of the opposite valve, a character that approximates the panopeæ to the solines. A large species is found in the hills at the foot of the Appenines in so high a state of preservation that it has been mistaken for a recent sea-shell. (*Mya glycimeris*, L.)

Another fossil species may perhaps be separated from it, which is completely closed at its anterior extremity. (*Panope de Faujas*, Mesnard, Lagr., Ann. du Mus. IX. xii.)

After all these various modifications of the myæ, we may place the

Pandora, *Brug.*,

In which one valve is much flatter than the other; the internal ligament is placed transversely, accompanied in front by a projecting tooth of the flattened valve. The posterior

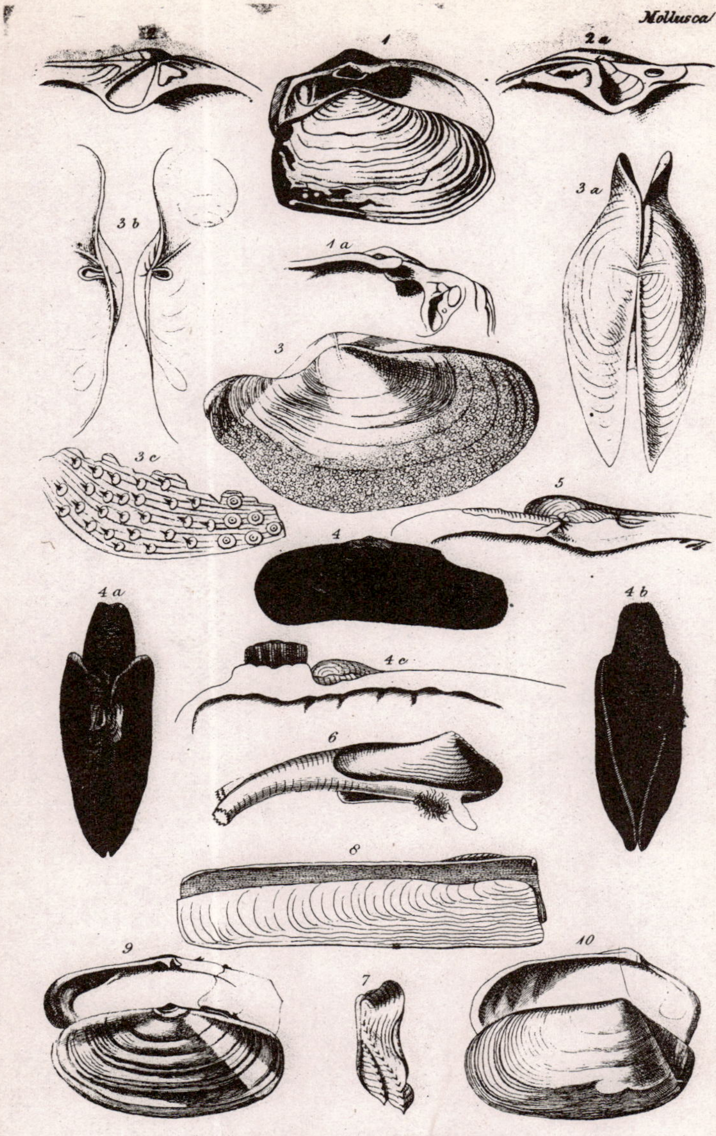

1 *Mya truncata.* 6 *Byssomia pholadis*

2 *Lutraria elliptica.* 7 *Hiatella arctica*

3 *Anatina hispidula.* 8 *Solen vagina.*

4 *Glycimeris siliqua.* 9 *Sanguinolaria livida.*

5 *Panopœa Aldrovandi.* 10 *Psammothea candida.*

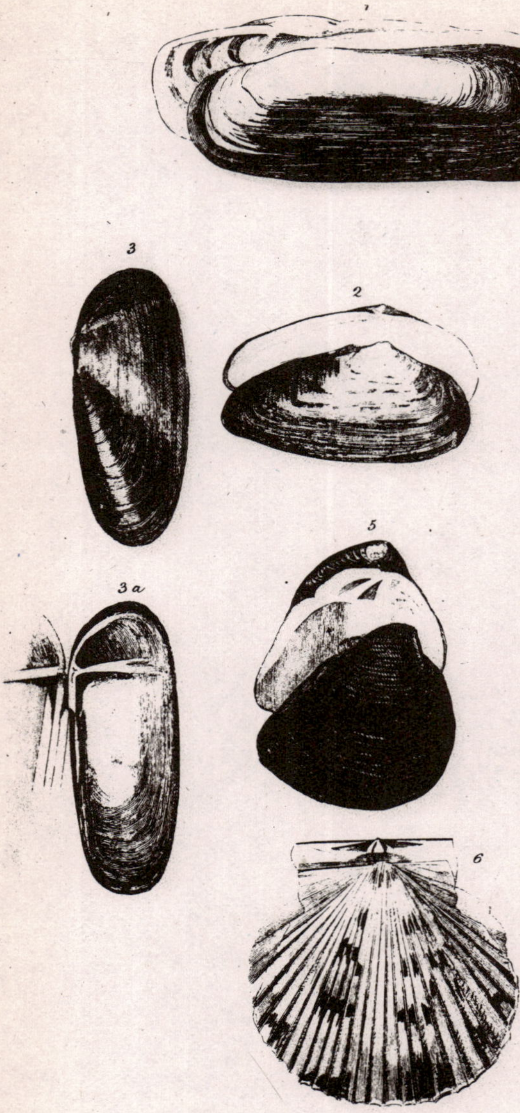

1 *Solen novaculina.* 4 *Solen tenuis.*

2 *Glycymeris apinensis.* 5 *Villorita cyprinoides.*

3 *Solen Sayii.* 6 *Pecten purpureus.*

side of the shell is elongated. The animal withdraws more completely into its shell than the preceding one, and its valves shut more closely; its habits, however, are the same.

But a single species is well known; it inhabits the seas of Europe, (*Tellina inæquivalvis.*)

Here also we find a group of some small and singular genera, such as

BYSSOMIA, *Cuv.*,

Where the oblong shell, which has no marked tooth, has the opening for the foot at about the middle of its edge, and opposite the summits. The byssomia also penetrate into stone, corals, &c.

A species which is provided with a byssus abounds in the Arctic Ocean. (*Mytilus pholadis*, Müll. Zool. Dan.)

HIATELLA, *Daud.*,

The shell gaping, to allow the passage of the foot, near the middle of its edges; but the tooth of the hinge is better marked than in the preceding genus. Ranges of salient spines are frequently observed on the hind part of the shell. They are found in sand, zoophytes, &c.

The North Sea produces a small species, (*Solen minutus*, Lin.)

SOLEN, *Lin.*,

The shell only bivalve, oblong or elongated, but the hinge always furnished with salient and well marked teeth, and the ligament external. In the

SOLEN, *Cuv.* (proper),

The shell is cylindrically elongated, and has two or three teeth in each valve near the anterior extremity, where the foot issues. The latter is conical, and enables the animal to bury itself in the sand, which it excavates with considerable rapidity on the approach of danger.

Several species are found along the coast of France.

We might distinguish those species in which the teeth approximate in the middle: some of them have still a long and narrow shell. (*Solen legumen*, Chemn. VI. v. 32- 34.)

In others it is wider and shorter; their foot is extremely thick. Two of the latter inhabit the Mediterranean. (*Solen strigilatus*, &c.) In

SANGUINOLARIA, *Lam.*,

The hinge is nearly the same as in the wide solens, and has two teeth in the middle of each valve, but the valves, which are oval, are much closer at the two extremities, where they merely gape, as in the mactræ. (*Solen Sanguinolentus*, Chemn.)

PSAMMOBIA, *Lam.*,

The psammobiæ differ from the sanguinolariæ in having but a single tooth in the middle of one valve, which penetrates between two on the opposite one, (*Tellina gari*, &c.)

PSAMMOTHEA, *Lam.*,

But a single tooth to each valve; otherwise resembling the psammobiæ. (*Psammothea violacea*, Lam., &c.)

PHOLAS, *Lin.*,

The pholades have two principal valves, broad and convex towards the mouth, narrow and elongated on the opposite side, and having a large oblique opening at each extremity; their hinge, like that of a true mya, is furnished with a plate projecting from one valve into another, and with an internal ligament running from that plate into a corresponding cavity. Their mantle is reflected externally upon the hinge, where it sometimes contains two or three supernumerary calcareous bodies. The foot issues through the aperture on the side next to the mouth, where it is widest, and from the opposite end project the two tubes, which are united, and capable of enlarging themselves in every direction.

The pholades inhabit canals which they excavate, some in ooze, and others in stone, like the lithodomi, petricolæ, &c. They are much sought for on account of their agreeable flavour.

Several species are found on the coast of France: such is the *Dail commun.* (*Pholas dactylus*, L., &c.)

TEREDO, *Lin.*,

The mantle extended in a tube much longer than the two small rhomboïdal valves, and terminated by two short tubes, the base of which is furnished on each side with a strong and moveable kind of operculum or palette. These acephala, while quite young, penetrate and establish their habitations in submerged pieces of wood, such as piles, ships' bottoms, &c., perforating and destroying them in every direction. It is thought, that in order to penetrate as fast as it increases in size, the teredo excavates the wood by means of its valves; but the tubes remain near the opening by which its entrance was effected, and through which, by the aid of its palette, it receives water and aliment. The gallery it inhabits is lined with a calcareous crust, which it exudes, and which forms a second kind of tubular shell. It is a noxious and destructive animal in the sea-ports of Europe.

Teredo navalis, L. This species, which is the most common, and is said to have been introduced into Europe from the torrid zone, has more than once threatened Holland with ruin by the destruction of its dikes. It is upwards of six inches in length, and has simple palettes.

Larger species inhabit hot countries, whose palettes are

articulated and ciliate. They should be remarked on account of the analogy which they establish with the cirropoda. Such is the *Teredo palmulatus, Lam.*

FISTULANA, *Brug.*

Separated from teredo: the external tube is entirely closed at its larger end, and is more or less like a bottle or club. The fistulanæ are sometimes found in submerged fragments of wood, or in fruits that had sunk in the water, sometimes simply enveloped in sand. The animal, like that of a teredo, has two small valves and as many palettes. Recent specimens are only obtained from the Indian Ocean, but they are found fossil in Europe. (*Teredo clava,* Gm., &c.) We should approximate to them the

GASTROCHÆNA, *Spengler,*

In which the shells are deprived of teeth, and their edges being wide apart, anteriorly, leave a large oblique opening, opposite to which there is a small hole in the mantle for the passage of the foot. The double tube, which can be retracted completely within the shell, is susceptible of being greatly elongated. It appears that they are certainly furnished with a calcareous tube.

In some of them, as in the mytili, the summits are at the anterior angle, (*Pholas hians*); in others they are placed near the middle.

They inhabit the interior of madrepores, which they perforate. Two genera of acephala, furnished with tubes like the teredines, have been detected among fossils, but the first of them, the

TEREDINA, *Lam.,*

Has a little spoon-shaped impression on the inside of each of its valves, and a small, free, shield-shaped piece on the hinge. (*Teredina personata,* Lam.) In the second,

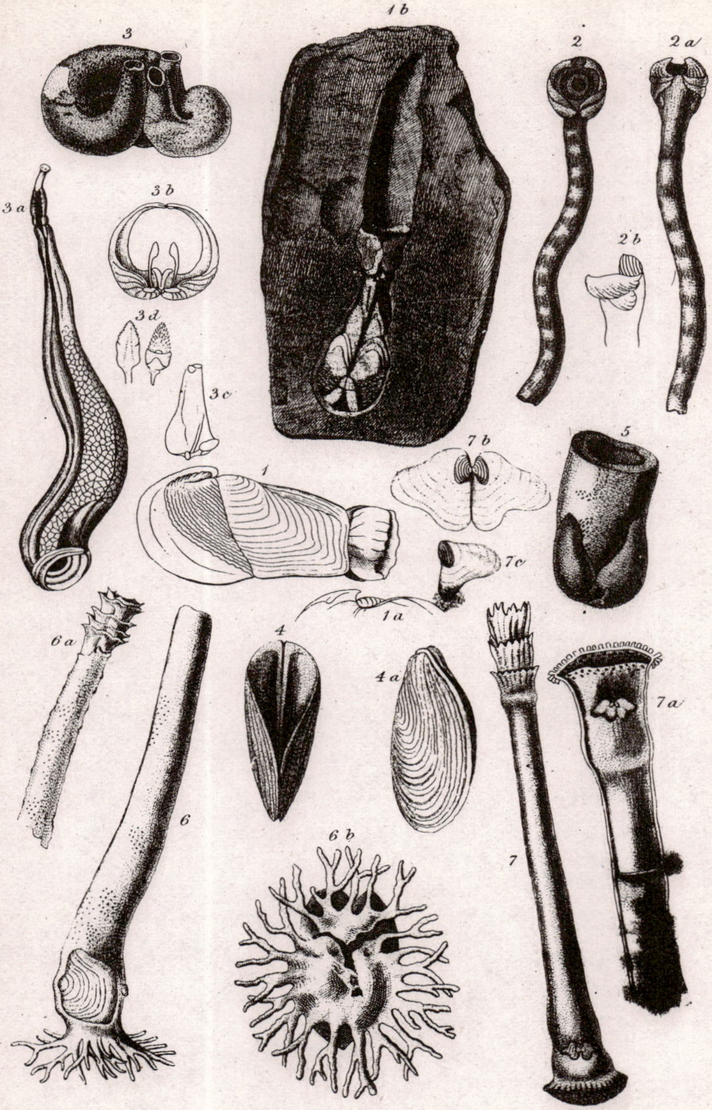

1 *Pholas striata.* 4 *Gastrochæna cuneiformis.*

2 *Teredo navalis.* 5 *Teredina personata.*

3 *Fistulana gregata* 6 *Clavagella coronata.*

7. *Aspergillum vaginiferum.*

Clavagella, *Lam.*,

One of the valves is clasped by the tube, leaving the other however free. (*Cl. echinata*, Lam.)

A single living species has been found in the madrepores of the Sicilian seas, which has been described by M. Audouin.

Some naturalists think we should also place in this family the

Aspergillium, *Lam.*,

The shell of which is formed of an elongated conical tube, closed at its widest extremity by a disk perforated with numerous small tubular holes; the little tubes of the outer range, being longest, form a kind of corolla about it. The reason for approximating them to the acephala with tubes is found in the fact that there is a double projection on one part of the cone which really resembles the two valves of the acephala. The affinity between these little tubes and those which envelope the tentacula of certain terebellæ, formerly caused this animal to be referred to the annelides.

The species most known (*Asper. javanens.* Mart. Conch. I. pl. i. fi. 7.) is seven or eight inches in length.

THE SECOND ORDER OF ACEPHALA.

Acephala Nuda

Are very few in number, and are sufficiently remote from the ordinary acephala to constitute a distinct class, should such a separation be deemed expedient. Their gills assume various forms, but are never divided into four leaflets; the shell is replaced by a cartilaginous substance, sometimes so thin that it is as flexible as a membrane.

We form two families of them; the first comprehends the genera whose individuals are isolated, and without any organic connexion one with the other, although they frequently live in society.

BIPHORA, *Brug.* THALIA, *Brown.* SALPA, and DAGYSA, *Gm.*,

Have the mantle and its cartilaginous envelope oval or cylindrical, and open at both ends. On the side of the anus the aperture is transverse, wide, and furnished with a valvule, which permits only the entrance of water, and not its exit; on the side of the mouth it is simply tubular. Some muscular bands embrace the mantle and contract the body. The animal moves by causing the water to enter through the posterior aperture, which has a valvule, and by making it issue through that on the side of the mouth, so that it is always pushed backwards, which has caused the posterior aperture to be taken by some naturalists for the true mouth. It also generally swims upon its back. Its gills form a single tube or riband, provided with regular vessels, placed obliquely in the middle of the tubular cavity of the mantle, in such a manner that it is constantly bathed by the water as it traverses that cavity. The heart, viscera, and liver are clustered up near the mouth, and towards the back, but the position of the ovary varies. The mantle and its envelope, when exposed to the sun, exhibit the colours of the rainbow, and are so diaphanous that the whole structure of the animal can be seen through them; in many they are furnished with perforated tubercles. The animal has been seen to come out from its envelope without appearing to suffer pain. The most curious circumstance respecting them is their remaining united for a

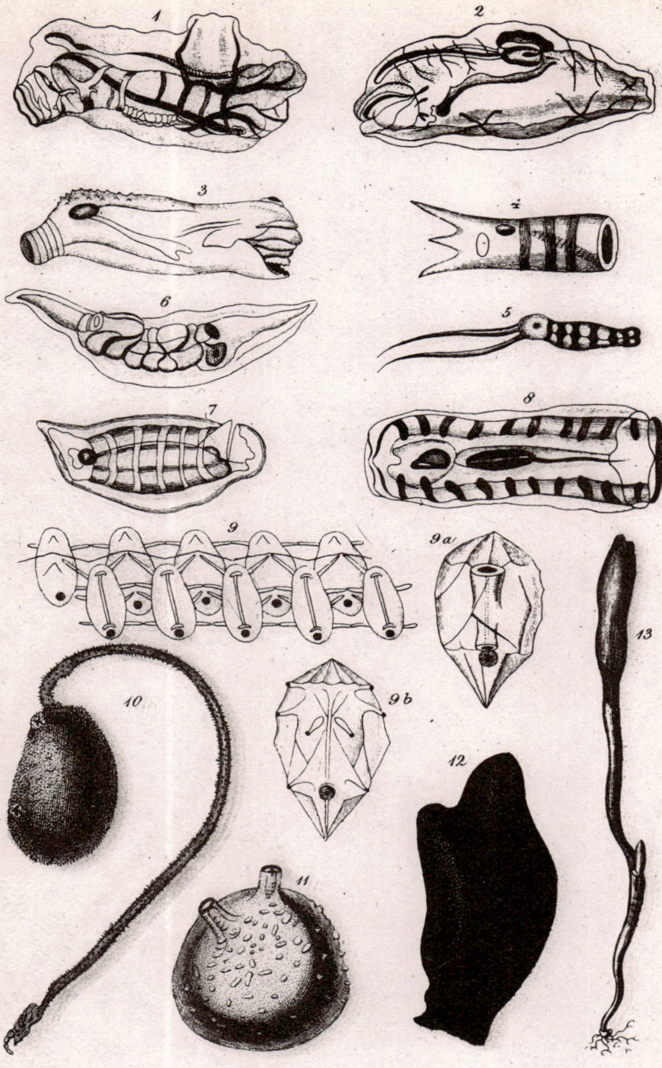

1 *Thalia cristata.*
2 *Salpa scutigera.*
3 *Sal. infundibuliformis.*
4 *Sal. tricuspis.*
5 *Sal. longicauda.*
6 *Sal. fusiformis.*
7 *Sal. zonaria.*
8 *Sal. Cylindrica.*
9 *Sal. Pyramidalis.*
10 *Boltenia ovifera.*
11 *Cynthia momus.*
12 *Phallusia nigra.*
13 *Clavellina borealis.*

long time, just as they were in the ovary, and thus swimming in long chains where the individuals are disposed in different ways, but each species always according to the same order.

M. de Chamisso assures us that he has verified a still more singular fact relative to these animals; it is, that the individuals which have thus issued from a multiplex ovary are not furnished with a similar one, but produce isolated young ones of various forms, which have an ovary like that which produced their parent, so that there is alternately a generation of a few isolated individuals, and another of numerous and aggregate ones, and that these two alternating generations do not resemble each other.

It is very certain that in some species little individuals have been observed adhering to the interior of large ones by a peculiar kind of sucker, which were different in form from those that contained them.

These animals are very abundant in the Mediterranean and the warmer portions of the ocean, and are frequently phosphorescent.

The THALIÆ, *Brown*, have a small crest or vertical fin near the posterior extremity of the back. (*Holothuria thalia*, Gm., &c.)

Of the SALPÆ, properly so called, some have a gelatinous dark-coloured plate in the substance of the mantle, and above the visceral mass, which may be the vestige of a shell. (*Salpa scutigera*, Cuv., &c.)

In others it is a simple prominence of the same nature as the rest of the mantle, but thicker. (*Salpa Tilesii*, Cuv., &c.)

Others, again, have neither plate nor prominence, but their mantle is extended by points, and of them

Some have a point at each extremity. (*Salpa maxima*, Forsk., &c.) Others have two at the extremity nearest the mouth. (*Salpa democratica*, Forsk., &c.); and even three, or more, as *Salpa tricuspis Id.*

Some have but a single one at this same extremity. (*Holothuria Zonaria*, Gm., &c.)

The greater number is simply oval or cylindrical. (*Salpa octofera*, Cuv., &c.) In the

ASCIDIA, *Lin.*, THEYTON of the ancients,

The mantle and its cartilaginous envelope, which is frequently very thick, resemble sacs every where closed, except at two orifices which correspond to the two tubes of several bivalves, one serving to admit water, and the other to give a passage to the fæces. The gills form a large sac, at the bottom of which are the mouth and the visceral mass. The envelope is much larger than the mantle, which is fibrous and vascular, and on which, between the two tubes, is one of the ganglions. These animals attach themselves to rocks and other bodies, and are deprived of all power of locomotion. The chief sign of vitality which they exhibit consists in the absorption and evacuation of water, through one of their orifices: when alarmed they eject it to a considerable distance. They abound in every sea, and some of them are eaten. The whole genus ASCIDIA, *Gm.* with some additional species.

Some species are remarkable for the long pedicle which supports them. (*Ascidia pedunculata*, Edw., &c.)

The second family of naked acephala,

AGGREGATA,

Consists of animals more or less analogous to the ascidiæ, but united in a common mass, so that they seem to communicate organically with each other, and in this respect to connect the mollusca with the zoophytes; but, independent of their peculiar organization, these animals, according to the observations of Messrs. Audouin and Milne Edwards, at first live and swim separately, only becoming united at a certain subsequent period, a fact which is in direct opposition to this opinion.

Their gills, as in the ascidiæ, form a large sac, traversed by the aliment before it arrives at the mouth; their principal ganglion is also situated between the mouth and the anus; a nearly similar disposition obtains with respect to the viscera and ovary.

Notwithstanding this, some of them, like the biphora, have an opening at each extremity; such is the

BOTRYLLUS, *Gærtn.*,

Of an oval form, fixed on various bodies, and united by tens or twelves like the rays of a star. The branchial orifices are at the external extremities of these rays, and the anus terminates in a common cavity, which is in the centre of the star. If an orifice be irritated, but a single animal contracts; if the centre be touched, they all contract. These very small animals attach themselves to ascidiæ, fuci, &c. (*Botryllus Stellatus*, Gærtn., &c.)

In some particular species three or four stars appeared to be piled one upon the other. (*Botryllus conglomeratus*, Gærtn.)

PYROSOMA, *Péron.*,

The pyrosomæ unite in great numbers, forming a large hollow cylinder, open at one end and closed at the other, which swims in the ocean by the combined contraction and dilatation of all the animals which compose it. The latter terminate in a point on the exterior, so that the whole external surface of the tube is bristled with them; the branchial orifices are pierced near these points, and the anus opens in the internal cavity of the cylinder. A pyrosoma may be thus compared to a great number of stars of botrylli strung together, the whole of which is moveable.

The Mediterranean and the ocean produce large species, the animals of which are arranged with but little regularity. They exhibit a phosphorescent appearance during the night. Several of the polyclina and aplidia of Savigny.

A smaller species is also known, where the animals are arranged in very regular rings. (*Pyrosoma Atlanticum*, Peron, &c.)

The remainder of these aggregated mollusca, like the ordinary ascidiæ, have the anus and branchial orifice approximated to the same extremity. The species known are all fixed, and till now they have been confounded with the alcyonia. The visceral bundle of each individual is more or less extended into the common cartilaginous or gelatinous mass, more or less narrowed or dilated in certain points; but each orifice always forms a little six-rayed star on the surface. We unite them all under the name of

POLYCLINUM.

Some of them are extended over bodies like fleshy crests.

Others project in a conical or globular mass. The *eucœlium*, Savig., the *distomi*, are arranged in the same manner.

Or expand into a disk, comparable to that of a flower or of an actinia. (The genus *diazona*, Sav.)

Or are elongated into cylindrical branches, supported by slender pedicles, &c. (The genus *sigillina*, Sav.)

Or form parallel cylinders. (The genus *synoicum*, Lam.)

Recent observations even seem to show that the ESCHARÆ, hitherto placed among the POLYPI, belong to this family of the mollusca.

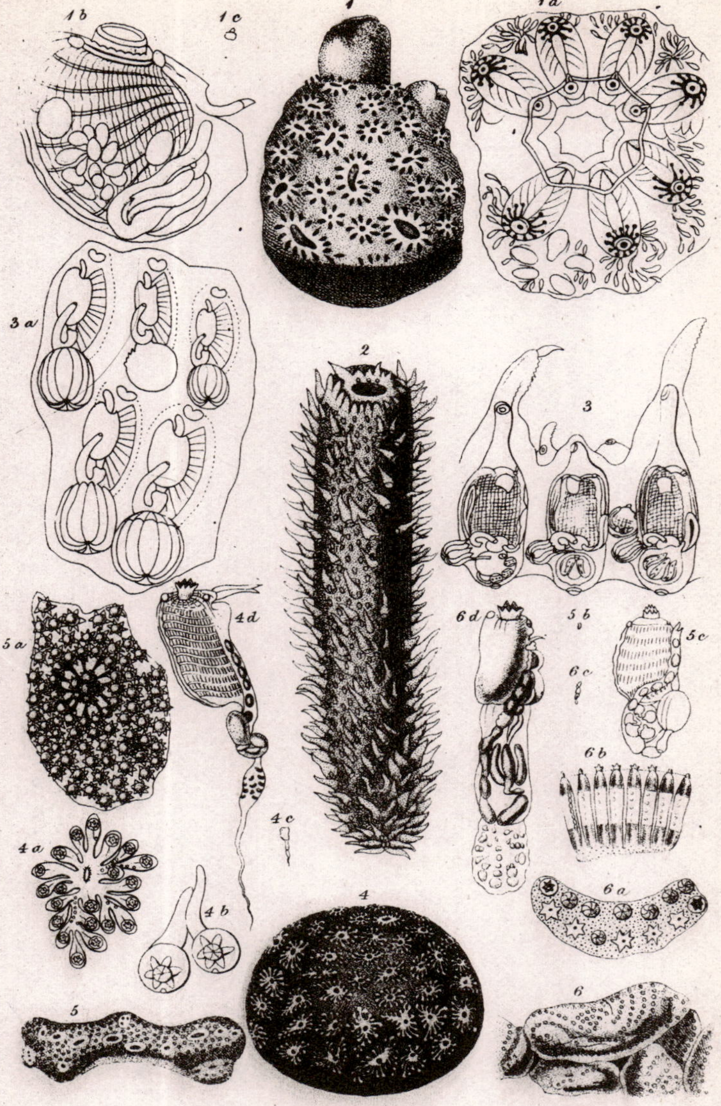

1 *Botryllus polycyclus.* 4 *Polyclinum constellatum.*
2 *Pyrosoma rufum.* 5 *Eucœlium hospitiolum.*
3 *Pyr. giganteum, details of.* 6 *Aplidium lobatum.*

THE FIFTH CLASS OF THE MOLLUSCA.

BRACHIOPODA.

THE mollusca brachiopoda, like the acephala, have a bilobed mantle, which is always open. Instead of feet, they are provided with two fleshy arms, furnished with numerous filaments, which they can protrude from and draw into the shell. The mouth is between the base of the arms. Neither their organs of generation, nor their nervous system, are well known.

All the brachiopoda are provided with bivalve shells, fixed and immovable. But three genera are known.

LINGULA, *Brug.*

Two equal, flat, oblong valves, the summits of which are at the extremity of one of the narrow sides, gaping at the other end, and attached between the two summits to a fleshy pedicle, which suspends them to the rocks; the arms become spirally convoluted previously to entering the shell. It appears that the gills consist of small leaflets, disposed around the internal face of each lobe of the mantle.

But a single species, *Lingula anatina*, Cuv. Ann. du Mus. I. vi.; Seb. III. xvi. 4, is known. It inhabits the Indian Ocean, and has thin, horny, and greenish valves.

TEREBRATULA, *Brug.*

Two unequal valves, united by a hinge; the summit of one, more salient than the other, is perforated to permit the passage of a fleshy pedicle, which attaches the shell to rocks, madrepores, other shells, &c. Internally, a small bony piece of frame-work is observed, sometimes sufficiently complex, composed of two branches that articulate with the imperforated valve, and that support two arms, edged all round with a long close fringe, between which, on the side next to the large valve, is a third simply membranous, and much longer appendage, usually spirally convoluted, and edged like the arms with a fine and close fringe. The mouth is a small vertical fissure between these three large appendages. The principal part of the body, situated near the hinge, contains the numerous muscles, which reach from one valve to the other, and between them are the viscera, which occupy but little space. The ovaries appear to be two ramified productions adhering to the parietes of each valve. I have not yet been able to ascertain exactly the position of the gills.

Numberless terebratulæ are found fossil or petrified in certain secondary strata of ancient formations. The living species are less numerous. (*Anomia scobinata*, &c.)

The shell of some is broader transversely, or longer in a direction perpendicular to the hinge, with an entire or emarginated contour, divided into two or several lobes; some of them are even triangular. The surface is either smooth, furrowed in radii, or veined; they are thick, and thin, and even diaphanous. In several of them, in lieu of the hole in the summit of the thin valve, there is a notch, and this notch is sometimes partly formed by two accessory pieces, &c. It is probable that when better known their animals will present generic differences. Already, in the

SPIRIFER, *Sowerb.*,

Two large cones have been recognized, formed of a spiral thread, which appear to have supported the animal. In

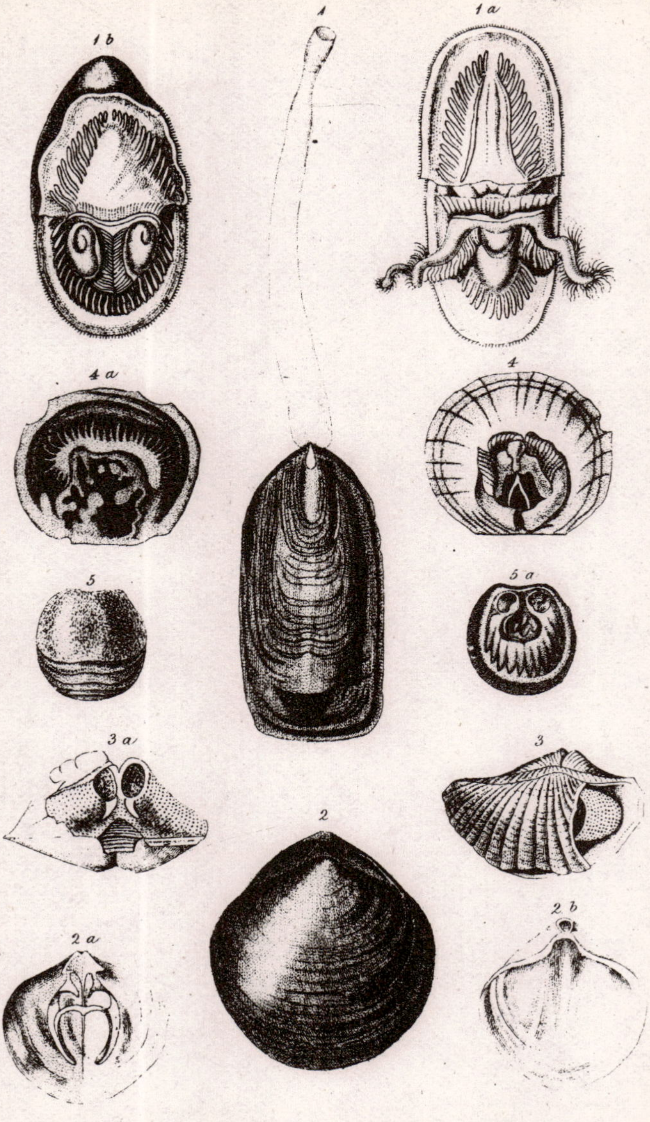

1 *Lingula anatina.* 3 *Spirifer trigonalis.*

2 *Terebratula Gaudichaudii.* 4 *Orbicula lævigata.*

Thecidea, *Defr.*,

This support seems to have been incorporated with the small valve. (*Thecidia Mediterranea*, Risso.)

Orbicula, *Cuv.*

The orbiculæ have two unequal valves, one of which, that is round and conical, when viewed by itself, resembles the shell of a patella; the other is flat, and fixed to a rock. The arms of the animal (*Criopus*, Poli), are ciliated and spirally recurved, like those of the lingulæ.

The seas of Europe produce a small species, *Patella anomala*, Müll. Zool. Dan. V. 26; *Anomia turbinata*, Poli, XXX. 15; Brett., Sowerb., Lin. Trans. XIII. pl. xxvi. f. 1.

The Discinæ, *Lam.*, are orbiculæ, the inferior valve of which is marked by a fissure. The

Crania, *Brug.*,

Should be approximated to the orbiculæ. The arms of the animal are also ciliated, but the shells have internally deep and round muscular impressions, that have caused it to be compared to the figure of a skull.

One of them inhabits the European seas, *Anomia craniolaris*, L., or *Crania personata*, Bret., Sowerb., Lin. Trans. XIII. pl. xxv. f. 3. Several are fossil, such as the *Cran. antiqua*, and the others, of which M. Hœninghaus has given an excellent monograph.

THE SIXTH CLASS OF THE MOLLUSCA.

CIRRHOPODA.

LEPAS and TRITON, *Lin.*

THE cirrhopoda, in several points of view, are intermediate between this division and that of the articulata: enveloped by a mantle and testaceous pieces which frequently resemble those seen in several of the acephala, their mouths are furnished with lateral jaws, and the abdomen with filaments named cirri, arranged in pairs, composed of a multitude of little ciliated articulations, and corresponding to a sort of feet or fins, similar to those observed under the tail of several of the crustacea; their heart is situated in the dorsal region, and the gills on the sides; the nervous system forms a series of ganglions in the lower part of the abdomen. These cirri, however, may be considered as analogous to the articulated appendages of certain species of teredo; while the ganglions, in some respects, are mere repetitions of the posterior ganglion of the bivalves. The position of these animals in the shell is such, that the mouth is at the bottom, and the cirri near the orifice. Between the last two cirri is a long fleshy tube, that has sometimes, but erroneously, been taken for their proboscis, and at the base of which, near the back, is the opening of the anus. Internally, we observe a stomach, inflated by a multitude of small cavities in its parietes, which appear to fulfil the functions of a liver, a simple intestine, a double ovary, and a double serpentine oviduct, supplying from its walls the prolific fluid, and prolonged through the fleshy tube to open at its extremity. These animals are always fixed. Linnæus comprised them all in one genus, LEPAS, which Bruguières divided into two, that have in their turn been subdivided.

ANATIFA, *Brug.*

A compressed mantle, open on one side, and suspended to a fleshy tube, varying greatly as to the number of testaceous pieces with which it is furnished; twelve pair of cirri, six on each side, those nearest to the mouth being the thickest and the shortest; the gills are elongated pyramidal appendages, that adhere to the external base of the whole of the cirri, or part of them.

The two principal valves of the most numerous species (PENTALASMIS, *Leach*), resemble those of a mytilus; two others seem to complete a part of the edge of the mytilus opposite to the summit, and a fifth azygous unites the posterior edge to that of the opposite valve. These five pieces cover the whole of the mantle. From the usual place of the ligament arises the fleshy pedicle; a strong transverse muscle unites the first two valves near their summit; the mouth of the animal is concealed behind it, and the posterior extremity of its body, with all the little articulated feet, is a little beyond it, between the first four valves.

The most common species of the European seas, *Lepas anatifera*, L., owes its specific appellation to the fable which represents it, as producing the *Barnacles*, a story founded on the rude resemblance that has been observed to exist between the pieces of this shell and a bird. The anatifæ adhere to rocks, piles, keels of vessels, &c. We may distinguish from them,

POLLICIPES, *Leach*,

Where, besides the five principal valves, there are several small ones near the pedicle (*Lepas pollicipes*, L., &c.), some of which, in certain species, are nearly as large as the former (*Lepas mitella*); frequently there is an azygous valve opposite to the ordinary one of the same description. In the

CINERAS, *Leach*,

The cartilaginous mantle contains but five small valves, which do not occupy the whole of its extent. (*Cineras vittata*, Leach.)

OTION, *Leach.*

The cartilaginous mantle contains but two very small valves, with three little grains that hardly merit that name, and has two tubular auriform appendages. (*Otion Cuvieri*, Leach.)

TETRALASMIS, *Cuv.*

But four symmetrical valves, which surround the aperture, two of them longer than the others. The animal is partly contained within the pedicle, which is large, and covered with hair. They are a kind of tubeless balani. (*Tetral. hirsutus.* Cuv.)

BALANUS, *Brug.*

The principal part of the shell of the balani consists of a testaceous tube attached to various bodies, the aperture of which is more or less closed by two or four valves. This tube is formed of various pieces which appear to be detached and separated, in proportion as the growth of the animal requires it. The gills, mouth, articulated tentacula, and anal tube, differ but little from those of the anatifæ. In

BALANUS,

Properly so called, the tubular portion is a truncated cone, formed of six projecting pieces, separated by as many depressed ones, three of which are narrower than the others.

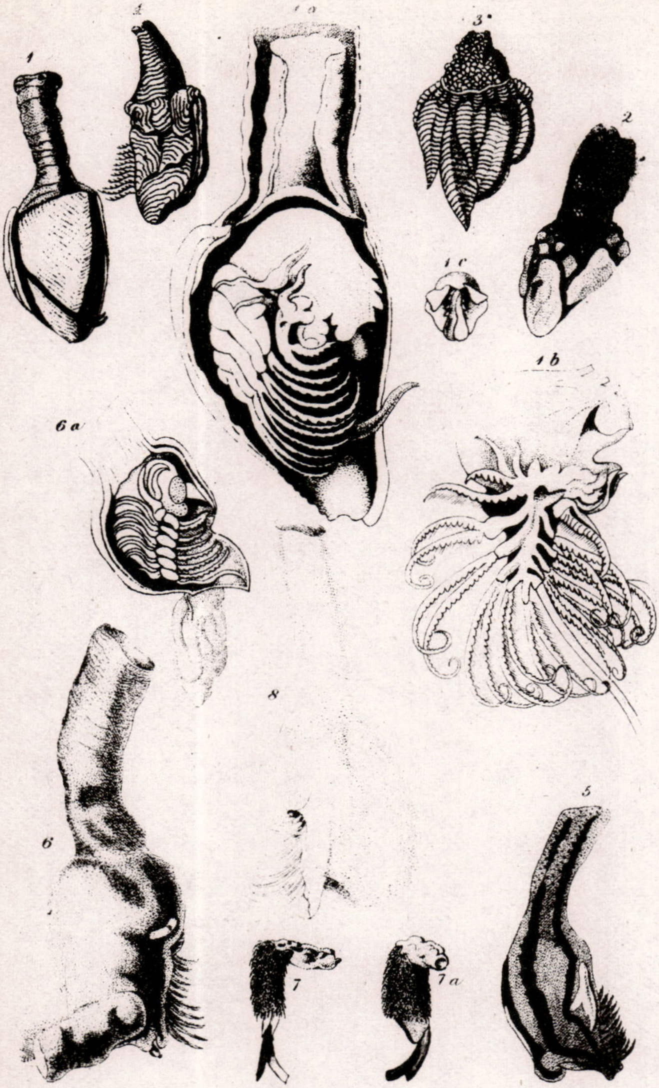

1. Anatifa lævis.
2 Pollicipes cornucopia.
3 Poll. mitella.
4 Poll. scalpellum.
5 Cineras vittata.
6 Otion Cuvierii.
7 Tetralesmis hirsutus.
8 Triton (Alepis) fasciculatus.

1 Balanus ovularis.
2 Animal of Bal. sulcatus.
3 Acasta spinosula.
4 Aca. Montagui.
5 Conia radiata.
6 Asemus porosus.
7 Pyrgoma cancellata.
8 The same after Savigny
9 Creusia spinosula.
10 Chthamalus stellatus, Poli?
11 The same after Blainville.
12 Ochthosia stroemii. Ranzani.
13 Coronula balænaris.
14 Tubicinella balænarum.
15 Diadema (Coronula Diadema) Lam.

Their base is usually formed of a calcareous lamina, and fixed to various bodies. The four valves of their operculum close the orifice exactly.

The rocks, shells, &c. of all our coasts are, in a manner, covered with a species of balanus, the *Lepas balanus*, L., Chemn. VIII. xcvii. 826. Naturalists have separated from it

The ACASTÆ, *Leach*, whose base is irregular, convex towards the exterior, and which does not become fixed. Most of them are found in sponge. (*Acasta Montagui*, Leach.)

The CONIÆ, *Blainv.;* the tube of which has but four salient pieces. (*Conia radiata*, Blainv.)

The ASAMÆ, *Ranzani*, where the tube has no decidedly salient pieces. (*Lepas porosus*, Gm.)

The PYRGOMÆ, *Savig.*, whose tubular part, forming a strongly depressed cone, has but a very small orifice, almost like the shell of a fissurella. (*Pyrgoma cancellata*, Leach.)

The OCTHOSIÆ, *Ranzani*, which have but three salient pieces in the tube, and only two valves to the operculum. (*Lepas strœmii*, Müll. Zool. Dan.)

The CREUSIÆ, *Leach*, with four salient pieces and two valves to the operculum. (*Creusia spinulosa*, Leach.)

M. de Lamarck, under the name of CORONULÆ, separates the very wide species, where the parietes of the cone are occupied by cells: so large that they resemble chambers. (*Lepas balænaris*, L., &c.) And under that of

TUBICINELLÆ, those in which the tubular portion is elevated, narrower near the base, and divided into annuli, which mark its growth. (The *Tubicinella*, Lam., Ann. du Mus. I. xxx. 1, 2.)

There are some species of these last two subgenera which affix themselves to the skin of the whales, and even penetrate into their blubber.

To the preceding subgenera must be added the

DIADEMA, *Ranz.*,

Where the tubular portion is almost spherical, and which has but two small valves hidden in the membrane which closes the operculum. The opercular valves would not effectually close the orifice without the membrane which unites them.

They also live on the balænæ, and otiones are frequently observed attached to their surface. (*Lepas Diadema*, Chemn. VIII. xcix. 843, 844.)

SUPPLEMENTARY TREATISE

ON THE

MOLLUSCA.

I. ON THE DIVISION IN GENERAL.

It is now five or six and twenty years since naturalists first so designated under the general name of Mollusca a numerous portion of the Animal Kingdom, including not only the true mollusca of Aristotle and Pliny, but also the testacea of these ancient writers.

This name, *mollusca*, comes from the Greek word μαλακα, in Latin, *mollia* (soft), because the majority of the animals to which this denomination has been applied are remarkable for the softness of their flesh, or more properly speaking, of their general envelope.

The science which treats of this portion of zoology has as yet received no peculiar name. *Conchology*, it must be remembered, is applicable only to the investigation and arrangement of the shells of such of these animals as have them.

M. de Blainville has proposed the name of *malacozoology*, or thus abbreviated, *malacology*, compounded of μαλακος, ξωον, and λογος; that is to say, a scientific discourse or treatise on the soft animals.

Aristotle, the most ancient and important of the natural

historians of antiquity, was the first writer who made use of the term *mollusca.* But under this name he comprehended but a portion of those animals which are at the present day classified under this type, giving that of *ostracodermata* (testaceous animals) to those which have a calcareous envelope, of greater or less hardness.

Pliny, and in general all the ancient Latin naturalists, have employed the same denominations, which they have translated into their own language by the words *mollia* and *testacea.* Ælian and the other Greek naturalists have followed Aristotle. Isidore de Seville, Wotton, Belon, Rondelet, have adopted the same denominations, as well as Gesner, Aldrovandus, and his abbreviator, Johnston.

Ray, the precursor of Linnæus, appears to have been the first who, having applied the name of worms to all white-blooded animals, or the invertebrata of modern naturalists, (the insects and crustacea excepted) has employed the names of molluscous worms and testaceous worms, which correspond, however, to the divisions of Aristotle, though the term worms, or *vermes,* is certainly most strangely inapplicable.

Adanson, perhaps the first who considered shells in a proper point of view, employed the term testacea (*coquillages* in French) in a classical manner; but under this name he merely comprised the species of mollusca which are invested with shells. Linnæus and his entire school have followed Ray. Pallas, after some very important observations on the subject, of which we shall speak hereafter, has shown for what animals the name of mollusca should be reserved.

The Baron Cuvier, a name which cannot now be pronounced without the deepest regret of every lover of science, appears to have been the first who fully appreciated the observations of Pallas, who carried them into operation, and united in one treatise all the animals indicated by the last-mentioned traveller and naturalist, comprehending them definitively under the classic name of mollusca, whether they were naked or invested with a shell of one or more pieces. This example was imitated by M. de Lamarck and almost all the French naturalists. Nevertheless, M. de Lamarck, in the last edition of his invertebrated animals, no longer employs the name of mollusca altogether in the same manner, and only applies it to a part of the mollusca of M. Cuvier, which nearly corresponds to his ancient division of the cephalous mollusca. M. Rafinesque, some years before, had designated this group under the denomination of *malacosia*.

M. de Blainville proposes the name of *malacozoaires* for the type which contains the true mollusca, and *malentozoaires* for the sub-type formed of the articulated mollusca.

Aristotle defined his mollusca properly so called, as animals which have no blood, the fleshy parts of which are external and the solid within ; the reverse of this constitutes his definition of the testacea. Pliny, and all the zoologists, at the revival of letters, have pretty nearly admitted the same definition.

Adanson understands by the word testacea, or *coquillages*, animals whose body is soft, without any sensible articulation, and covered altogether, or in part, with a stony crust, called a shell, to which it is attached by one or more muscles.

This is the definition of Linnæus: MOLLUSCA—*A. simplicia, nuda, absque testa, artubus instructa.* TESTACEA—*A. simplicia, domo, sæpius calcareo, obtecta.*

Bruguières, in separating the mollusca from the insects, gives them as common characters, the being without bones, without stigmata, without feet, or without articulations. He distinguishes the mollusca, properly so called, because they are naked, from his testacea, which are contained in a shell of one or more valves.

M. Cuvier defines them according to their anatomical characters: animals without vertebræ or articulated skeleton, nervous system not collected into a spinal marrow, but merely into a certain number of medullary masses, dispersed over different parts of the body, the principal of which, termed the brain, is situated on the œsophagus, which it encircles with a nervous collar; blood cold, whitish or blueish; circulation double; hearts often numerous.

M. de Lamarck admits nearly the same definition: oviparous animals, with soft body, not articulated in its parts, and having a variable and muscular mantle; respiration by diversified gills; a brain, some ganglia, and nerves to give sensibility and irritability to the organs; but no spinal marrow; conglomerate glands; a shell enveloping or enveloped, and sometimes none.

M. de Blainville proposes the following:

Animals with the body and its appendages soft, not articulated, enveloped in a skin or muscular dermis (mantle), of a variable form, in or on which is most frequently developed a calcareous part (the shell), of one or two pieces; circulation complete, with white blood, heart essentially aortic, and superior to the intestinal canal, except in the brachiocephala; respiration aquatic or aerial; nervous system composed of a cerebriform ganglion, sub-œsophageal, communicating with the ganglia of the different functions; those of locomotion being lateral.

As to the place of the mollusca in the series of animals, Aristotle separates his two groups by the crustacea. Aldrovandus, Johnston, Ray, Linnæus, his whole school, and M. Dumeril, place them after the insects; Cuvier, Lamarck, and their followers, place them at the head of the invertebrated animals.

In considering these animals as constructed on a peculiar plan, and forming a distinct type, they might even be approximated to man, the top of the animal scale, as may the insects, but in a different direction. Nevertheless, we think that if, in truth, the structure of the first genera possesses something of that of vertebrated animals, in the rudiments of skeleton which envelope the brain, yet the last are considerably degraded, and pass quickly into the lowest type of animals, so that we may place them in a line parallel to the animals articulated externally, and as it were passing to the actinozoaires, by the ascidiæ, &c.

Although nearly to these latter times, as we shall see in noticing the history of this part of zoology, these animals have been very much neglected, for the almost exclusive examination and study of their envelopes or shells, they do not the less present a very considerable degree of interest in many respects, so as to induce us to smooth the difficulties involved in their investigation. Thus the anatomist, and more especially the physiologist, will find that their study throws no inconsiderable light upon certain general questions of the first interest and importance: they will find, for instance, the organ of hearing reduced, so to speak, to its most simple expression, to that which is absolutely essential to it, in the *sepia* and the neighbouring genera;—they will find the principal organ by which the blood is moved divided into several parts, and sometimes presenting the singular disposition of appearing to be traversed by the intestinal canal. In studying the organs of generation in these animals, they may form more correct notions of hermaphrodism, complete or incomplete; thus, in a great number the organs, both male and female, exist in the same individual, though such individual cannot fecundate itself; so that the species is necessarily composed of two individual hermaphrodites, while in others, in which the female organs alone are perceptible, generation takes place by means of a single individual.

Geology will derive not less advantage from the minute study of the shells or envelopes of the mollusca, to assist in the determination of the identity of the different strata of the earth. It will discover, in the innumerable quantity of these animals, succeeding each other from generation to generation, in the depths of the sea, one of the evident causes of the increase of continents.

But man can find, in the knowledge of the mollusca, applications still more direct to his well-being in society, either in the advantages or disadvantages which he may derive from them. Thus a considerable number of species are proper to supply him with food; the sepia, and especially the calmar, or ink-fish, are in great request in Greece, and even in several parts of Italy; the larger species of snails, many buccinæ, and some cognate genera, are in tolerable estimation in several countries, and were so much so among the ancient Romans, that Pliny has not disdained to recount the names of those who thought proper to collect them in depôts, and to provide them with abundant nourishment, to fatten them, and render them more succulent. Oysters and muscles also constitute a prime object of commercial speculation in our own days.

Although not very numerous, the advantages which the mollusca present to us are nevertheless much greater than the injuries we receive from them; and it is probable that we cannot stigmatize among them an animal truly hurtful, except the *teredo* (ship-worm), which, attacking the wood of our vessels and dikes, often occasions very considerable mischief. The knowledge, therefore, of its habits and manners, is of the first necessity in those countries which are infected by it. Slugs and snails are also strongly and justly dreaded enemies in our gardens.

It may neither be unuseful nor uninteresting to give a brief sketch here of the history of this part of zoology.

All the ancient authors, such as Aristotle, Pliny, and their abridgers, appear to have been very little acquainted with these animals: they place them among those which they designated *exsanguia*, a division which corresponds altogether with the white-blooded animals of Linnæus, and the invertebrated animals of modern naturalists; not that they conceived these animals to have no blood, but they termed them so merely in comparison with the red-blooded animals. They contented themselves with dividing them into two sections, the *mollusca* and *testacea*, in which they were followed by the naturalists at the revival of literature, who, however, added but little to the facts already detailed by the ancients. But soon, the easy collection of the envelopes of these animals, often of the most extraordinary beauty, having become an object of curiosity, and even of rivalship, among the rich, the study of the animal itself was forgotten in an exclusive attention to the shell. Thus arose that part of natural history properly called conchology, on which we have so many magnificent works, mere objects of luxury, almost in all countries. On this department we shall bestow a few general remarks in the proper place In vain did our celebrated countryman, Lister, before him Fabius Columna, and after, Willis, Heyde, Swammerdam, &c., give the anatomy of many molluscous animals; no attention was paid to establish their classification, on their external organization, on their form, and still less on their internal structure. It is true that Linnæus, in the earlier editions of his *Systemæ Naturæ*, speaks of the animal of his testacea, before he exposes the characters of the genera; but he confines himself to citing the name of his mollusca, with which they have the greatest relation, and the genus is really established on nothing but the form of the shell.

The great majority of the naturalists of the last age followed the example of this great man, as we shall see by and by; but some French naturalists began to see the necessity of recurring to the animals to arrive at a proper classification of the shells. Thus, in 1743, Daubenton read to the Academy of Sciences, of which he was not yet a member, a memoir on the methodical distribution of these envelopes, in which, after having proved that the knowledge of them may suffice, he nevertheless remarks that that of the animals is indispensable to form a complete system of conchology, and a natural distribution of shells. We do not, however, find that he carried this principle into execution; at least there is no mention of his having done so, in the extract given from his memoir by the secretary of the Academy.

In 1766, Guettard, a member of the same society, was the first who put in practice what Daubenton merely hinted; for in a very detailed article, inserted in the Acts of the Academy, the covert object of which appears to have been a criticism of some observations of Buffon, at the commencement of his description of the ass, on species and its distinction,—not only does he establish on indubitable principles the necessity, in the classification of shells, of having recourse to the animal which they contain, and of which they form a part, but he characterizes a certain number of genera, at least among the univalves, extremely well.

Although in this memoir Guettard tells us that the genera of bivalves ought also to be susceptible of being equally characterized from the animal, he confesses that his observations on them have been too few to enable him to make the trial; but he very well discriminates how far the division of shelled mollusca into terrestrial, fresh-water, and marine, is exact. He likewise pays great attention to the presence or absence of the operculum.

These new observations of Guettard doubtless determined d'Argenville, in the second edition of his Conchology, in 1757, to add a great number of figures, unfortunately very bad, of animals, under the name of *zoomorphoses*, but without serving any purpose in the characters of his genera of shells, which had been justly criticised by Guettard, in the memoir just noticed.

In the same year, 1757, Adanson, who grouped animals in families on no arbitrary system, but according as they exhibited the greatest number of relations among themselves, made a more extended application of these principles to the conchiferous mollusca, which he designates under the classific name of *coquillages*, in the first and only volume which he published of his Voyage to Senegal. He studies with care, distinguishes, and denominates in a suitable manner, all the external parts, both of the animals and their shells. He then occupies himself in arranging those which he had observed in Senegal into a great number of systems, or tables of relation, considering, for instance, in the shell of the helix, the summit, the aperture, the opercle, the periosteum; in the conchs, the valves, according as they are equal or unequal, &c.

Then passing to the animals, and always admitting his original division, into helices and conchiferous mollusca, he considers, in the first, the tentacula or horns, the eyes, their absence, existence, or situation in the head, the mouth, either with two jaws and without proboscis, or with proboscis and without jaws, &c. In the conchs he considers the mantle, the trachea, &c. He then describes and figures the species of shells, which he observed in Senegal.

We may remark, that this work of Adanson's, if not the first in which the principle of classification, by which the animal is regarded as well as the shell, is established, is at least that in which the means of applying this principle are to be found. It must be confessed, however, that Adanson has not always employed with success the excellent materials which he had prepared in so suitable a manner. In fact, the distinction of his genera is very far from being complete, especially in the conchifera: his approximations are not always very natural, in a variety of cases. To him, however, we are indebted for a knowledge of the numerous relations between *pholas* and *teredo;* but of him, likewise, has science to complain, for the erroneous approximation of *oscabrio* and *patella*.

Another French naturalist, to whom the science of malacology is indebted for the use of the same principle, first broached by Guettard, and so well supported by Adanson, is Geoffroy, the physician of Paris. We find, in effect, in his little treatise on the terrestrial and fresh-water shells of the environs of Paris, published in 1766, the description of the animals to which they belong; and the characters of the few genera which this book contains are equally derived from the animal and the shell. He speaks but of five genera of univalves, among which there is but one new one, *ancilium*, adopted by all modern zoologists. Although he has established pretty nearly the same genera as Guettard, *cochlea*, *buccinum*, *planorbis*, and *nerita*, he has not been equally happy in their circumscription: for instance, he has confounded *physis* with *planorbis*, and in his genus nerita he has placed terrestrial and aquatic cyclostomata, &c. As to the two only genera of bivalves which he establishes, those of *camus* and *mytilus*, he places in the first the *cyclas fluviatilis*, in the second an anodon and a unio.

Müller, the celebrated author of the *Danish Fauna*, was the first foreign zoologist who adopted the same principle in his history of terrestrial and fresh-water worms. But in general his system of classification, though more complete than that of Geoffroy, since it extends to all the conchyliferous animals, is yet by no means natural, and is much inferior to that of Adanson.

About the same period, we begin to discern certain important changes in the distribution of molluscous animals, in the *Systema Naturæ* of Linnæus.

In the first nine editions, Linnæus does not appear to have yet employed the denomination of mollusca, the naked species being placed in his order zoophytes, class vermes, and the conchyliferous, in his third order of the same class, under the name of testacea. Though he does not yet distinguish his different genera but by a very small number of characters derived from the shell, he nevertheless cites the naked animal, which he supposed to belong to it, and which he had placed in his zoophytes, but that evidently in an accessory manner.

But in the tenth edition we find considerable augmentations, and still more in the twelfth, which may be considered as having received the finishing hand of its celebrated author. The class vermes is there divided into five sections, the second of which has the name of mollusca, and contains eight genera of the true mollusca, *ascidia*, *limax*, *aplysia*, *doris*, *tethis*, *sepia*, *clio*, and *scyllæa*. The third is almost entirely consecrated to the testacea, divided into multivalves, bivalves, and univalves.

In the characters of their genera, however, Linnæus always confines himself to the citation of an analogous naked molluscum; so that if the work of Adanson had any influence over the last editions of the *Systema Naturæ*, it was only in the more numerous division of the genera of shells, and in their better circumscription. But it had little real effect on the part relating to animals. Thus we find among the mollusca of Linnæus some which are articulated and others which are radiated animals. In his testacea also there are several inconvenient approximations.

Though the French zoologists whom we have mentioned may be considered the founders of a scientific arrangement of these animals, yet they paid no attention except to the external parts of the animal body inhabiting the shell, and moreover, took no notice whatever of the naked mollusca.

Nevertheless, the impulse given to his age throughout Europe, by the system of Linnæus, and in France by the writings of Buffon, occasioned many naturalists to publish the anatomy and description of a considerable number of molluscous animals. This was done by Bohatsch, Baster, Forskahl, Fabricius, Müller, &c. Besides, the application to the different parts of zoology, of the principles so happily imagined for botany by Bernard de Jussieu and Adanson, changed in some measure the mode of considering the classification of animals. Wishing to arrange them in such a manner as less to interrupt the natural relations, zoologists felt the necessity of a knowledge of their internal structure, and Pallas may be regarded as the chief of this new school, which the French philosophers have supported with so much success, and which is now fast propagating throughout all enlightened Europe.

It was in his *Miscellanea Zoologica*, published in 1766, that Pallas exhibited, as it were, the germ of those ameliorations of which the methodical arrangement of the *malacozoaria*, was susceptible. He proves that Linnæus, in the disposition of his molluscous worms, has departed very considerably from nature: that his sub-division of testacea, considering the shell and not the animal, could not be preserved, and that in general he was totally wrong in separating these two orders. Accordingly, he proposes to unite in the univalves, as forming a natural order, not only the univalve testacea but also the limaces, (and under this name he comprehends doris, tethys, and scyllæa,) as well as the sepia, and perhaps, adds he, the medusa; but this is evidently wrong. In the second order, he thinks, should be placed all the bivalve testacea (joining with them the teredo), of which the ascidia appears to him to be the analogue, or to speak more properly, the naked type.

Notwithstanding this, Bruguières, an author to whom conchology is greatly indebted, has almost completely imitated the arrangement of Linnæus. His genera of testacea are, however, more numerous, and better defined.

Gmelin, who was too little of a zoologist to profit by the labours of the writers we have named, has scarcely made any change in this department, in his edition, the thirteenth, of the *Systema Naturæ*.

An Italian physician, M. Poli, was the first to establish the genera of mollusca, according to the animal alone, without paying any attention to the shell. In 1791 appeared the first volume of his magnificent work, on the testacea of the two Sicilies. It appears that he took into consideration all the molluscous animals, whether naked or testaceous, and divides them into three orders: 1. *Mollusca brachiata*, characterized by having several arms, in the manner of the hydra; 2. *Mollusca reptantia*, creeping after the manner of snails, by means of one broad foot, and always having a head and eyes; and 3. *Mollusca subsilientia*, provided with a long foot, fixed to rocks, or not, and constantly devoid of head and eyes; this contains the bivalves and multivalves.

During the ten or twelve years of the revolutionary whirlwind which agitated Europe but few works appeared on this department of zoology; indeed, it appears to have lain almost uncultivated, if we except some facts recorded in journals, and the establishment of some new genera. Thus the science remained stationary until, in 1798, our illustrious author, feeling like those we have before mentioned, that the methodical subdivision of the mollusca, like that of all other animals, should rest upon the study of organization, proposed his new classification. He thought in the first instance that the whole division of the malacozoaria should rise a degree in the animal series, and precede the entomozoaria, or animals articulated externally; a second innovation was to unite definitively, as Pallas had done, under the classific name of mollusca, the molluscous worms of Linnæus, to his testaceous worms, that is, to consider the existence or absence of the shell, as but a very secondary consideration. He therefore made a distinct class of this great group, which he still named white blooded animals, and which were soon to be known by the denomination of invertebrata, characterized them in a clear and precise manner, as well as the other three, those of insects, worms, and zoophytes, which he admits among the animals without internal articulated skeleton; then taking into consideration the form of the mollusca, he divides them into three sections, cephalopods, gasteropods, and acephala. In the first he placed not only the sepia of Linnæus, but also the argonauta. But it would be superfluous to enlarge further here on the labours of Cuvier.

M. de Lamarck had chiefly confined his speculations on this subject to a consideration of the shells, but in his work on invertebrated animals, published in 1801, he has pretty nearly followed the example of M. Cuvier. He imitates him at first in this, that the class of the mollusca is put at the head of the invertebrated tribes. But subsequently he departs from his plan pretty often. Thus his first division of mollusca into two orders rests on the presence or absence of the head, a division which is merely implied in the systems of Poli and Cuvier.

In spite of this evident march towards perfection in the classification of the mollusca, some persons even in France thought proper to stick to the system of Linnæus, improved by Bruguières: such, for example, was M. Bosc. In his supplements to Buffon, although he felt the value of the innovations which had been made, he nevertheless adopted the division of the molluscous *worms* for the naked mollusca, and of testaceous worms for the conchyliferous species; and in each of these divisions he exactly followed Bruguières, adopting, however, the new generic division of Cuvier and De Lamarck. M. Bosc, however, who has often had occasion to study living mollusca, has introduced many new facts into their history, and has also established some genera.

The first work which collected all these recent labours was the Natural History of the Mollusca just commenced by Denys de Montfort, and executed almost altogether by M. de Roissy, a work which constitutes a part of the edition of Buffon by Sonnini, and which developed in a suitable manner the system of Cuvier.

In 1809 M. de Lamarck, obliged by his place of Professor of the natural history of invertebrated animals, to follow the progress of the science, and to put together the new facts which it had acquired, proposed a new distribution of those animals, in his work entitled Philosophie Zoologique. Dividing the animal kingdom into six degrees of organization, he places in the fourth, ascending from the lowest, and in the third, taking an inverse course, the animals with which we are now occupied. But he divides them into two classes, one to which he leaves the name of mollusca, while to the other he gives the new designation of *cirrhipoda.* In this new system, though considerably improved, M. de Lamarck had still established some unnatural approximations, and in 1812 he found it necessary to make some further alterations in his general classification of these animals. On this occasion we may observe, that in the prodromus of his course M. de Lamarck divides the animal kingdom into three primary sections, 1. *apathetic animals ;* 2. *sensible animals,* (these two divisions composing the invertebrata); and 3. *intelligent* or *vertebrated animals.* The philosophic propriety of this division is more than questionable.

Towards the end of 1814 M. de Blainville published his first notions on the methodical arrangement of the *malacozoaria,* in which he particularly established the necessary relation between the shell and the organs of respiration. He also drew from this the new character of the symmetry or non-symmetry of those organs for the establishment of his orders.

To carry this history further would be tedious: suffice it to say, that this branch of zoology has been much indebted to the subsequent labours of the French naturalists we have already cited, and to others of the same nation; to some zealous philosophers of Germany, such as Oken and Dr. Goldfuss; to some of the United States of America, as Mr. Say; and last, though not least, to some distinguished men amongst ourselves, of whom we shall merely mention Dr. Leach and Mr. Gray. To enter into an analysis of their systems and improvements would be wholly beside our present purpose.

We shall now briefly treat first of the form and organization of the animals of this class.

The form of the body of the molluscous animals is extremely variable, though it presents the constant negative character of being never articulated. Thus, though most usually oval, more or less elongated, convex above, plane underneath, as in doris, limax, &c., yet it is sometimes equally oval and convex, above and underneath, as in sepia, elongated and subcylindrical, as in certain loligines, globular, as in octopus, it is often compressed more or less strongly on the sides, as in scyllæa, and particularly in all the lamellibranch acephala. It may also be very much elongated and claviform, as in the teredo and neighbouring genera. In many cephala a large portion of the body is rolled, like the shell, into a spiral more or less elevated, and of different forms. In fine, the form may be so irregular that the animal scarcely appears symmetrical externally, as is the case with the ascidiæ and neighbouring genera, and even in the biphoræ.

A tolerable number of these animals presents a very clear separation between the head and the rest of the body, as in octopus. Sometimes, however, it is much less marked, as in doris, &c.; and finally, in an entire class, consequently named acephalous, this separation no longer takes place, and no head properly so called exists.

The distinction of neck, breast, abdomen, and tail, is still less evident; the body forming but a simple mass, or sometimes subdivided in a vertical direction, but never in a longitudinal.

The body is but rarely provided with locomotive appendages properly so called, but sometimes presents cutaneous expansions, more or less extended, which serve to locomotion. It is only in some that the disposition of the appendages assumes a form a little analogous to that which takes place among the entomozoaria.

The skin which envelopes the body of the malacozoaria presents a peculiar character in its softness, its sponginess, and especially in the manner in which the dermis is confounded with the subjacent muscular fibre, so that it is contractile in all points and in all directions. This dermis, as to the rest, may either be tuberculous, or very smooth; the vascular net-work there, is moreover, very considerable. The colouring pigmentum is often very lively. It is also probable that the nervous stratum may be tolerably complete, from the great quantity of nerves which repair thither. As for the epidermis, it is often reduced to a nullity.

If we may judge from the great quantity of mucosity which is spread in general over the surface of the skin of the mollusca, we should believe that the mucous cryptæ there were very numerous; but it is often very difficult to demonstrate their presence. We find parts, however, where the mucous pores are evident, as at the thickened edge of the mantle, which constitutes the collar of the conchyliferous cephala, and probably at the place which often forms numerous folds in the bottom of the respiratory cavity, towards the anus, and which have been designated under the name of mucous folds. In effect, there issues from these places in the skin much more mucus than from all the others.

We never observe genuine hairs in any animal of this type;

sometimes, however, the epidermic mucous part of the shell is prolonged, as it were, externally, and is rounded or flatted, so as to present a pilose aspect, as may be seen in certain species of helix, and of bivalve shells.

In the oscabrions, this disposition is still more marked upon the skin itself, and sometimes we find in certain species bundles of corneo-calcareous hairs on each side of the body.

As it happens pretty often that the skin of the mollusca is larger than necessary to surround the body exactly, or the mass of the viscera, and that the folds which it forms seem to envelope it, as our body might be in a mantle or cloak, this name (pallium) has been generalized to designate the skin of the mollusca, though in reality this disposition does not always exist.

The general disposition of the mantle of the mollusca presents so great a number of differences, that it would be almost tiresome to enumerate them; we shall therefore confine ourselves to the principal ones. In octopus, sepia, and loligo, it forms a sort of purse or very thick sheath, open at the lower circumference of the neck, and it is through this aperture that the water penetrates into the branchial cavity which it constitutes. In the conchyliferous acephala, the part of the skin which covers the viscera is excessively slender; it thickens gradually towards the edges of the mantle, and forms round the pedicle which joins the foot to the visceral mass a sort of ring, more slender behind, much thicker in front, and to which the name of *collar* is often given. It is in the thick part of this free edge of the mantle that are found in the greatest abundance the mucous pores which produce the shell, and it is into the middle of these edges that the head and foot of the animal re-enter when it is desirous of a complete shelter in its shell. The extent and form of the aperture of the mantle are always in relation with the bulk of the pedicle of the feet; accordingly, very much contracted in the buccina and the neighbouring genera, which constitute the family of the siphonobranches; and even in those of the pulmobranch family, where it really merits the name of collar, it is very long and very narrow, in the cones, the olives, the porcelaines, where it is constituted of two lobes, more or less unequal, and which may sometimes greatly pass the aperture of the shell, and be curved over it, so as to envelop it totally. Finally, the aperture of the mantle may be oval or circular, as in the symmetrical or unsymmetrical cervico-branches. In the naked or almost naked cephalous mollusca, the mantle being very thick in its whole extent, or a very little more so on its edges, or otherwise covered with tubercles, as in doris, peronia, tritonia, and even in limax, the projecting edges nevertheless pass the foot, so as to resemble a species of large buckler.

In the lamellibranch acephalous mollusca, whose body is usually very much compressed, the mantle constantly very slender, if we except towards the edges, is divided into two great lateral lobes, equal, or nearly so, which fall back on each side of the body, compress it between them, and often very much exceed it. This is an arrangement pretty much analogous to that of the porcelaines; and it is here that this part of the envelope really deserves the name of mantle. Always united in a greater or less extent along the dorsal line, the lobes of the mantle of the lamellibranches may be separated in all the rest of their extent, as in the oysters; half separated, as in unio, cardium, venus; are well joined, so as to constitute a sheath, opening only in front and behind, as in solens and many other genera; or finally, form a sac, pierced only with two posterior apertures, approximated as in the ascidiæ, or more or less distant, as in the biphores, in which the mantle, in its external stratum, becomes almost cartilaginous.

The edges of the aperture of the mantle in the cephalous mollusca are often simple; that is to say, without elongations, without lobules, or digitation, or tentacular cirrhi, as in sepia and the neighbouring genera; but it often happens that the upper edge advances a little to form a sort of shelter for the head, as in onchidia, and even in limax, or that it is considerably prolonged by the addition of a thick muscular appendage, in the form of a cornet, open below, but constituting a complete tube, more or less elongated, and serving as an introduction for the water into the branchial cavity. This is observable in all the siphonobranches, in which the aperture of the shell is emarginated or siphonated.

We find a small number of species of mollusca of this class, in which the lateral edges of the mantle are lobate or digitate; but there are a few more which have them furnished with fringes or tentacular cirrhi. The cervicobranches, and especially the patellæ and haliotides, are the species which more particularly present this character.

But it is especially in the class of the acephala that the marginal cirrhi of the mantle acquire the greatest development, both in size and number. In the limæ, for example, they are almost small cylindrical tentacula, forming a quadruple cordon round the edges of the mantle. In pecten, the cirrhi, which are also large and numerous, are intermixed with small oval plates, iridescent, in the form of eyes, regularly intervalled, and the use of which is completely unknown.

In this same class of animals, the edges of the mantle pretty frequently present lobules, or digitations, more or less marked; and in the species in which the labial lobes are more or less completely united, they are so behind, by means of one or two muscular tubes, entirely contractile, distant or not, short, or very much elongated, whose orifices are often furnished with cirrhi, and assume an almost radiated arrangement. These tubes seem, one, or the ventral, for the introduction of food, the other, or the dorsal, for the ejection of excrement. In the biphoræ, where they are so much separated that they seem to be at the two extremities of the body, one of them, the dorsal, is provided with a valvular apparatus.

But a more singular character of the skin, in a great number of these animals, is, that in a part of its thickness, and most frequently between the vascular net-work and the pigmentum, a mucous matter is deposited, mixed with a greater or less quantity of cretaceous substance, the accumulation of which, when dried up, produces a protecting body, or in other words, a shell.

We shall say something hereafter concerning the forms of shells, and those of their different parts, so as to draw from them the characters of this accessory branch of zoology. Our business at present with these bodies is as to the relations of their structure, chemical composition, the manner in which they originate, grow, and are modified with age; and finally, of their connexion with the animal.

A true shell is always composed of mucoso-calcareous strata or laminæ, applied one inside the other, the oldest and smallest being outside, and the newest and largest being the last inside. This is evidently observable in the foliated shells, such as oysters, especially, when by exposure to the heat, or by long action of the air, the mucous matter which connected not only the molecules of each lamina, but also those of the two superposed ones, has been removed. The edges of the composing laminæ, which are seen at the external face of the shell, constitute what are named the striæ of augmentation.

This structure, the best known of all, is the foliated structure. But there is another which differs from it, by the composing strata being much better connected, and their calcareous molecules more approximated together: such is that of the shells of pecten and patella. Accordingly, these shells may be heated very considerably without breaking, which is the reason why the first of them is often used as a sort of dish.

Sometimes, when the calcareous molecules are depositing themselves in the formation of one of the composing laminæ, they correspond or place themselves one above the other in all those which compose the shell, and from this results the fibrous structure, in which the shell is more easily broken in the direction of the fibres than in that of the laminæ. This is often very observable in the shell of the pinna.

We find some shells in which these two structures may alternate; that is, one part of their thickness is simply foliated, and the other fibrous. This is the *fibro-lamellary* structure.

A structure much approaching to this is that which we remark in nacre-shells, or those which produce mother-of-pearl, whether univalve or bivalve. The nacreous part appears to be always lamellated, and the other to be fibrous, or more or less oblique.

When a shell is arrived to the degree of size of which it is capable, the dermis of the animal appears to produce a greater quantity of calcareous matter and less mucous matter, and the molecules which compose it are no longer deposited by laminæ, or regular strata. They are very much crowded, heaped together, and assume a *vitreous* structure, which is polished more and more with age, by the rubbing of the parts of the mantle, which may be remarked in all the univalve shells, at their internal surface, and especially near the aperture, as in galea, for example; but it is seen still better in the porcelaines, and some neighbouring genera, where, in consequence, the animal being provided with two large lateral lobes to its mantle, the shell is almost every where enveloped by it.

It is with this matter that the holes are filled, which accident may have made in the shell. The posterior part of the spire of those which are turriculated is also thus filled, which forces the animal to abandon it, and even the tubes or calcareous tunnels, which are formed by certain acephalous bivalve mollusca at particular periods of their life, are also filled in the same manner. In fine, it is by this vitreous deposited substance that the aperture of a tolerable number of univalve shells is narrowed, and that it often assumes quite another form than that which it had before the adult age of the animal.

This part of the shell of the mollusca is so far remarkable that it is very brittle in all directions, somewhat after the manner of glâss; this explains what is named by naturalists, the decollation of the spire in many cephalous mollusca.

It is very rarely that the shell is coloured in its composing strata; in fact, in the far greater number of cases it is white: but, on the contrary, it is sometimes coloured in some parts of its internal surface, and almost always at the exterior.

Every shell which is completely dermal is never coloured, which may be easily supposed, as the pigment remains at that part of the skin which covers it.

The coloration which is sometimes remarked at the internal face, which seldom occurs, as would appear, except in the bivalves, belongs to the matter of deposition, and appears to be produced by an impregnation which extends by almost imperceptible degrees in surface and in depth. It is therefore probable that it is owing to some humour of the animal, produced in an organ whose contact with the shell tinges it with the colour of this humour. This at least appears certain as to the yellow or brown colour which is sometimes seen in the univalve shells. It is unquestionably owing to the contact of the liver. That of the janthina is in the same case;

it is a true colouring of impregnation, which appears to proceed from the depurating organ.

As for the nacreous or iridescent coloration which is still more frequently remarked at the interior of univalve and bivalve shells, the experiments of Mr. Brewster (of which more hereafter) put it out of doubt that it is owing to the mechanical disposition of the molecules, and not to a really colouring matter.

The coloration of the external surface of the shells is altogether different, and in reality does not belong to them. It is always extremely superficial, and produced by the coloured pigmentum of the edge of the skin. These are the coloured molecules which are deposited above the calcareous deposition, and which are of another nature, since they disappear in the course of time and by the action of heat. Accordingly the colour is so much the more lively, as the animal is younger, and as the produced part of the shell is more new. We are indebted to Reaumur for experiments which prove that it is only the anterior border of the mantle, which thus produces coloured molecules. In effect, it is quite certain that the new piece which is formed to fill a hole made in another part of the shell than its edge, is constantly white. We see, besides, that in the helix nemoralis, on which he made his experiments, and whose robe is agreeably zoned with black on a yellow ground, the part of the collar which corresponds to the black zones presents a tint of this colour, so that if we break a portion of the edge of the shell the piece which is reproduced is black opposite the black part of the border of the mantle, and yellowish on the rest. Although we have no direct proofs that such is the case with all the other shells which are coloured by zones decurrent from the summit to the base, analogy permits us to conclude that it is likely to be so; it must be confessed, however, that in the species whose coloration is by oval, square, or irregular spots, and especially by transverse bands in the direction of the striæ of augmentation, this analogy becomes less evident, unless we admit with Bruguières, that there is a change, a displacement, irregularly or not, in the parts of the border of the mantle which produce the coloured deposition, phenomena of which it is much more difficult to give an explanation, and which it would be necessary to submit to new observations.

We have said just now that the coloration of the shells is constantly superficial; there is, however, one group in which, notwithstanding the existence of this superficial colouring, there is another deeper and not visible, and always very different not only in its kind but in its form. This group consists of the porcelains and some olives. Bruguières has perfectly explained this fact. During the course of a pretty long life these animals are invested, as we have seen above, with a very thin shell, its edges not denticulated, the spire visible, &c., and which is especially coloured at its superficies, as are the majority of shells. This coloration, owing to the edges of the mantle, takes place by degrees, along with the growth of the shell; but at a later period, perhaps when the animal is adult, the cutaneous appendages, which from each side of the body rise over the back of the shell, as the animal crawls, deposit a calcareous ebony-coloured matter, which thickens by degrees, and at the same time a colouring matter, which constantly presents a disposition totally different from that of the first. We must then admit that the upper face of these cutaneous lobes presents spaces where the pigment is produced which colours the cutaneous matter thence exhaled; and, as in the development of these lobes, it is rare that these spaces fall precisely in the places of the first depositions, we may conceive how this new coloration not only never takes place by decurrent bands, but is always exhibited in rather irregular spots.

Light unquestionably has a very powerful influence on the coloration of shells, since those which are altogether interior, or deposited in some large lodge of the dermis, are always white, in the same manner as those of animals which live constantly in holes from which they never come forth. But another proof of this fact is, that in certain bivalve shells, which are always fixed somewhere, more or less horizontally, the fixed valve is constantly white, while the upper one is often coloured in a very lively manner. The spondyli, and a considerable number of pectines present examples of this. We must then admit here that one lobe of the mantle, not receiving the exciting action of the light, does not produce any coloured pigmentum, while the reverse is the case with the other; or, to express ourselves with more precision, the pigmentum is coloured by this action alone; so that if one of those shells should be designedly reversed there would be a reverse in the coloration of the valves, as actually does take place with the sides of certain pleuronectes.

In general the coloration of shells is so much the more lively as the animals from which they proceed are more exposed to the action of the light. The helices, terrestrial animals, are, in fact, those whose shell varies the most in colour; the tubicolæ, among the bivalves, have, on the contrary, their shell constantly white. Olivi, who has made some researches on this subject, has equally remarked, that the shells which are enveloped by sponges or alcyones, or which live in sand, or even in places which are constantly shaded, are much paler than those which are constantly uncovered in places greatly exposed. The same shell too is often more coloured in its uncovered parts than in those which are concealed.

We find almost all kinds of colour on the external surface of shells, most commonly however brown and fawn-colour, but green less frequently. There are a great number of systems or arrangements of coloration, sometimes uniform, sometimes pitted, or spotted, sometimes striped longitudinally or transversely.

Finally, a last part which enters into the composition of shells is the epidermis, which covers the colouring pigment, and which is sometimes named *epiphlosis*. It is evidently in the epidermis of the skin that the shell is deposited. This epidermis is formed of a mucous or corneous matter dried up; sometimes producing a stratum more or less thick and smooth at the surface of the shell, and sometimes rising in laminæ, or filiform productions, flatted or conical, and sometimes elongated, so as to resemble kinds of hairs. In the bivalves this part is of the same nature as the ligament, and it sometimes envelopes the valves altogether, as in solen. It is this part which first begins to be formed in the growth of a shell, univalve or bivalve, and whether it is to remain with an epidermis or not.

After what we have said concerning the structure of the shell of the mollusca, it is certain that it is chemically composed of two substances; 1. of an animal mucous matter, more or less abundant, according to the age of the molluscum, the part of the shell analyzed, and its structure; 2. of a calcareous salt, but which varies in quantity according to the age of the conchyliferous mollusca. Although the analysis of shells given by chemists is very incomplete, inasmuch as it is exercised on all their parts at once, without distinction of age, we can nevertheless recognize that the differences in the results are pretty much in relation with the differences of structure.

The species which contain in general the largest portion of animal matter appear to be those which possess the fibrous and nacreous structure. According to Mr. Hatchett they are formed of subcarbonate of lime and coagulated albumen. The mother-of-pearl itself, in 100 parts is composed of 66 of the first and 34 of the second.

The shells of oysters contain much less animal matter, and this matter resembles more a gelatinous substance. M. Vauquelin has found there, besides the organic matter, some subcarbonate and phosphate of lime, subcarbonate of magnesia and oxide of iron.

The shell of the patellæ, which presents a very close lamellated structure, approximates still more in its chemical composition to those whose structure in general is vitreous. The latter, according to Mr. Hatchett, who names them *porcelaine* shells, contain but a very small quantity of azotic matter; we find there, on the contrary, much subcarbonate of lime, but without any traces of phosphate or sulphate of the same basis.

After what we have now said, it is evident that the shell of the molluscous animals, a mucoso-cretaceous substance, is not a hardening of the skin by the deposition of calcareous molecules in the meshes of a cellular tissue, but truly a deposition of mucoso-calcareous matter, not however excreted at the superficies of the skin, but between two of its parts, the vascular net-work and the epidermis, and sometimes even in the dermis itself; and, in fact, it is connected organically with the rest of the animal, and especially with the muscular or contractile fibre, while a simple calcareous tube, like that which exists in the tubicolæ, is in reality but a deposition, an exhalation altogether external, and accordingly is not, properly speaking, attached to any part of the animal. It is this point of relation of the animal with the shell which produces the impressions of variable form which are remarked in different parts of the shell, and especially in the bivalves. This necessary relation then does not permit us to suppose that a conchyliferous molluscous animal, whose shell had been removed, could reproduce it, still less that it could quit it of its own accord, as Bruguières supposed in the case of the porcelaine shells. Neither does it permit us to adopt the notion of Klein and

Bonnet, that the shell grows by intus-susception. In fact, the experiments of Reaumur, in which he has shown that a hole made in the shell, or in a part of its spire, or even at the edge, is not filled by the circumference, but at once and independently of the shell itself, have placed the matter beyond the reach of doubt.

The form of this shell, and even the predominance of animal over mineral matter, must then be in relation with the form of the skin, or mantle, and with the age of the animal. Accordingly the tubular, spiny, lamellary elongations which we often remark at the surface of a shell, are only elongated productions, lobes, strips as it were of the mantle; in the same manner as the sinuses, the emarginations are produced by the habitual but intermittent production of some organ, such as the respiratory tube, the head itself, the oviduct, &c.; but properly to comprehend their formation, we should follow the developments of a conchyliferous molluscum, from the moment of its appearance in the egg from which it has issued up to the completion of its growth, and from that point to its death.

Every molluscous animal, however large and disproportionate to its body the shell may become in the course of time, has once presented an inverse disproportion, that is, the shell which is perceptible very early in the egg, has been at first much smaller than the body, and was consequently very far from being able to contain it, as may be seen in the helicolimax. It has equally commenced by being almost entirely membranaceous; originally then its free edges were really in the skin itself, since they did not yet attain the limits of the mantle. By the addition of new interior strata, and the increase of the quantity of calcareous molecules, the shell is thickened and solidified, but at the same time it has grown in such a manner that the edges of its aperture have attained to the limits of the mantle, at first only in the state of repose or retraction. However, the animal has issued from the egg pretty nearly at this epoch, and its growth has continued. For the purpose of seeking its food, and placing itself in such circumstances as may be generally necessary to its development, it has been obliged to extend the different parts of its mantle, and especially the lobules, the strips, the digitations with which it is provided, and which are always more large proportionally, and even more numerous in youth than at the period of decrepitude, when they have a tendency to disappear. It is then that the edges of the aperture of the shell are extended and pass those of the retracted mantle, that the deposition of new strata augments incessantly, and so much the more that the animal, from some circumstance, is obliged to contract and retract itself still more. The shell has become a shelter, a protecting organ, so much the better and so much the more complete as the animal has more approached to the *summum* of development of which it is susceptible. If the edges of the mantle were simple those of the shell are so likewise; if, on the contrary, they were prolonged in any direction to facilitate some function, the edges of the shell have followed these prolongations, and similar prolongations result in the envelope. We must admit, however, that the prolongations of the mantle had the organization necessary for excreting along with the mucous matter which the skin of the mollusca always throws out, a sufficient quantity of cretaceous matter. Without this it would be impossible to explain why, among the siphonobranches, there are species whose cutaneous tube has produced a tube to the shell, as in the siphonostomata, and merely an emargination, as in the entomostomata. It is thus we may explain not only the formation of the siphon, and of the spines which it has, but also that of the points or desections, more or less numerous, of the right edge of the aperture of a shell, &c. As a general proposition, it is certain that the spines, tubercles, and prickles of a shell, however solid they may be, have at first been channelled;

those whose canal or scissure is within, and these are the most numerous, have been produced by digitations of the mantle; those in which the scissure is external, as the horn of the ***purpura monoceros***, and the spines of the corslet of the ***venus dionea***, appear, on the contrary, to have been made by the concavity of an appendage of the mantle, which projects externally.

But these lobules, these desections of the mantle, have not existed, as it would appear, at all periods of the active life of the animal, and then the shell could not have been furnished with corresponding desections. This may be seen very well in the pterocères and neighbouring genera, whose shell, in youth, much resembles that of a cone. We must then conclude that in these genera the right lateral lobe of the mantle is dilated and widened, and sometimes digitated in rather an irregular manner, with age, and it is then that the shell presents the wing or the digitations which characterize it. We must also necessarily admit that this disposition of the mantle diminishes, by little and little, at the period of old age, since the digitations of the shell, at first evidently canaliculated, are filled and solidified completely, and the right lobe of the mantle presents no trace of division at the places corresponding to those digitations of the shell now become solid.

In a tolerably great number of mollusca, it appears that during the period of growth their activity suffers no interruption, which probably is to be attributed to the constant reunion of favourable circumstances, more especially in temperature and nourishment; and then the growth of the shell, more or less slow, is however uniform, until it has attained the *summum* of its development. But there are also many others of them in which, from the intermission of favourable circumstances, the animal, being forced to diminish the intensity of its vital activity at certain periods of the year, or of its life, the shell presents indications of these periodical intermissions in the swelling or thickening of the right edge of the aperture in the univalves, or of the entire free edge in the bivalves, which is to be noticed at very different intervals in the extent of the spiral cone; while between these intervals the shell is more smooth and slender. Are these intermissions themselves determined by that of the activity of the digestive organs, or by that of the generative organs? This is a point not easy to ascertain or to decide, but which might nevertheless be plausibly referred to the state of either of these organic systems. We may conceive, in fact, that during the period of generative activity, the vital congestion being carried to the organs of generation, would proportionally diminish that of the skin, and of the cretaceous excretion, and then that the growth of the shell would go on as usual, from whence are the intermediate spaces at the swelled or thickened parts; but when this congestion had ceased, it would be directed towards the skin, whence an accumulation of calcareous matter at the edge of the aperture, which would produce simple or ramified swellings, according to the simplicity or the subdivision of the edges of the productive mantle. The rarity or the frequency of these subdivisions would determine the number and distance of the swellings, sometimes very close and crowded, as in scalaria, lyra, and certain species of venus, or widely intervalled, as in tripterous and dipterous murices, and triton, where these swellings in the growth of the spire are regularly disposed, three in number, one on each side, and one medio-dorsal; or two in number, symmetrical, one on each side, which gives to the shell generally considered a flatted form; or two in number, not symmetrical. But we must remark that these swellings are always formed of vitreous and not of lamellary substance.

When the animal has arrived at the term of its growth, and within limits of bulk sufficiently variable, its shell is always terminated by a swelling or thickening in the species of which we have been just speaking. But even in those in which the intermissions of the growth are not so sensible, and are marked only by simple striæ, the termination of the growth is very often indicated by a swelling more or less thick, sometimes simple, sometimes denticulate, and which is equally formed of a vitreous substance. It is also at this epoch, that in the univalves the vitreous substance of the interior depôt, and even exterior, as in the porcelaines, increases, thickens, appears as it were to extravasate, and tends to diminish the aperture, the form of which it very often changes, as may be seen in the true cassides and certain helices, so as sometimes to join the two edges, and form a sort of continuous peristoma. The orifice of a univalve shell is also often modified by the formation of teeth, not only at the internal side of the right edge, but again upon the left edge, and the columella itself. These teeth are evidently produced by the cannelures of the mantle which accompany the pedicle, that joins the foot of the animal to the tortuous part of its body.

The explication of the formation of sinuses, notches, emarginations, is similar to that of tubercles, canals, swellings, and varices, with this difference, that these solutions of continuity in the edge of univalves or bivalves, are owing to some moveable projecting part, which is alternately put forth and withdrawn. Thus, in the univalve shell, the anterior emargination of the aperture is, as we have already had occasion to observe, owing to the tube formed by the edge of the mantle. The sinus which is sometimes remarked in the anterior part of the right edge, as in pterocera and strombus, results from the passage of the head; the medial or sub-posterior notch of the same edge, and which is found in pleurotoma, and in many other genera, is referrible to the exsertion of the female organ of generation or the oviduct. In all probability this does not exist except in female subjects. It is at least certain that this disposition has not been hitherto found except in the diœcious species. As to the sinus, sometimes extending to the length of a spur, and forming a sort of gutter, it appears owing to a prolongation or fold of the mantle, and perhaps also to the organ of generation.

Another consideration to which the examination of shells gives rise, and of which it will be proper to say something, is that of the muscular impression produced, as we shall see farther on, by the communication or adherence of the muscular fibre with the shell. This adherence, so strong in the living state, is nevertheless very little so after death. Does it consist in a simple application? This appears very probable. Whatever it may be, the traces of it which remain upon the shell are always more or less evident, and form very fine striæ, more or less parallel or concentric. In the univalves there is almost always but a single muscular impression produced by the dorsal bundle of the columella, and which perfectly well indicates its form. Little or not at all visible in the spirivalves, in consequence of its depth, it becomes so in the species where the last turbination is very large, as in concholepas, haliotides, and the argonauts. But it is especially so in the patelloïd species, or those in which the shell is not rolled: its form is then almost always like a horse-shoe, open in front, for the passage of the head of the animal, and with branches more or less unequal. On some species of the non-symmetrical patellæ ot Linnæus, the right branch of the horse-shoe is divided into two by a smooth space, or canal, of no great depth, through which, without doubt, the water proceeds to the gills. Some other species of true patellæ have their muscular impression, as it were, lobate, or strangulated from space to space; and finally, some species not symmetrical, have really two distinct impressions, the horse-shoe being interrupted behind.

The shell of the acephalous mollusca presents, on the contrary, several muscular impressions, much more frequently than one. They are deeper, and result as well from the attachment of the ligamentous fibres as of the muscles.

We shall see by and by that the first, which have so much analogy with the epidermis, have not less with the dried muscular fibres of the byssus. Accordingly, the impressions which they leave upon the shell are absolutely of the same aspect; but two sorts have been observed, one external, or extero-internal, more or less elongated, occupying the dorsal part of the valves behind, and very rarely in front of the summits; the other, entirely or almost altogether internal, usually rounded under the summits, as in mactra, crassatella, &c.

The impressions produced by the muscular fibres are much more numerous. They may be divided into those of the adductor muscles, of the retractor muscles of the foot, of the attachment of the edges of the mantle, and finally of the attachment of the tubes.

The impression of the adductor muscles is sometimes simple, or unique, central or not, as may be seen in the ostracea or sub-ostracea, and in the pholades. It is sometimes subdivided, as in anomia.

It appears again to be unique in the mytilacea, but on considering it attentively, we see, altogether in front, a very small impression, which is the commencement of the double muscular impression, that is found in almost all the lamellibranch acephala, one part of which is buccal and the other anal. The form, the proportion, and even the position of these two impressions, vary very much, and furnish good characters for conchology.

The impressions of the retractor muscles of the foot are always much smaller, and are frequently confounded, especially the posterior ones, with those of the adductor muscles, where they form a sinuosity. They are numerous in the *mytilacea;* in the *conchacea*, the anterior one, alone distinct, ascends over the hinge.

The impression of the edges of the mantle, and that of the

attachment of the tubes, constitute what we name the *abdominal impression;* the one descending from the anterior adductor muscle follows the direction of the shell, in a variable breadth, and at a variable distance, and arrives at or passes the impression of the attachment of the tubes, which forms an excavation, or a sinuosity more or less deep, and open behind.

When a shell is finally come to its greatest degree of development in extent, the changes which it undergoes, always in relation with those of the animal, which tends to contract itself, especially in the lobes of its mantle, consist of little more than its augmentation in thickness, not by the augmentation of the strata which compose it, but by that of the vitreous matter, and in its increase of weight, by the diminution of the proportion of organic to inorganic matter in its composition. The external strata lose more and more the filiform productions, and the little epidermis which they might have had. The colours grow pale, are effaced, and disappear; the striæ, the tubercles, and even the varices, grow blunt, wear out, and become more and more levelled; the shell is covered with earthy cretaceous deposits, and with animals which excavate lodges there; the spiny and tuberculous elongations are filled and solidified. On the contrary, the ordinary sinuses grow hollow and become larger. Some cavities of this kind are even developed, especially in the female individuals, in places where there were none during the greater part of life, so as to form *pleurotomata* in a great number of genera. The aperture is narrowed, the posterior extremity of the cavity is filled or partitioned by the successive advancement of the animal, and the death of the latter, the necessary consequence of life, determines that of the shell. This shell then loses by little and little the animal matter which it contained, and finishes by being nothing but a composition of carbonate of lime, and consequently often becomes very friable. The insensible movement produced by the laws of attraction between the molecules, brings them to unite under an inorganic form, and to crystallize; and then the testaceous spoils of the mollusca have a greater or less tendency to disappear, and to form calcareous masses by their agglomerations, and especially by that of their pieces or detritus, which constitutes the formations of the shelly limestone.

According to all that we have now said, the shells present very considerable differences, according to the age of the animal to which they belong; and these differences sometimes prevail on the form of the aperture, and especially on that of the right edge of the univalve shells.

They also present differences according to the sexes in the dioecious groups, that is, where the male sex characterizes one individual and the female another, as we shall see by and by.

We shall now cast a glance on another production of the skin, the use of which is to render the protecting apparatus still more complete, and which is designated under the name of *operculum,* because it serves to close more or less completely the aperture of the shell, even at its orifice, or more or less deeply. Some authors, and among others Adanson, have regarded it as the analogue of one of the valves of a bivalve shell; but this is clearly incorrect, for its position, in relation to the body of the animal, indicates no analogy of the kind. The two valves of a bivalve are placed one on each side of its body, except, perhaps, in the palliobranches, while in the operculated mollusca the shell alone, dependent on the mantle, constantly occupies its dorsal face, and the opercle has never any connexion but with the upper dorsal face of the foot, sometimes at the angle of its junction with the pedicle of the body, rarely at its posterior extremity, and most frequently in its middle part. It is evidently the production of the skin which covers the foot. This production is, without doubt, an excretion of calcareous or corneous matter; but how a plane, oval, or circular surface produces a substance which rolls itself into a spire, and often in a very regular manner, and sometimes forming a great number of turbinations, is a question to which it is extremely difficult to reply, and more especially so as the subject has been by no means sufficiently studied. From this, however, may be derived good characters for genera and families, for the opercle differs not only in its point of attraction, in its size, relatively with that of the orifice of the shell, but also in its form, its chemical nature, and its mode of adherence.

We have already observed its principal differences with respect to its point of attachment.

As to its size, it is often sufficiently developed to close the aperture of the shell even at its orifice, as in all the cyclostomata, being applied almost on the edges. But sometimes it is much less so, and does not close it but when it has been more or less sunk in the spiral cavity. This is the case with almost all the siphonobranches. Finally, it also happens that it is almost rudimental; that is to say, that it can only close a very small part of the aperture of the shell, as in some purpuræ, in strombus, and more particularly in the cones.

This facility with which the opercle can enter more or less into the aperture of a univalve shell, has a necessary influence on its general form. In fact, when it remains at the orifice itself, applied in the little widening formed by the peristoma, it has constantly the form of its aperture; accordingly, almost circular in the cyclostomata, it is elliptical in the ellipsostomata, semi-circular in the hemi-cyclostomata, or nerites, &c. In the species where it sinks into the spiral cavity it again pretty nearly presents the form of its orifice, but it is much smaller. Finally, in those where it is only rudimentary, there are no longer any relations between its form and that of the aperture of the shell.

As to its specific form, it also varies in a manner determinate for each very natural group: oval or rounded in the shells of the siphonostomata, where it is always corneous; it is not formed into a spiral, but on one side we see the striæ of augmentation, which have commenced towards one of the extremities, and on the other a space more or less ovaliform, ornamented with subregular striæ, at the middle of a border or smooth swelling, much broader on one side than the other.

Another form is that of the calcareous or corneous operculum of the *anentomostomata.* It presents, in fact, constantly, a spiral enrolment on one and the same plane, more or less visible on the two faces, and constantly so on the internal one. But the summit of the spire varies much in its degree of eccentricity; sometimes it is even altogether central, as in the corneous opercle of the trochi, which is formed of nine or ten spiral turbinations; it is much less so in that of turbo; and finally, in nerites it is completely lateral. The presence or absence of colour on the external face of this kind of opercle, the disposition of the striæ which adorn it, the furrows with which the internal or adherent face is often marked, may also furnish excellent characters to confirm the distinction of genera and species. Unhappily, this part of the organization of the mollusca has been too much neglected.

This same property of entering or not with the animal into the shell, appears also to have some influence on the chemical, corneous, or calcareous nature, and on the thickness of the operculum. In fact, in the first case, it is constantly corneous, and most frequently slender and flexible, especially on the edges, while in the second it is often calcareous and very thick. It may, however, be simply corneous; accordingly, we sometimes find in the same natural genus of the anentomostomata some species which have a corneous operculum, and others a calcareous operculum, which is never observed among the siphonostomata and entomostomata.

Finally, the last relation under which the operculum can vary, is that of adherence. All the calcareous opercula, and even a part of the corneous, appear to adhere through all their internal or inferior surface, so as to leave nothing free but their circumference; while the corneous opercula of all the entomostomata are fixed to the skin only by a small part of the same surface, at their base, and are free in all the rest. This may be perfectly well seen in murex, buccinum, purpura, &c. In the hemicyclostomata, the adherence to the foot is made by means of one or two processes of the anterior or right edge, and the operculum seems to be articulated with the internal edge of the shell.

We must take care to distinguish the piece of the shelly envelope of which we have just spoken from the epiphragma, because, if there is any relation of the usage, which is completely to close the aperture of the shell, there is none of structure, nor even of position in relation to the animal. The epiphragma, or temporary operculum, is, in fact, but an aggregation of dried calcareous molecules produced by the edges of the mantle, or the collar of certain species of helix, when they have completely withdrawn their head and foot within the mantle. The stratum, more or less thick, which results, adheres in no wise to the animal, and several may be successively formed, in proportion as the unfavorable circumstances, such as cold, great drought, or absence of nourishment, which forced it to re-enter the shell, are more prolonged.

After this digression, into which we have been obliged to enter, respecting the shell as a dependence on the skin, or seat of touch, we shall now proceed to the examination of the apparatus of this sense, and successively of the rest.

The apparatus of the sense of touch in the mollusca consists in the tentacula, or the tentaculary cirrhi, with which the edges of the mantle may be furnished, and of which we have already spoken. We may also include certain tentaculary appendages, sometimes in the form of a fringed membrane, as in the Janthini, or even true flatted tentacula, as in certain turbines, monodontes, and nerites, which are on each side of the pedicle of the foot. Frequently, when these appendages are broader, they answer the purposes of swimming.

This sense becomes less delicate in the species whose external envelope, being always uncovered, is more or less tuberculous, and appears to be almost lost in those whose envelope is more or less solidified.

The organ of taste, when it exists, is doubtless, as in superior animals, situated at the lower part of the buccal cavity, where may be frequently remarked a lingual swelling; but it must be owned that the skin which invests this part does not appear to differ much from that which is at the orifice of the mouth itself, and in many other portions of the body. We shall presently see, however, that this skin is furnished with little corneous hooks, arranged symmetrically, which have some analogy with those which are observed at the superficies of the tongue of certain mammifera, and that it receives a great number of nerves.

The acephalous mollusca have no trace of this swelling.

The seat of the sense of smelling, which appears to exist only in the cephalophorous mollusca, is not, perhaps, as yet sufficiently determined; and in fact the nature of the skin of the mollusca, resembling in general, in its structure, the olfactory membrane of vertebrated animals, many persons have thought that the mollusca might have the power of smelling in all parts of their skin. Others having assumed, as a principle, that an odorant molecule must be suspended in a gaseous vehicle, have supposed that it was only the species which breathe the atmospheric air that could smell, and that consequently the seat of the function must be the edge of the respiratory orifice? But this is to deny without any reason the existence of this sense in the aquatic species, who questionless must smell as well as the others. Another opinion has obtained with some naturalists, and one which is sanctioned by the authority of M. de Blainville, that the extremities of the true tentacula form the organ of olfaction. In fact, the skin there is softer, more smooth, and more delicate than in any other part, and the nerve which repairs thither is more considerable.

That certain of the cephala do smell, may be inferred from the fact that slugs and snails seek after particular plants in perfect darkness.

The organ of vision has not given rise to so many opinions, because in its structure the relation of cause and effect is much more evident. It is wanting in all the articulated mollusca, and in all the acephalous. It is, on the contrary, almost certain that it exists in all the cephalophorous races, the hipponices perhaps excepted. But it is susceptible of very different degrees of development.

The eyes of these animals are never more than two in number, disposed very symmetrically, one on each side of the head, or of the anterior part of the body, in case the head is not very distinct.

Certain fibrous, vascular, and nervous coats may be recognized in the structure of these eyes; but the cornea belongs only to the skin. We also see there some humours and a crystalline very distinct. Sometimes there are even small muscles which can move them a little in a sort of orbit, or protecting cavity, as takes place in sepia and other neighbouring genera. But in general, these eyes would be immoveable, if they were not pretty frequently more or less pedicled; that is to say, carried at the extremity of a sort of tentaculum, analogous to that of olfaction, as may especially be observed in the family of the limacines, and by which their eyes can be directed by the animal in a great number of ways; or on the olfactory appendage itself, in a point of its extent, more or less elevated, as in the buccina, strombus, &c. In the instances where they are sessile, their position varies much, in relation to the true tentacula, for they may be anterior to them, posterior, exterior, or interior, all which positions furnish good characters in zoology.

The organ of hearing presents much fewer differences in the mollusca, and is found no farther than in the brachiocephala and a few other genera, where it is reduced to a small sac, hollowed at the lower lateral part of the cephalic cartilage, and which has not even any immediate communication with the exterior.

We shall now advert to the apparatus of locomotion.

We have observed, in treating of the structure of the skin in the mollusca, that the contractile fibre is not often distinct from the dermis itself, properly so called ; hence it happens that all the points of this skin are susceptible of contraction in all directions. It is, in fact, a thing perfectly certain, that all the external parts of a molluscum, and even the gills, can execute a number of vibratory movements; but that would produce little more than a sort of partial locomotion. The general locomotion is determined by a true distinct muscular fibre, visible at the internal face of the skin, and disposed in bundles, having a determinate form and direction. It sometimes has its point of rest on the solidified part of the skin, but nevertheless, it is very seldom that this part can, in reality, serve for the purposes of locomotion, except in the articulated mollusca.

The disposition, the number, and even the form of the muscles, in this type of animals, are necessarily in relation with their general form. Thus, wherever there is a well-marked separation between the head and trunk, there are superior lateral and inferior muscles, as is observed in the brachiocephalous mollusca; but in all the rest of the trunk this distinction does not take place. In the oscabriones it is different, each articulation of the back having its particular muscles; but they can scarcely be considered as genuine mollusca.

The mantle which envelopes the body of the mollusca,

although the muscular stratum which doubles it does not form distinct muscles, presents differences only in the thickness of this stratum in different points of its circumference. Thus, sometimes, this thickness is pretty nearly the same, whence results a sort of sac, as in sepia, &c.; but more frequently it is much greater at the lower part of the body, where even the fibres, though longitudinal, experience frequent intersections, and a sort of muscular disk, more or less thick, is thus formed, to which the name of *foot* is given.

In a number of cases, this sort of foot extends through the whole length of the body, or rather of the mass of the viscera which is above, and forms a kind of sole, somewhat varying in form, by the assistance of which the animal creeps, and which occupies its entire belly. From this conformation comes the denominations of *repentia*, or gasteropods, which have been given to the limaces, and neighbouring genera.

In some species the visceral mass comes forth, as it were, beyond the mass of the body, being rolled more or less spirally; and the foot, containing no more viscera in its posterior part, where it is free, appears to be no further attached to the body, except before or behind the head, at the part which might be regarded as the neck; from which arrangement these mollusca have obtained the name of trachelipods. Most of the cephalous conchyliferous mollusca are thus formed, especially when the shell is strongly spiral and rolled.

The proportional size, and even the form of this foot, vary much, sometimes even in genera pretty much approximating to each other. Thus it is almost circular in the patella, oval and very thick in the haliotides, rounded in front, and slender on the sides in the turbines, buccina, &c. It is auriculated on each side in some. In a great number of genera of the order cyclostomata, it is cut by a transverse furrow at its anterior edge. The foot of the auricula pecten is divided into two heels by a broad transverse furrow.

All the species of cephalous mollusca are eminently gasteropods. It is not thus with the conchyliferous species: they are not necessarily trachelipods, though this be most commonly the case. But what they present in common is, that the shell is in muscular communication with this foot, so that the latter can be drawn in more or less deeply, as well as the head itself, by a muscular bundle called the muscle of the columella, because, in the spiral shells, it is to that part it is attached. The disposition of this bundle of muscles differs much, according to the form of the foot, and especially according to that of the shell. Thus, when the latter is simply covering, and not spiral, as in the patelloïdes, the bundle in its dorsal attachment forms a sort of horse-shoe, open in front or behind, and the termination of the foot extends almost through its entire circumference. In those whose shell is a turriculated spiral, the common bundle is pointed at the columella, and is carried more or less obliquely from the hinder part to the front, and from top to bottom, towards the middle of the foot, so that by its contraction it almost bends the latter in two in withdrawing it within the shell. In the species where the rolling is lateral, as in the porcelaines, it is, on the contrary, a broad muscular band, which is longitudinally inserted at the columella, and which terminates at the foot, in such a manner as to bend the latter in its length, to draw it within the shell.

We may also regard as making a part of this muscular bundle the more or less considerable muscles, which are carried forward to proceed to the tentacular and ocular appendages, where these organs can be withdrawn internally, as is the case with the limacines. They penetrate, in fact, into these tentacula, and proceed as far as their extremity, surrounding the nerve, which also repairs to the organ.

In the species which have these sorts of tentacula, and have not a crustaceous shell, and whose foot has consequently no retractor muscles, we do not the less find the retractor muscles of the tentacula. But their origin takes place above the muscular partition, which separates the visceral cavity from the pulmonary cavity, pretty nearly at the same point as in the conchyliferous species.

It is also nearly from the same point that proceeds the retractor muscle of the excitator organ when it exists.

Finally, it is also from the muscular bundle of the columella that springs the retractor muscle of the operculum, where this part exists, and to which is attached the siphon of such species as are provided with one.

We have observed, a little farther back, that some cephalous mollusca are provided on each side with locomotory appendages, pretty considerable, as the sepiæ, calmars, and pteropods generally speaking. In this case these appendages have levator or depressor muscles, which are carried from the back or the belly to their roots. But when the appendages do not really serve for locomotion, they are formed of a contractile dermis, in which it is not possible to distinguish true muscles.

The acephalous mollusca present a disposition of locomotive organs considerably different, and which appears still more so when the passage from the cephala to the acephala has not been very precisely observed, as in all the mollusca; in general, all the parts of their envelope, whether branchial or not, are really contractile. But we sometimes remark besides some distinct muscular fibres, which, from the edges of the mantle, more or less thick, proceed to fix themselves to the shell at a little distance from its circumference, so as to be capable of drawing those edges in more or less; and more rarely we perceive small slender muscles, which proceeding from the adductor muscles, of which we are about to speak, direct themselves into the different points of each lobe of the mantle. In the cases where the latter has only this last sort of muscle the shell does not present any submarginal impression, and the mantle is considerably retractile; but in the contrary case an impression is very perceptible, in the form of a strip, which follows more or less regularly the edge of the shell, descending from the anterior muscle, and which often forms behind a large flexuosity drawn inwards, and which indicates tolerably well the size of the posterior and tubular elongations of the mantle. In this last disposition the mantle has no part contractile but that which is between its edge and this line of insertion.

We often find, besides, that the middle of the abdomen is occupied by a muscular mass more or less thick, polymorphous, and which, besides its intrinsic contractile fibres, has also its extrinsic muscles. This mass has received the name of *foot*, as that which occupies the lower part of the gasteropods, of a form and size extremely variable, to such a degree that there sometimes exists no trace of it, as in the oysters, it is attached more or less in front, which depends upon the habitual position of the animal; but, besides, it may be carried in different directions by true muscles, which divided into one or a greater number of bundles, direct themselves to different points of the shell, and especially in front, behind, and in the intermediate space, as is observed in the *mussels*, anodon, &c. This extensible foot sometimes resembles a sort of cupping-glass, as in the *nuculæ*, sometimes a sort of tongue, as in the mussels where it is channelled behind; sometimes it is like a hatchet, as in *Venus*, sometimes like a sort of human hand, as in camus, and finally, like a sort of whip, as in the loripedes, &c.

Besides these muscles of the foot, which have evidently some analogy with those which we have seen serve for its retraction in certain gasteropod mollusca, as the patella, there are some others directed transversely, that is, from one side to the other of the animal, and each extremity of which is attached to one of the valves in such a manner as by their contraction to approximate one to the other. These are the adductor muscles; sometimes they form but a single mass drawn together in the middle of the valves, at other times the mass tends to subdivide into two or three; finally, in a great number of cases there are two very distinct, one anterior and the other posterior, the form and proportion of which, moreover, are variable. It is the insertions of these muscles in the valves of the shell which form what are named the impressions or the muscular attachments, in the same manner as it is that of the edges of the mantle which forms at the inferior and posterior edges of the shell a line more or less broad, more or less sinuous, or retracted backwards, which we have already noticed in treating of the shell, and the consideration of which is not without importance in conchology.

In the locomotory apparatus of the bivalve mollusca we ought also to consider the mode in which the pieces of the shell unite, or their system of articulation, and especially the ligaments, which serve not only to retain them in a determinate relation, but also to act as antagonists of the adductor muscles. On the first we shall make a few remarks when we come to speak a little more on conchology. As to the ligaments, we may add here, that, formed of a corneous substance, evidently epidermic, they are composed of transverse fibres which pass from one valve to the other, just as do the contractile fibres of the adductor muscles, and which have much resemblance to those which constitute the dried part of the true byssus, and still more of the tendinous foot of the tridacne. The ligaments observed in the shell of the acephalous mollusca may be distinguished into epidermic, external and internal. The epidermic ligament is that which is formed by the epidermis itself of the valves, which is continued in passing from one to the other, as in solens, &c.

Perhaps we should consider in the same way the ligament of the arca and some kindred genera. The external ligament is always more thick, more gibbous, and more elastic. It always occupies the back of the shell, just behind the summits. Finally, the internal ligament, simple or multiple, is that which is more inside than the line of articulation. Its fibres are usually short and straight. The mactra, crassatella, pectines, &c. present us with examples of this kind of ligament.

One of the most remarkable peculiarities in the acephalous mollusca is, that in many species, a greater or less number of the fibres of he adductor muscles may be attached and agglutinated to foreign bodies, so as to serve for an external point of attachment for the animal. It is this which constitutes the byssus in the pinna, and the mytilacea, and the tendinous foot of the tridacne, and certain species of arca. This byssus is not really formed, as some authors have asserted, of a mucosity secreted by a gland, and spun (like the silken productions of insects) in a groove in the foot. It is nothing but an assemblage of muscular fibres, dried up in a portion of their extent, still contractile and living at their origin, and which were so throughout their whole length, at the period in which they became attached.

Another singularity, not less worthy of remark than the preceding, is the observation of the march, or change of place of the adductor muscles, in proportion as the animal grows, as well as its shell. In fact, if in a very young shell the muscle is subcentral, it must necessarily happen that, to retain this position when the shell is twice as large, it must have moved as the shell grew. It is admitted, that in aspiri-valve shells the muscle seems to descend with the animal in like manner as in a bivalve shell the subcentral muscle advances, not that it detaches itself all at once, but that an anterior rank of fibres is detached at the same time that a posterior rank is produced. But if it was exactly this, we ought to find at an advanced period of the animal and its shell a part of the impression which would be without muscular fibres, and this has not been observed, even in oysters and other bivalve shells, where it would be necessary besides that it must exist in an inverse direction for every muscle. It therefore may be preferable to admit that the muscles grow like all the rest of the organization, throughout their whole circumference, but especially on the side where the shell grows the most, as behind, which is the case with oysters; perhaps it might also be supposed that the muscle altogether is at a certain point detached when the new stratum of growth is formed, and that it is thus that its apparent movement takes place, for the thickness increases equally at the place of the attachment of the muscles, where the strata are, however, in general more crowded, which causes this impression very frequently to form a sinking in, and in the fossil state occasions this part to be preserved longer than the rest

The apparatus of locomotion is yet more different in the balani and anatifæ. In the first the mantle is very slender, and presents no muscle but at its posterior or open extremity, for the movements of the pieces of the operculum. In the second it presents, besides, this singularity, that at its cephalic or inferior extremity, in consequence of the position of the animal, it is prolonged into a fibro-contractile flexible tube, which attaches the animal in a fixed manner to submarine bodies. There is, moreover, an adductor muscle between the two principal valves of the shell.

As to the muscles of the animal itself, or of the trunk and its appendages, their disposition evidently begins to approximate to that of the entomozoaria.

The oscabriones have also in the assemblage of their locomotory apparatus something of the true mollusca and something of the last mentioned class. In fact, the whole inferior part of the body is occupied by a sort of foot very analogous to that of the patellæ, phyllidii, &c., while the back, in its conchyliferous part, presents as many double pairs of oblique muscles, one at the right and the other at the left, as there are testaceous pieces.

The faculty in the molluscum of changing its relations with external bodies being generally in a direct proportion with the quantity of the sensibility, it is evident that the locomotion of the malacozoaria cannot be very active, must be confined in its range, and often does not exist at all.

The brachiocephala *(cephalopods)*, being those mollusca whose sensorial faculties are most extended, are also those which move with the greatest quickness, and in all directions. The acephala, and especially the last of them, such as the ascidiæ, are exactly in the opposite extreme, and in fact they live fixed upon immerged bodies.

We remark, however, among the mollusca, many kinds of locomotion : a certain number swim by the aid of fins, or sorts of appendages, in pairs, such as the calamary, the sepia, the pteropods in general, and many monopleurobranchia, pretty nearly in the same manner, as do the fish with their pectoral fins. These organs even enable them sometimes to issue from the water and shoot into the air, to different degrees of distance. This is a certain fact with respect to the calamary. The same may be said of certain species of bivalves, which thus employ the valves of their shell as a sort of wings, with which they take their point of support upon the water.

Another sort of motion is that which is performed by an odd middle fin, or by a very compressed foot, and consequently by alternate movements to right and left, as may be seen in the firolæ and carinariæ. But in this case motion appears never to take place but in a reversed situation ; that is to say, the back being under and the belly uppermost.

There is a third method, more singular, and which is to be met with in the first species of the type and in the last. It is executed by the contraction of the envelope, which thus expels the fluid with which it has been filled in its dilatation, from which results a movement of translation, often very quick. The sepia, the calamary, or loligo, and the biphoræ, move in this manner.

Some mollusca sail upon the surface of the waters, impelled by the current, or by the wind, some by the assistance of a sort of hydrostatic bladder, as the janthinæ and others, by unfurling a sort of sail, formed by the salient edge of the mantle, or by some widened appendage, while they row with others, as is related of the argonautæ. In the first instance, it appears that the animal is always at the surface of the water, for it cannot make the bladder re-enter, which is subcartilaginous. In the second, the octopus can, it is said, at will, develope its sail and oars, or refold them within the shell, which serves it as a boat, and then sink more or less deeply. But after all, may not some degree of doubt yet attach to the truth of this ingenious manœuvre?

Perhaps it is the octopi only which perform a sort of walk, by means of the long appendages which crown their head; but then they have the mouth below and the trunk uppermost. It appears also that they can roll over and over at the bottom of the sea with great velocity, and without fixing themselves by their tentacula.

Some species of auriculæ, and even the terrestrial cyclostomata, also make sorts of steps, by taking a resting point on the anterior part of the foot, or on the advanced muzzle, and approximating to it the posterior part or centre foot all at once.

A much greater number crawl on the surface of the ground, either on land or in the waters, by means of the foot or muscular disk with which their belly is provided. But this sort of reptation has no resemblance to that of reptiles : it is rather a kind of sliding of the foot, produced by extremely fine undulations of all the little longitudinal fasciculi which compose it, and which proceed in succession from the first to the last, each being alternately the point of resting, or the fixed point for the following. By this mode of locomotion, the animal touches one after another all the eminences, all the anfractuosities of the ground on which it moves; its advance is therefore in general very slow. Nevertheless, the species whose foot is broad, thick, extended, and has no shell to drag along, especially where the distinct contractile fibres have an evident fascicular direction, as in the limaces, helices, &c., can get on with a much greater degree of rapidity than one would be led to suppose from their first aspect. Others, on the contrary, such as the patellæ and the haliotides, though possessed of a broad foot, creep so slowly, and so rarely change place, that some persons have erroneously believed that they never do so. They are capable, besides, of adhering to a fixed point with very great force, through the viscosity of their foot and the vacancy which it can make, either altogether or in little fossettes.

The hipponyces and some others remain fixed to the bodies on which they have fallen when born. Accordingly, their foot is scarcely muscular, and much resembles the horse-shoe muscle of the back, which serves as an attachment for the shell.

The scyllæ, whose foot is extremely narrow, and, as it were, canaliculate, cannot move but along the stalks and pedicles of plants, and always in a gliding manner.

A tolerable number of species can also creep along the surface of the water, taking as a *point d'appui* a light stratum of this fluid. But then they are obliged to do it in an inverted position ; that is to say, the shell being under, and the inferior surface of the foot above. This is observable in the limnææ, the planorbes, the paludinæ, glaucus, doris, thethys, &c. The theory of this movement is in other respects precisely the same as that of the reptation of ordinary gasteropods.

The movements of the acephalous mollusca are often confined to the trifling opening of the valves of their shell, and to their complete occlusion. The first is the natural position, or the repose of the animal; and in fact it is only then that it can receive the water which brings it nutriment, especially when its mantle is not provided with external tubes. It is produced by the disposition of the ligament of the hinge, the perpendicular fibres of which, at each valve, are drawn or compressed according to their position without or within the point of support, when an attempt is made to cause the two valves to touch. Their closure is, on the contrary, entirely active; that is to say, owing to the contraction of the fibres of the adductor muscles, which are the antagonists of the ligament. Willis, and more lately, Dr. Leach, have considered, that in oysters, a part of the central adductor muscle was formed of elastic substance, an antagonist of the other part which should be alone contractile. Some doubt seems to attach to this notion.

The family of the palliobranchia contains many genera, in which, instead of a ligament, the two valves of the shell are united at their summit by a long elastic tube, which is fixed to submarine bodies, and which might even perhaps be a little contractile. Nevertheless, the animal has no other movement than that of opening and closing the shell, like the other acephala.

In the species fixed immediately by the shell, or by a tube,—such are the only motions permitted,—there is then not the slightest degree of transference of situation. In all the others there is one, though to very different degrees. Thus many species are almost in the situation of those of which we have just spoken; that is to say, they are fixed, but it is with a certain degree of mobility. Such are they whose attachment is made by dry muscular fibres or byssus, as some species of pectines, the limæ, the crenatulæ, and especially the mussels, and the pinnæ. In this case, it appears that the filaments of attachment are fixed to solid bodies, by means of the canaliculate foot with which these animals are provided, and which in fact appears very extensible and very long. They cannot detach themselves from their position, but if they have been thus detached accidentally, they can re-attach themselves to their post.

The arcæ, and even the tridacne, can also fix themselves to solid bodies, by a sort of agglutination of their foot, something like the byssiferous species, but *en masse*, and not fibre by fibre. Accordingly, it may happen that from the growth of the animal it becomes naturally detached. This is at least presumable from the observation that the shell of the tridacne loses, in becoming more bulky, the great præcordial aperture which it had when small, and through which the muscular bundle passes.

In most cases, the acephalous mollusca not being adherent, can change place. They move by the assistance of their foot. Some, however, are confined to a movement of ascent or descent in the hole which they inhabit, whether it be excavated in a stone, in the sand, or in mud. Their foot, attached more anteriorly than in the other species, comes out more or less, is elongated, and takes its point of support on the bottom of the lodge. This is the case with all the pyloridæ, and perhaps a little with the tubicolæ.

All the other bivalve mollusca, although again they frequently live more or less buried in mud or sand, can issue from their retreats at will, and consequently completely possess the power of motion. Some of them move by leaping, almost as if they were impelled forward by a spring. For this purpose their foot, very much extended, is bent lengthwise, and then suddenly straightened. It is this singular mode of locomotion which caused M. Poli to extend, though evidently erroneously, the denomination of *subsilentia* to all the acephala; for if most of the animals of the family of the conchs can leap in this manner, the submytilacea, the arcacea, &c. cannot, and appear in reality to crawl with their foot; much less can a similar power belong to the species which have but a rudiment of this organ, or to those which do not possess it at all.

The oscabriones move by crawling with their abdominal foot pretty much in the same way as do the patellæ. As for the nematopods, no species belonging to them possesses the faculty of changing place altogether. The appendages of their caudiform abdomen can issue from the shell, and move in the water, but, as far as it appears, only to determine the current of that fluid into the interior of the mantle of the animal, and to seize the little animals that pass within reach.

The composition of the mollusca is complete, that is, it is formed of organs of digestion, of respiration, and of circulation. We shall first treat of those of digestion.

The mouth is always anterior in the mollusca, although it be not constantly terminal or visible. It is sometimes altogether inferior, as in doris, onchidion, scyllæa, oscabrio, &c. In fact, it does not appear that it is ever situated superiorly.

Its form is extremely variable, depending on the disposition of the lips, which is different in the different groups. Thus in sepia and the neighbouring genera it is a sort of circular veil, sometimes double, pierced in its middle and fringed in its circumference. In the polybranches, cyclobranches, inferobranches, and even in many cervicobranches, they form a thick semicircular pad, in the inferior middle of which is the mouth, and which is prolonged sometimes on each side into a sort of appendage, which forms the labial tentaculum. In many species of doris, in tritonia, &c. the anterior edge of this labial pad is dilated, fringed, and forms a membranaceous veil of greater or less extent; at other times we find that the lips are prolonged into a sort of cupping-glass, in the bottom of which is the proboscis, as in the cones; finally, we often remark that they are prolonged in the same way, but acquiring a tolerable degree of thickness, and there results a proboscidiform muzzle, as in a great number of species of the family of cyclostomata, an organ susceptible of contraction or elongation, but without ever being able to re-enter into the buccal cavity, which distinguishes it from the true proboscis, of which we shall presently speak.

Within these contractile lips, and sometimes provided with some small specific muscles, are often found corneous or calcareo-corneous organs, to which the name of jaws has been erroneously applied; these in fact are genuine teeth produced from the skin, which they cover, and the structure and mode of formation of which are altogether analogous.

We rarely find two of these teeth acting vertically one upon another, as in sepia, or horizontally, as in tritonia. In such cases they are surrounded at their bases by a thick circular muscle, which presses them strongly one against the other, after they have been separated by the action of the levator muscles of the upper, and the depressor of the lower.

In a much greater number of cases there is but one upper tooth in the form of a comb, curved and denticulated on the edge. It is then pretty nearly immoveable, and the tongue, of which we shall presently speak, acts upon it. This is what may be seen in all the animals of the family limacines, in that of lymnæa, of auricula, and even of patella.

In a much greater number, again, there is no trace of true marginal teeth, as in all the cephalous mollusca, provided with a proboscis, and in the entire class of the acephala.

At the lower face of the buccal cavity there often exists a swelling more or less considerable, which has been compared with some justice to that which forms the tongue in the boned animals. This swelling is regular and symmetrical, and receives a considerable number of nerves. Its upper surface is most usually furnished with very small corneous hooks, the point of which is directed backwards, and which are disposed in a very symmetrical manner. These are again dependences, productions of the skin, but which cannot be compared in consequence of their disposition and place to the marginal teeth.

This sort of tongue is never exsertile in front but along with the whole buccal mass, but it is sometimes prolonged in a singular manner behind, into the interior of the visceral cavity, being contorted like the spring of a watch. In general its disposition differs according to that of the teeth.

In the species which have two teeth opposed, as in sepia, &c., the lingual plate is not very projecting nor very mobile.

In those which have an upper tooth the lingual swelling is thicker, more mobile, but much shorter, and is easily carried forward by the motion of the adductor muscles. This may be seen in the limaces, helices, bulimus, limnææ, &c.

In the species which have no teeth whatever to the buccal orifice we find that the tongue forms a long narrow band, which is prolonged backwards into the abdominal cavity, rolling spirally. Its surface is bristled by a number of little hooks, bicuspid or tricuspid, directed backwards, and whose solidity or resistance, proceeds always decreasing from the base to the point, where they are soft and very little apparent. This singular kind of tongue is found in the porcelains, cones, patellæ, and even in the oscabriones.

Finally, in a great number of species which have no teeth, properly speaking, we observe that by a singular disposition of the œsophagus it may be prolonged externally or re-enter into the buccal cavity under the form of a cylindrico-conical organ, to which the name of proboscis has been given. Beside the muscular dermis which composes this organ, and which is capable of being elongated or shortened, according as the longitudinal or transverse fibres act, we find at its base some extrinsic muscles which facilitate this action, some in drawing it backwards and others in carrying it forward.

In the species which are provided with this sort of proboscis there has not been seen any lingual swelling, properly so called, and consequently no corneous hooks. But this swelling is pretty frequently replaced by a double group of hooks placed at the right and left, and which are more or less deeply sunk into the proboscis, so that they do not become marginal but when it is strongly turned back. This takes place in buccinum and neighbouring genera. In the *vis maculata*, which has also a very long proboscis, there is no trace of these hooks.

Other species have been remarked whose palate is armed with a plate of corneous teeth, like the tongue; such are many monopleurobranches, and among others bulla and umbracula.

No acephalous molluscum presents any traces of teeth, nor any lingual swelling whatever. But the aperture of the mouth, of variable form, though ordinarily very large, and almost always inferior, is accompanied with two lips, most frequently simple, sometimes fringed, which are prolonged at their angles into labial or tentacular appendages. These appendages, of a triangular form and very variable size, are striated, especially at their internal face, so as a little to resemble gills, with which their connexion is always tolerably intimate. They are almost always very soft, and directed backwards. In the nucula they are, on the contrary, stiff, and directed towards the mouth, so as to represent sorts of jaws.

The salivary apparatus, deficient in all the acephala, present in most of the cephalophora, is ordinarily simple, that is to say, formed on each side of a single salivary gland, which commencing more or less behind, on the sides of the intestinal canal, or even placed free in the visceral cavity, traverses the nervous ring, to open into a part of the buccal aperture. Sometimes the salivary apparatus is composed of two glands on each side, one disposed like that which we have just described, and the other filiform, and often extended very far along the intestinal canal. The cones have one very singularly formed, unequal, situated in the visceral cavity, and of which the excretory duct, very long and reentering, proceeds to open at the base of the tongue.

The union of organs of which we have just spoken constitutes a mass more or less considerable, ordinarily oval, which is sometimes perceptible through the skin, but most frequently indistinct. This buccal mass is surrounded by a great number of muscles, which can draw it forward, carry it back, and sometimes make the lower part act upon the upper.

No indication of this is found in the acephala; it is, on the contrary, very strong in most of the cephalophora, especially where there is a true mastication.

It is almost always at the superior and posterior part of this mass that the intestinal canal, properly so called, commences with an œsophagus, whose diameter is always more narrow than its own.

The intestinal canal of the mollusca, taken generally, is composed of an internal mucous membrane, most usually forming longitudinal folds, and of a muscular stratum more or less distinct, but evidently contractile in all its points. Its extent, its stomachal swellings, its direction, and its circumvolutions seem to present a great number of variations.

Thus we sometimes find an œsophagus long and narrow as far as the stomach, or an œsophagus very broad, very large, as in most of the phytophagous mollusca. We also see, though more rarely, a sort of distinct crop, as in some brachio-cephala. Its direction, sometimes almost medial, as at its origin, is often from right to left, so as to unite itself to the stomach on this last side.

The stomachal swelling, often simple and indistinct, is on the contrary, in a great number of species, divided into several pouches or lodges. Sometimes even one of these pouches has its parietes comprised between two muscles very thick, almost as is the case in the gizzard of birds. In peronia and limnæa the conformation is similar. We find also in many species, and among others in the monopleurobranches, that the internal membrane of the stomach is armed with corneo-calcareous productions, very analogous in their structure and composition to teeth, which occurs even in the shell.

The stomach of the acephalous mollusca has not its parietes distinct. Of a form usually irregular, it appears hollowed in the tissue itself of the liver, which envelopes it on all sides, and pours in the bile through numerous apertures or sinuses, very large, in which very singular bodies are observable, whose use and mode of formation are completely unknown. They are named *crystalline* stylets, because they are usually in the form of stylets, the point of which is in the canals, and they are a little transparent.

In the cephalous mollusca, where nothing similar has been yet remarked, the liver never completely envelopes the stomach, neither does it adhere to it. It is even most frequently carried back into the remotest part of the visceral mass, and to the very point of the spire. It is composed of lobes and lobules, the last of which are in the form of hollow globules. From each of these lobules springs one of the minute biliary ducts, which successively uniting, constitute one or three or four thick canals, opening freely into the stomach itself, or sometimes into the commencement of the intestine. This structure of the liver often permits it to be inflated with the greatest facility. This is most especially evident in the brachiocephala.

The liver sometimes appears more considerable in the phytophagous mollusca than in the zoophagous.

The intestine, properly so called, varies more than the stomach in its diameter, the number and form of its circumvolutions, in its direction, and in the point of its termination.

Most usually its circumvolutions between the lobes of the liver, from which it is at times rather difficult to separate them, are in the posterior part of the body of the animal. It is disengaged from them pretty often, proceeding in a middle line underneath and in front, or above and behind, but often also proceeding from left to right, or in front towards the anterior and right side of the animal, where the anus exists.

The acephala offer perhaps less variations in the extent, the circumvolutions, and especially in the mode of termination of the intestine; in fact, after having formed a sort of handle, more or less large, in the liver, and sometimes a cœcal sinus at the root of the foot, it re-ascends towards the back of the animal, places itself in the middle line, and directs itself from front to rear, where it is terminated in the cavity of the mantle by a free elongation, more or less considerable, at the extremity of which is the anus.

The position of the anus in this class of mollusca is thus almost constantly the same, and it is pretty generally pedicled. The same is not the case with the cephalous mollusca; in these the anus, sometimes medial, inferior and anterior, as in the brachiocephala, is sometimes medial, posterior, superior or inferior, as in doris and peronia. Finally, in most cases it is placed at the right, sometimes altogether in front, as in limax, or altogether behind, as in onchidia. When it is at the left the animal and its shell are so. The haliotides and ancile, nevertheless, have it on this side, and are rolled from left to right.

The mode of nutrition in the malacozoaria is in general much less known than that of their locomotion.

A very small number can seize their prey before they introduce it into the buccal cavity. To effect this, the singular appendages with which their head is provided interlace and attach themselves very closely, by the aid of the sort of cupping-glasses with which they are furnished, to the living animal which is about to be devoured.

The mollusca whose buccal orifice is furnished with teeth appear to be able to seize and chew their food with them. When there is but one tooth above it serves as a resting point on which the lingual swelling acts in its anterior part, which is very observable in the limaces, helices, and approximating genera.

We are not equally well acquainted with the mode of action of the proboscis in such mollusca as are provided with one. It is believed, however, that the teeth with which it is often armed at its extremity, when it is sufficiently unrolled, may serve to make a hole in the shell of other mollusca, through which this proboscis may be introduced to tear or suck the soft parts ; but this is very far from being indubitably ascertained.

The mollusca feed on all kinds of substances, animal or vegetable, in all states, living or dead, fresh or putrified ; but each species, and even each genus, but less certainly each family, confines itself to one or other kind of these aliments.

All the known cryptodibranchia are nourished upon living animals, which they tear, and perhaps break, but which probably they do not masticate.

The siphonobranchia appear also to be all carnivorous ; but it is probable that they rarely swallow their prey entire, but that they suck it, draw it into their proboscis, armed or unarmed, but that they do not chew it, being unprovided with organs adapted for true mastication.

The asiphonobranchia appear in general to be less carnivorous; perhaps they are not so at all, or take indifferently animal or vegetable nourishment, in a state of putrefaction. They seem, in fact, to make use of their proboscidiform unarmed muzzle rather to swallow rotten vegetable matter, than to masticate it. This is certain, at least, in the terrestrial cyclostomata.

The pulmobranchia, on the contrary, are most generally phytophagous, and they chew or cut the substance of which they make their nutriment in small pieces, which they swallow by little and little. We have already seen that their mouth is armed always with an upper cutting tooth, and denticulated on the opposite side of the lingual mass. It is said, however, that the testacella swallows earth-worms whole, drawing them by little and little into the intestinal canal.

The chismobranchia, the monopleurobranchia, are probably in the same situation as the asiphonobranchia, since they have no teeth to the mouth.

The pteropods also appear not likely to chew their prey, but to suck it, or take it in a state of decomposition, for the same reason.

We may say as much of the cyclobranchia, inferobranchia, and even of the polybranchia, although in this last order there are some genera, such as those of tritonia and scyllæa, in which there are two jaws acting laterally, like the blades of scissors, and which consequently ought at least to cut their nutriment.

As to the nucleobranchia, it appears that they feed on small animals. The cervicobranchia are perhaps in the same way, but it is more probable that their aliment is composed of substances in decomposition.

In the entire class of the acephalophora, this must be still more necessary, since the mouth of these animals, soft in all its parts, could not exercise the least action on bodies of the slightest solidity. Accordingly, it is probable that they feed upon animal and perhaps vegetable particles, the result of the decomposition of the beings of either of these kingdoms, and which are drawn along with the fluid which enters into the cavity of the mouth for respiration. It might also happen that their nutriment is composed of the innumerable animalculæ which the microscope enables us to observe in the water where these animals live, and which are of an extreme softness. The nuculæ may perhaps feed on substances of greater solidity, as might be supposed from the disposition of their labial appendages.

According to the nature of the aliment, and the state in which they seize it, it is evident that the means which the mollusca employ to take it must be very different.

Species, such as the brachiocephala, and even the testacellæ, which feed upon living fugitive prey, are obliged either to pursue it when they have the means, as the sepiæ and loligines, or to wait in ambuscade and cast themselves suddenly upon it. This is the case with the octopi, and probably with the testacella.

Those which, on the contrary, feed on living animals, but immoveable, attach themselves upon them, pierce their envelopes, of whatever nature they may be, by the assistance of the hooks with which their proboscis is armed, and consequently have but little difficulty in finding their prey, which is generally incapable of all sort of motion.

The mollusca which feed on animal or vegetable substances in a state of decomposition, seek them, without doubt, guided essentially by the smell, and have need of no great efforts to obtain them.

There are even some among those which, like the greater portion of the limacinæ, compose their aliment of living vegetable substances, more or less solid. They have only to seek them out, and cut them into small pieces.

Finally, for the species whose nutriment consists of molecules already disunited, or of microscopical bodies suspended in the fluids in which they live, there is no necessity of research or prehension. It is sufficient for the animal to produce in the water an almost circulatory movement of this fluid which should bring with it the nutritive substance, and probably to swallow this substance and the vehicle at once.

The mode in which the species with a spiral tongue use that organ is totally unknown.

We know but little more concerning the method by which the acephala take their food. It would appear that it must be almost in a molecular state, suspended in the water, and carried by the buccal appendages to the mouth; for there is no indication of masticatory and salivary apparatus.

The palliobranchia, by the aid of their long labial appendages, should be better able to seize their food, since they can put them forth out of their shell, and agitate them in all directions. The ascidiæ and biphoræ, having no appearance of such apparatus to the mouth, must be placed in a situation altogether opposite.

Deglutition, at least in the cephala, must take place as in animals of a higher grade.

The organs of respiration are pretty nearly known in all the true malacozoaria and in all the malentozoaria; but they vary considerably, not only with respect to form and situation, but even as regards structure.

In fact, in this last respect, although in most mollusca the part of the exterior envelope, modified to form the organ of respiration, be disposed like gills—that is, in such a manner that it shall be an organ capable of immersion in the ambient fluid—it sometimes happens that there is a contrary disposition, and that it forms a sort of pouch or cavity, into which the ambient fluid shall penetrate, which constitutes an aërial or pulmonary organ; and then the afferential and efferential vessels invest the internal face of this cavity. This disposition takes place in the different species of mollusca which live habitually in the air; but these mollusca may really appertain to different families. The greatest number, however, belong to those of the limacinæ and limnææ; but there are some also in the family of the cyclostomata, and that of the cyclobranches, and even, according to M. de Blainville, in that of the cervicobranches; for it would seem that the true patellæ respire by a lung, and not by gills.

The form of the organs of respiration varies still more. In fact, in the aërial mollusca it is always a cavity more or less inclining towards oval; but in the aquatic, the organ may be simple or multiple: it may be formed of species of ramified arbuscula, as in the tritonia; of little tufts, as in scyllæa; of laminæ, or shred-like strips, as in the carolinæ and eolides; of triangular pyramids, very large, one on each side, as in sepia, &c.; or very small and numerous, as in phyllidia and oscabrio, the latter of which, nevertheless, so much differ from the former; of sorts of combs more or less elongated, as in most spirivalve cephala, in the dismembered genera of the patellæ; and of large semi-circular plates, as in most of the acephala; or, finally, of a reticulated tissue, as in the ascidiæ; or of a long fringe, as in the biphoræ.

The situation of the respiratory organ presents, perhaps, still greater variations than its form. Thus, in a great number of species, it is exterior, and can then be constituted only by gills. This is observed in all those genera which, from this circumstance, have been named by M. Dumeril *dermobranches*, and by M. Cuvier *nudibranches*. It even occurs in the inferobranches. This disposition would be still more evident in the pteropods, if it were certain that the gills formed a net-work at the surface of the natatory appendages. In all the others, the respiratory organ is more or less interior, but more in the pulmonary than in the other genera, where it may be almost exterior, as in certain monopleurobranches and cervicobranches. In the brachiocephala, the gills are contained in the sac formed by the mantle.

In all the acephala, the gills are between the mantle which conceals them and the body.

The place which the organ that we are now examining occupies also varies in a notable manner. Thus it is sometimes at the upper and posterior part of the body, as in doris, peronia, and even in the testacellæ; it is at other times on each side of the back, as in scyllæa, the eolides, the tritoniæ; in other species it passes underneath, all around the edge of the mantle, between the foot and this last, as in phyllidia, umbracula, and even a little so in oscabrio. But rarely the respiratory organ is on each side of the body, in the sac formed by the mantle, as in the brachiocephala, or only on the right side, as in all the monopleurobranches. In fine, most generally it is at the anterior and upper part of the origin of the back, and of the back itself, that we observe the respiratory organ, as in most cephalous mollusca, whether pulmonary or branchiferous, and even in the dentales.

In all the conchiferous acephalous mollusca, on each side of the body, between it and the mantle, are two large semi-lunar lobes, which are generally regarded as the gills of these animals.

In the order of naked acephala, the respiratory organ is in a sort of tube, which conducts from the posterior part of the body to the mouth.

As to the structure of the gills of the cephalous mollusca, it reminds us of that of these organs in the fish. Whether it be sorts of triangular laminæ, ranged like the teeth of a comb upon a common axis, or sorts of tubercles irregularly collected, in the manner of granulations, the skin which constitutes them is considerably attenuated, although it preserves its contractile quality. Its principal afferential vessels may be very well injected, whose ramifications, often very fine, proceed to unite in a principal efferential trunk, which is so directed as to come forth from the organ in an inverse direction from that of the afferential vessel. The projection of these branchial combs, or of these tubes, is sometimes not inconsiderable.

The gills of the acephalophora are composed, in most cases, of two pairs, more or less unequal, of semi-lunar plates, placed vertically between the abdomen and the lobes of the mantle, and applied one against the other. Separated or united more or less in the extent of their lower edge, that of one side is joined to its correspondent on the other, through a part more or less considerable of its superior or dorsal edge. But it is attached on the sides of the belly by its anterior extremity, the other being often free. Each of these four gills is itself formed of two plates, which leave between them a free space, divided into a great number of vertical lodges, open at the dorsal edge, by numerous triangular partitions. These plates are constituted by two strata of parallel vessels, vertical, and united by other transverse vessels. One of these strata is formed by the ramifications of the branchial artery, and the other by those of the vein. These ramifications unite in two thick trunks, which border the back of the branchial plate, and which are in communication, one with the auricle, on its side, and the other with the venous system of the rest of the body.

In the lingulæ and approximating genera, it appears that the gills are a little different in their structure and in their position, since they are in the form of a comb, applied to the internal face of each lobe of the mantle.

In the naked acephala, the branchial apparatus is still more anomalous. In fact, in the ascidiæ, it is formed by a net-work, with quadrangular meshes, which invests the cavity of the tube as far as the mouth; and in the biphoræ it is a kind of narrow plate, almost free, and directed obliquely.

The nematopods have their respiratory apparatus approximating to that of the entomozoariæ, if it be certain that it is formed by small triangular plates, attached to the root of the first pair of appendages, as is the opinion of Baron Cuvier. We may also conceive that the filaments which bristle these appendages might take place of them, and then these appendages might be considered as gills.

The oscabriones have their respiratory system formed almost as in the phyllidiæ, of small triangular plates placed under the edges of the mantle.

According to the structure, the form, and even the position of the respiratory organ, the apparatus by means of which the ambient fluid is brought in contact with the modified cutaneous envelope, must of necessity be different.

There was no need of any when the gills were external, either on the back or under the edges of the mantle.

When, on the contrary, they have become internal, some particular modification was necessary in the edges of the cavity which contains them, and even in the shell which covers and protects them. Thus in a great number of the pectinibranch species the anterior edge of the mantle is prolonged into a tube, while others have but a sort of inferior auricle in place of this tube, or present but a broad cleft, which conducts into the branchial cavity. The pulmonary species have only a hole pierced into the thickened edge of the mantle.

In almost all the acephala the water arrives at the gills through an aperture formed by the two lobes of the mantle, which are often prolonged behind by the addition of a long contractile tube, distinct, or united to that of the anus, as before observed.

The theory of the function of respiration, independently of the organs, appears pretty nearly the same as in the more elevated types of animals. We know, in fact, that the mollusca absorb the oxygen of the air in which they are retained. But is this done alone by the organ of respiration? That is by no means probable, the general envelope being by its nature so absorbent; but as this organ contains a much greater number of vessels than any other part, the aerial absorption should be much stronger there.

We also know that the species which are provided with a pulmonary cavity die in a short time after they have been retained a certain depth under the water, without allowing them to re-ascend to its surface; and that, on the contrary, the species with gills cannot live for any time in the atmospheric air, especially when those gills are uncovered. When they are internal, the animal can survive for a long time, in consequence of the water by which they are moistened, and which does not quickiy evaporate.

The mechanism by which the ambient fluid is brought in contact with the elaborating fluid, or the blood, is in general rather simple.

In the species whose gills are exterior, as tritonia, scyllæa, phyllidia, &c., it is sufficient for the animal to swim for the purpose of respiring.

Those, on the contrary, which have the respiratory organs formed by the parietes of a cavity, as the pulmobranchia, or contained in the cavity, as almost all the other cephalous mollusca, the ambient fluid (air or water) is introduced or expelled by the dilatation or contraction of the cavity, and of its simple or tubular orifice; and these two effects are facilitated in all the species, and especially in such as are provided with a shell, by the extension or contraction given to the anterior part of the body where the apparatus exists, and by its advancement into the broadest part of the shell. But in no case is there any regularity in the inspiration and expiration: they do not even exist (as independent motions) in the brachiocephala, where the water introduced into the cavity of the mantle occupied by the gills serves at the same time for locomotion.

The acephalous malacozoaria, which are all aquatic, present

pretty nearly all of them the same mode of respiration. The labial appendages with which the mouth is provided appear, by their continual movements, to produce a sort of current in the water where the animal is plunged. This is very well distinguished, especially in the species in which the posterior extremity of the mantle is prolonged into two tubes, more or less long; the water enters by the lower, and issues out by the upper. It is thus in the ascidiæ, and perhaps also in the biphoræ. It is at the time of the passage of the fluid into the branchial cavity that the effects of respiration take place.

In all the malacozoaria the circulating system is complete, although the re-entering or absorbing centripetal system is only formed by veins.

These vessels have their parietes extremely thin, and often so much confounded with the tissue of the parts, especially in the dermoïd envelopes, that it is often rather difficult to perceive them; and sometimes they really exist only through the internal membrane, as in the bivalves. They sometimes present this peculiarity, that they are pierced with gaping orifices, tolerably large, at least in the visceral cavity, as may be well observed in the aplysiæ. We also find that they sometimes bristle with sorts of little spongy bodies, also proceeding into the visceral cavity.

As in the animals of more elevated types, the veins spring, without doubt, in part, from the continuity of the arteries, and in part from the tissue itself of the organs, but they never constitute more than two systems, one which returns from the whole body, and the other from the special organ of respiration; that is to say, there is no system of the vena portæ. The venous radicles of the general system of the body, after having been successively united in trunks more and more thick, distinguished, however, for some time into those of the viscera, and those of the sensible contractile envelope, arrive towards the respiratory organ, and according as it is simple or complex, symmetrical or not symmetrical, act in somewhat a different way. In fact, in the first case all the veins of the body unite into a single gross trunk, which most usually, without the intervention of a muscular swelling or heart, is subsequently changed into a pulmonary or branchial artery. In the second case, on the contrary, the veins unite into two principal trunks, which are subdivided into as many branchial arteries as there are gills. At the point of this transformation there is never any true heart or organ of impulsion, but in all the branchiocephala, and even in a small number of cephalous species, and probably in the acephala, we find in this place a venous sinus, to which the name of heart has been sometimes given, but which probably cannot be so designated, as it possesses no muscularity. Be this as it may, the branchial or pulmonary artery, simple or multiple, is ramified in a manner more or less regular, according to the form of the respiratory organ in the modified skin which constitutes it.

It is from the capillary extremities of the branchial artery, subdivided in the respiratory organ, that the second venous system originates. After that the ramifications, disposed like those of the arteries, have been successively united in branches more and more thick, there results from them finally a gross trunk, which issues from the respiratory organ, and repairs into an aortic heart situated in a different manner, according to the position and symmetry of the gills.

The heart of the mollusca, in the greater number of cases situated in the back, above the intestinal canal, unless perhaps in the brachiocephala, where it is inferior, is placed, as in osseous animals, at an equal distance from each respiratory organ when there are two, or obliquely to the left, and rarely to the right, when the organ is odd. This heart is not contained in a true pericardium, but in a muscular lodge of the diaphragm kind, which separates the visceral cavity from that of the gills. As to the rest, it is formed of an auricle, sometimes double when the gills are symmetrical and lateral, as in the brachiocephala and conchiferous acephala, and of a ventricle.

The auricle, of a very variable form, ordinarily oval, sometimes triangular, has its parietes very slender. We may observe, however, at the interior some muscular cords which traverse it. It does not appear that there is any valvule at the entrance of the branchial or pulmonary vein into this auricle.

Its communication with the ventricle is made by a sort of pedicle or contraction, often pretty long, as in the calmars, for example, and by means of a narrow orifice usually transverse, situated between two folds of the internal face of the ventricle, but without valvules, properly so called, something in the way that the small intestine opens into the cœcum in the human species.

The ventricle, in general much thicker, is also of a form and direction very variable; its parietes are always thicker than those of the auricle, and we can very well distinguish the transverse muscular bundles which form it, between two of which is the auriculo-ventricular orifice.

It is from its point, or from one of the extremities of its greatest diameter, that the arterial or centrifugal system proceeds, most usually by a single trunk, but sometimes also by two, as may be well observed in the calmars.

The arteries of the mollusca have evidently their parietes more thick and more resistant than the veins. They possess great elasticity; and in the largest of these animals, they appear of a gelatinous tissue, without any trace of fibres.

Their distribution is too variable to speak generally concerning it; nevertheless, most commonly there are two principal trunks, one anterior and the other posterior. The first furnishes branches to the head and to its different parts, to the œsophagus, and even to the anterior organs of generation;

while the second, which has more resemblance to the cœliac tripod, sends its ramifications to the stomach, to the rest of the intestine, to the liver, and to the secretory organs of generation.

In the acephalous mollusca the circulatory apparatus presents some differences from that of the cephalous. The veins of each gill are united in a lateral auricle, placed on each side, and after a contraction often very sensible, each of the two auricles opens into a ventricle, which is situated in the medio-dorsal line. This ventricle is usually fusiform; but what it exhibits most remarkable is, that it appears to be traversed by the rectum, because, in its breadth it recurves around this intestine, so that the two extremities of its transverse diameter appear to touch. From the root of this ventricle spring two aortæ: a posterior one, smaller, which passes under the rectum, and gives out branches to the posterior parts of the body, and an anterior much more considerable, which proceeds as far as the anterior adductor muscle, furnishes branches to the stomach, to the liver, to the foot, and to the other surrounding parts, recurves below by an anastomosing branch, which follows the edge of the mantle, to unite itself to a similar branch of the posterior aorta, and forms a large arch, of which the lower branches go to the tentacula of the edge of the mantle, while the others, more considerable, remount and distribute themselves to all its parts.

The venous radicles of the belly, and of all the anterior parts of the body, unite in two gross trunks, which proceed from the hepatic region below the rectum, and after having received, by many radicles, two veins, which have followed the edge of each lobe of the mantle, they open at the anterior extremity of a sort of auricle or venous reservoir, placed longitudinally under the heart in the dorsal line. This reservoir receives by its posterior extremity two other veins, tolerably thick, which have taken up the blood of the posterior parts of the body, and even of the edges of the mantle.

This medial sinus, which is surrounded by a brown organ, of which we shall speak further on, when we come to treat of urinary depuration, appears also to receive a considerable number of vessels from it, or perhaps those vessels may originate from the former and proceed to distribute themselves to this organ, while a much greater number unite in the branchial arteries. These are two in number, one on each side. They are considerable, and placed longitudinally along the upper edge of the branchial plates, thicker in the middle, they diminish in diameter, and finish in a point at the extremity in proportion as they have furnished branches to the gills. Those branches form two planes, one for the internal face of the external leaflet, and the other for the external face of the internal leaflet of the gills, descend vertically, diminishing as far as the edge of the leaflet, and furnish numerous longitudinal anastomosing branches, so that from them results a net-work with square meshes. From this same net-work spring, by a contrary disposition, the branchial veins, whose net-work occupies on each leaflet the face opposite to the arterial net-work, and they unite in as many thick longitudinal veins as there are branchial plates, at least in front, where they are perfectly separated at the superior edge; for behind there are but three, the medial being common to the two internal plates, which are united. The external branchial vein changes into a sort of sinus or long auricle, with which the external vein communicates by several venous pedicles; and this auricle itself, after having been narrowed, opens into the ventricle.

The palliobranch species appear to have the auricles still more distinct than the ordinary acephala, and it is probable that it is this which has caused it to be admitted that they had two hearts.

The oysters have also the heart placed differently, and not occupying the back of the animal, but the anterior part of the central muscle.

In the naked acephala it occupies pretty nearly the same place as in the conchiferous, but it is perhaps less symmetrical.

It is completely so in the nematopods, and especially in the oscabriones, whose ventricle occupies the posterior part of the back, along with a large auricle, very symmetrical on each side.

We shall now speak briefly of the urinary and generative processes.

The apparatus of urinary depuration, in general very simple, appears to exist in all the malacozoaria, that have been sufficiently examined. It always accompanies the termination of the intestinal canal. In the cephala it is found sometimes described under the names of the organ of *glue*, or the *calcareous sac*, and in the acephala under that of pulmonary organ.

In the first it consists of a single secretory organ, not symmetrical, of very variable form, often situated near the organ of respiration, making a projection into the interior of the cavity which contains it. From it springs an excretory canal, which, after a passage more or less long, often accompanying the rectum, terminates externally by a rounded sessile orifice, at a small distance from the anus.

In the class of the acephala the depurator organ is double, or at least symmetrical. It is situated equally on each side of the rectum underneath it, in front of the posterior adductor muscle, and behind the connexion of the branchial lobes. Its colour is usually of a deep green. Its form is more or less cylindrical, its structure is cellular or vascular, and it receives a great number of arterial, and more especially of venous vessels. It appears that the excretory canal is not sufficiently known. The organ, however, is contained in a pouch opening by a very small orifice into the upper and anterior part of the branchial cavity, according to M. Bojanus, which has led him to think that this organ is a true lung, those which have been hitherto regarded as gills being, in his opinion, the organs for the deposition of the eggs. According to many this organ is behind, and under the rectum, which appears to be more in analogy.

The apparatus of generation, which is more or less completely known in all the animals of this type, is often very complicated, and at other times reduced to the greatest possible simplicity. In fact, it is sometimes composed of the female part only, as in all the acephala and some cephala, which causes all the individuals of one species to be similar. There are found a tolerably great number of mollusca in which the sexes are distinct, but both existing in the same individual, from which it results that they again are all similar. Finally, there are also many in which the sexes are separated, thus producing in the same species the distinction of male and female individuals.

The female apparatus of generation, in the cases where it exists alone, is formed but of one or two secretory organs or ovaries, situated a little differently in the acephala from the cephala. From this ovary there proceeds a canal or oviduct, which, after being for some time swelled in a part of its extent, directs itself forward or backward, and terminates on one side or the other, but more usually on the right than on the left.

In the acephala the ovary may be single in its origin, and capable, by its successive enlargements, of extending itself into all parts of the mantle. It appears always to be prolonged by two distinct oviducts, which are placed, being directed from front to rear, on each side of the abdomen, where they terminate by a rounded orifice, at the extremity of an elongation, of greater or less extent, situated between the second branchial plate and the body. Before this termination the oviduct contains at a certain period a milky white liquor. Some authors, and among others MM. Bojanus and Mery, add to this essential part of the generative organ of the bivalves the organs which we have already described under the name of gills, and which they regard as depositaries for the eggs.

In the monœcious cephala, such as the patellæ, haliotides, &c., the ovary is always single, and on one side. It is the same with the oviduct, which is constantly directed from behind to front, sometimes at the left, but more frequently at the right side, where it terminates by a very short tube in the respiratory cavity.

The disposition which constitutes the union of two distinct sexes in the same individual obtains only among the cephalophorous malacozoaria, and only in a certain number.

The female part is formed by a single ovary, situated posteriorly in the liver; from it proceeds a first oviduct, which originates by ramifications, like those of the biliary canal in the liver. Very fine at first, its bulk increases; it bends, runs into heaps more or less compact, approaches the male part, becomes intimately united with it, and finally opens into a second oviduct, much wider, with thick folded parietes, secreting an abundant viscous matter, and which is sometimes designated under the name of matrix. Near its mediate or immediate termination at the exterior is remarked that of a canal more or less long, proceeding from an oval or spherical bladder, contained in the great visceral cavity, and the use of which is entirely unknown. Might it not be a sort of prostate?

The male part is also composed of a secretory organ or testicle, situated most usually on the left side, and in front of the ovary. The deferential canal which springs from it, after an intimate union with the first oviduct, follows the passage of the second, against which it is cemented more or less closely, forming sometimes a sort of epidydymis by its numerous folds, then is changed into a cylindrical canal with thick muscular parietes. This is directed towards a species of excitatory organ, at the base of which it terminates in the majority of cases. This organ is only a sort of long hollow tentaculum, contractile in all its points, of a form extremely variable even in species of the same genera, and which most generally retracted into the interior of the visceral cavity, by the aid of a retractor muscle, may also, from the disposition of muscular annular fibres, be turned outwards, like the finger of a glove.

At the point where the deferential canal terminates we sometimes find a heap of cylindroïd organs, or kind of hollow cœca, variable in number, and which successively united at their base, finish by opening through a single orifice. These have been called the *vesiculæ seminales.*

Finally, in a certain number of hermaphrodite mollusca we remark another organ again neighbouring to the external termination of the male apparatus. It consists of a muscular mucous pouch, in the form of a bladder, which produces and contains in its interior a labial corneo-cretaceous body, in the form of a dart, or poignard ; this body can issue forth through the orifice of the pouch, which is situated near that of the rest of the male apparatus.

In spite of the intimate connexion which subsists between the two parts of the genital apparatus of these hermaphrodite mollusca in their passage, they may terminate at distances more or less considerable from one another, although always on the right side. Thus in some species the termination of the female organ is altogether behind, and that of the male organ in front, more or less near the tentaculum on that side, as in the veronicellæ and onchidia. In some others the separation is less great; many have them united in the same external tubercle, as doris, tritonia, &c. Finally, in all the pulmobranch species these terminations are made in a sort of common vestibule at the root of the right tentaculum, so that in a state of inaction but a single orifice is seen at the exterior; but in the act of coupling the vestibulary pouch is reversed, and the two terminations become apparent.

The third disposition of the genital apparatus of the mollusca constitutes the division or isolation of each sex in a distinct individual, whence result females and males. As for the rest, each apparatus is pretty nearly conformed according to the preceding disposition. We find, however, perhaps more frequently in the female sex, the swelling of the second oviduct performing the office of matrix, and in the male sex we remark that the vesiculæ seminales are sometimes replaced by a single swelling situated towards the end of the deferential canal. Finally, another difference is that the excitatory organ, when it exists, never appears to be completely retractile, but only contractile, so that it is always more or less visible at the right and anterior side of the animal, sometimes being recurved into the branchial cavity.

With respect to the products of the organs of generation, that of the male sex, when it exists, appears to be always a fluid of a viscous white, but in general it is little known. We are not even acquainted positively with the mode in which it is ejected, whether at once or by little and little, nor into what part of the female organ it penetrates. The product of the female sex is much more known, and always constitutes a true egg, composed of envelopes of a vitelline mass, and a germ placed upon this mass, which no doubt makes part of it.

The form of the eggs of the mollusca does not fail to exhibit a good number of varieties; sometimes being spherical, as in those of the limaces; sometimes ovaliform, as in a great number of species, or even more or less longitudinally pedicled, as those of many buccina.

Those envelopes which may be called *adventitious* or casual, usually viscous at first, to determine the adherence of the egg, pass subsequently to a corneous or concrete mucous state, and sometimes even to a cretaceous state, so that they resemble tolerably well the calcareous envelope of the egg of a bird. This may be observed in many terrestrial mollusca.

The proper envelopes are but little known, but it is probable that they do not differ much from those of the eggs of animals of a higher rank.

We are not much better acquainted with the form and disposition of the germ, except when it may be sufficiently developed to resemble the parents which gave it birth. We only see that this germ is contained at first in a lodge or superficial excavation of a true vitellus, which communicates as usual with the intestinal canal, perhaps even altogether near the mouth. This vitellus is evidently a mucous or gelatinous matter concrescible by alcohol, translucid, and not very thick in the fresh state.

The development of the germ in the interior of the egg of the mollusca is so complete that the little animal which comes from it almost entirely resembles its parents. Accordingly it often happens that this development takes place in some part of the mother, and that both in the cephala and the acephala. They, therefore, in such cases bring forth their young living, and these mollusca are termed viviparous. All the acephala appear to be in this predicament.

The disposition of the eggs, laid by the malacozoaria, externally is also very variable. Thus sometimes they are placed and attached one by one on submarine bodies, as in a great number of cephalous mollusca, but at other times they are joined together so as to form masses more or less considerable, and which more or less resemble clusters of grapes, especially when the eggs are of a black colour, as those of sepia. Often they are united by a gelatinous substance, in which they are plunged, like those of the limnææ, planorbes, and aplysiæ. At other times many of these eggs are enclosed in sorts of corneous envelopes, piled like pods one after the other, an arrangement which is observable in some species of fusus.

We next proceed to the nervous system of the mollusca, or that portion of the organization on which the irritability of the animal and its susceptibility to external impressions, entirely depends.

This apparatus of the nerves, as may be seen in the general characters of the type, presents a disposition somewhat singular. It is, however, always composed of a central portion, or brain, situated above the intestinal canal; of ganglions, for the special organs of the senses, where these exist, as well as for the apparatus of locomotion; of some visceral ganglia; and finally, of conducting threads or nerves, the structure of which is sometimes singular, in that they have a fibrous envelope larger than the nervous cord, so as to permit, as some say, their injection, which appears very doubtful.

The general disposition, and especially the proportion of the parts of the nervous system, are very different in the two classes of the malacozoaria, and especially in the sub-type of the malentozoaria.

In the cephalous mollusca the brain, composed of two similar parts, more or less bulky, more or less united by a sort of commissure, is sometimes contained in a sort of cranium or cartilaginous box, which serves as a support to the contractile fibre. But in a great number of cases it is scarcely covered with cellular tissue, and it is placed at the origin of the œsophagus, behind the buccal mass, so that it follows their movements.

With this brain communicates the ganglion of the organ of vision, which is always placed immediately behind the bulb of the eye, as well as that of hearing, when there is any, and from it proceed the different nerves, which repair to the tentacula as well as to the lips.

Beside the communication more or less close which there is above the œsophagus, between the two parts of the brain, there is another inferior one, which passes under the œsophagus, and constitutes a sort of ring or collar surrounding it.

The nervous system of the sensitive and locomotive apparatus is formed only by a single ganglion situated on each side, sometimes tolerably distant from the brain, with which it always communicates by a cord, but most frequently so near this organ that it really appears to form a part of it. In both cases it gives out the more or less numerous threads, which repair to all the parts of the musculo-cutaneous envelope, and especially to those which serve essentially to general locomotion, as for instance, to the foot of the gasteropods and trachelipods, to the sac of the brachiocephala, to the wings of the pteropods, &c.

The visceral ganglia appear to be but two in number; the one which appertains especially to the male excitatory organ is usually situated near the orifice through which that organ issues, and it furnishes threads to the organ as well as that by which it communicates with the brain. The other visceral ganglion is more constant in its situation; it is usually placed towards the stomachal swelling, and the nervous threads which it furnishes are equally of two sorts, those which proceed to the intestinal canal and those which ascend to communicate with the brain, through the medium of the œsophageal ring.

It is unnecessary to observe that the development of the different parts of this nervous system is proportional to that of the organs to which they appertain, and that consequently it is much greater in the brachiocephala, which stand at the head of the class, than in the patellæ, &c., which close it.

This observation is equally applicable to the nervous system of the acephalous malacozoaria; in fact, it is so very little developed in them that its very existence was for a long time unperceived. Their brain is no more than a double ganglion, or rather a sort of flatted cord, situated always above the œsophagus. It appears that there are no threads which should form round the latter a true ring, as in the cephalous mollusca. From this sort of brain there proceed two long cords, but they go much farther behind, and proceed to establish the communication between this organ and the ganglion of locomotion which is found underneath the adductor and posterior muscle, and which, in fact, receives threads from them, as do the mantle and the tubes, when there are any.

In the common *mussel* (*Mytilus edulis*), the nervous system seems to be more evident than in any other species of acephalous mollusca. It is composed of three pairs of ganglia; the first, the most anterior, is certainly placed under the œsophagus, or rather under the anterior retractor muscle of the foot, in part covered by the posterior edge of the union of the second pair of labial tentacula. The lesser ganglia of which it is composed are of triangular form, and of a white opake colour: they furnish, first, a very fine transverse net, which serves as a commissure between them; second, a thicker branch, which is distributed to the anterior adductor muscle, and to the labial appendages; and third, from their hinder termination, a very thick thread, which proceeds externally, is applied on the membrane of the liver, traverses obliquely the anterior retractor muscle of the foot, follows the sides of the abdomen above the termination of the ovary, and proceeds to join the posterior ganglion.

The second pair of ganglia, which alone can be regarded as

a little superior to the intestinal canal, is placed above the anterior retractor muscle of the foot, applied immediately upon it, below the liver, against which it is cemented. It is a double ganglion, or divided into two lateral parts by a medial furrow, of a softer consistence, and more pulpy appearance than the other two pairs. There is seen to issue forth in front a very fine thread, which perhaps proceeds to join the anterior ganglion; a fact, however, that M. de Blainville, from whom we derive this information, was not able precisely to ascertain. Behind, there is another thread, which repairs to the muscles of the abdomen.

The third pair of ganglia is altogether behind, below, and a little external to the anterior part of the posterior adductor muscle; that of one side is separated from that of the other by the entire thickness of the muscle. They furnish, first, a very fine thread as a transverse commissure; second, behind, a thicker thread, which penetrates into the muscle itself; third, from their external and posterior angle, two threads, which go backwards, probably to the edges of the mantle; finally, from their anterior and external angle, the thick cord, by which they anastomose with the anterior ganglion.

In the oscabriones the nervous system approaches more to what it is in the cephalophorous malacozoaria, with this difference, that the two lateral locomotive ganglia are replaced by two sorts of cords, which follow the sides of the back, and furnish threads to each kind of articulations.

In the nematopods, or balani, we find almost completely the disposition which exists in the entomozoaria, or articulated division of animals, the nervous system of locomotion having passed below the intestinal canal, and being composed of as many small ganglia as there are articulations in the caudiform part of the body.

At the end of a series of reproductions more or less repeated, the molluscous animal tends to its decomposition, or in other words, to death. We are completely ignorant of the duration of its natural life, but it is probable that it is tolerably long, if at least we may judge by the duration of its growth, and because it passes its life under circumstances of no great variety. Nevertheless, we possess no positive data on this subject, and it must be confessed that it is exceedingly difficult to procure any.

As to the duration of the shell, and the changes which it is susceptible of experiencing from the action of the air, and in the bosom of the earth, all this depends much upon its structure, its solidity, its bulk, and some accessory circumstances.

If it is exposed to the action of the air, and to the vicissitudes of temperature and humidity, it at first loses its colours, which are very quickly changed (the ferrugineous resist the longest), and it usually becomes of a tarnished white; the animal matter is destroyed, and disappears by little and little. The composing laminæ being no longer connected, exfoliate, expecially from the alternation of heat and cold; and soon, by this continual action, the laminæ themselves become resolved into a sort of calcareous dust, which is drawn along by the currents of the water.

The particular structure of the shell, its age, and even its bulk and thickness, facilitate or arrest more or less its terreous decomposition.

If, on the contrary, the dead shells are, by particular circumstances, sunk in the sand, or in the mud, in which they have lived, and where they have been encrusted with a cretaceous deposition, which takes place in greater or less quantities in all fresh and salt-water, but especially in the first; or, finally, if by the action of currents they are accumulated together, whether broken or not, in some localities of seas or lakes, as in these different cases they are placed in shelter from the vicissitudes of temperature and humidity, their decomposition is infinitely more slow, and their colours are preserved for a much longer time. The corneous fibres of the ligaments are sometimes preserved for a great number of ages, and still more their laminated or fibrous structure, to such a degree, that they have often lost none of the parts which serve as generic, and even as specific characters. Finally, when the colours have disappeared, as well as the animal gluten, they arrive at a point, when white, and grating to the tongue, they may resist the inroads of time for such a number of years as it is impossible to calculate. However, in the long run, the pressure, determined by the new depositions which cover them, tends to break them, by approximating the molecules together. The diminution and disappearance of the animal matter, which retained the inorganic substance in forms, as it were, accidental to it, and determined by the living principle, facilitate the tendency which these molecules have to approximate, according to the simple laws of the inorganic kingdom. The shell then tends to disappear altogether, by the successive removal of the calcareous molecules which constitute it; but as its cavity was filled by the pressure in all directions of the terreous or argillaceous molecules which surrounded it, when the true testa has disappeared, it is as it were represented and prolonged for a lengthened period, by what is named its mould, which exhibits all the forms of its cavity. It is equally possible to conceive, what indeed happens, that the calcareous molecules, although having obeyed the laws of crystallization, do themselves preserve the form of the shell. The structure, it is true, is lost, but not the form. This is what constitutes a spathified shell, and prolongs, as it would appear, almost for an indefinite period, the proof of the existence of the organized being, through a series of ages, until at last it is converted, as it were, by continual pressure, by the molecular movement of the parts which surround it, into the rock itself, which it contributes to form.

The diseases of the mollusca are doubtless not very numerous, but assuredly they are not very well known, at least as far as the animal itself is concerned. Should we regard as such the peculiar alteration exhibited by oysters when they pass to a state of greenness? Nothing can be less certain than this. Nevertheless, observing that the oysters which pass into this kind of state live in stagnant water, and that they are in general smaller and less fleshy than the others, we may be allowed to think that the particular vibrio, to which they owe their green colour, does not sufficiently nourish them, and that the water, half fresh, and seldom renewed, in which they are, does not sufficiently excite their organic activity.

The diseases of the shell are probably more numerous, and better known. The first is the separation or breaking of the point of the spire. This is observed in many species of univalves, and among others in some of the genus *bulima*. Although this takes place only with shells of the turriculated form, nevertheless it cannot be this circumstance alone which determines this fracture, since most shells of this form exhibit no such appearance. It is more probably owing to the circumstance of the animal growing very fast, and quickly abandoning the commencement of the spire, while the vitreous matter, deposited to fill the deserted cavity, is more fragile and less lamellated.

A kind of change, which is remarked at the summits or hooks of a great number of fresh-water bivalve shells, which compose the genera unio and anodon, has perhaps some analogy with what we have just mentioned concerning the univalves. This, however, is not certain; and in fact many authors have supposed that this kind of caries, which appears to eat in an irregular manner, not only into the summit, but even into other parts of the unio, and that often pretty deeply, was owing to the destructive action of the animals which feed on the mollusca. Be this as it may, we know that this caries augments in breadth and depth with age, and that the uniones of all countries exhibit this singularity.

The anomalies or deformities of shells are of two sorts: the one is sufficiently explicable, the other not so.

We may rank in the first kind the relative thickness which individuals of one and the same species may attain to in their growth; and, in fact, we find some individuals in certain genera, which, although complete, are much smaller than others. This, without doubt, is owing to a difference in the quantity of nutriment, either in the same, or in a different locality, as may be seen among the hexapod insects. Accordingly, we cannot admit the notion of Bruguières, that this difference necessitates the change of skin in the animal, something like that of the epidermis in insects.

We must also consider in the same light the double swellings which are formed in certain univalve individuals, after that, having come to their full growth, the normal swelling has been produced. This, no doubt, is referrible to an over excitation of the vital powers, determined by some local circumstance.

We must equally place in the same class the artificial form which certain slender bivalve shells may assume, and the inferior valve of which adheres in its full extent. Not only does this assume the form of the body on which it is applied, but the upper valve follows the form of the lower. This observation, made upon the anomiæ, is owing to M. Defrance, and is explained by the circumstance, that the upper valve must follow the form of the body, which has itself been modified by that of the lower valve moulded on the foreign body.

A phenomenon pretty nearly inexplicable is the degree of elevation in the spire of the univalves. In fact, we know that the same species presents, under this relation, some differences which, although contained within limits sufficiently bounded, are not the less very evident. But it sometimes happens that they depart considerably from the limit determined for a species, in that the turbinations of the spire become more apart, are elongated in a vertical direction, and are very far from touching, which gives to the shell the appearance of a ladder, and has given a denomination to the individuals of this anomalous kind derived from this resemblance. An example of this may be found in some species of helix.

But the most inexplicable monstrosity of shells, and even of the molluscous animals themselves, is that in which there is an inversion in the position of the viscera, and consequently in their termination, which, instead of taking place at the right side, takes place at the left. The shell having followed this inversion, is then turbinated from right to left, and constitutes the variety to which the denomination of *sinister* has been given. There are some genera, in which this inversion is much more common, and even to the point of serving as a character. Such are *physis* and *planorbis*. In many other genera we find examples of it, but more rarely; and finally, there are some in which it has not been observed, such as the porcelaines and cones.

The intelligence of the mollusca, at first pretty evident in the higher species, such as the octopi, which employ stratagems to reach and seize their living prey, decreases very rapidly, and arrives at its minimum in those, all of whose movements are confined to the opening and closing the valves of their shell, as the oyster, and which receive their nutriment in the form of dissociated molecules, and already nearly in a fluid state; the habits and manners, therefore, of this class, do not afford much matter of interest, when compared with those of the other classes.

The mollusca are found in habitations of all kinds. In fact, there are some which appear to live constantly under ground, as the testacellæ; but this is rare. More of them live in the air, at the surface of the earth, as the limaces, helices, &c. Some are to a certain point amphibious; that is to say, they can breathe the air, by means of their organ of respiration, and yet live in the water, which they quit but seldom, as for example, the limnææ and planorbis. Finally, most of the malacozoaria live constantly in fresh or salt water, current or stagnant, which is the case with all the acephalophora without distinction. The waters of the Dead Sea, though so strongly bituminous, contain living conchyliferous mollusca. Some are also found in thermal waters; for example, the *turbo thermalis*, a paludine species no doubt, lives in those of Abano, the temperature of which is 40° of Reaumur, while the *Clio borealis* appears to be unable to quit the polar seas.

There are some characters which indicate this difference of media inhabited by the mollusca. This is certainly the case with the aquatic and terrestrial species, since the organ of respiration has a peculiar structure.

But this cannot be observed of those species which are entirely aquatic, whose gills present no difference to explain why they should act in fresh water or in salt. Would the shell itself alone furnish any characteristic differences which would indicate the nature of the sojourn of the animal? Not by considering it merely in itself, but up to a certain point it will, by comparing the shells of marine animals with the fresh-water or terrestrial, as we shall see a little farther on, when we come to add a few remarks on the subject of conchology.

It has been a question, whether the species which are found habitually in salt-water can live in the fresh-water, and *vice versâ*. This question, to which a great importance is attached in geology, would be resolved in the affirmative, if we were content with analogy alone; for we know indubitably that certain fishes quit the waters of the sea to live in those of rivers, and others quit the latter for the former, as for instance eels, and that almost suddenly. Why should not the mollusca do the same? However, there is no very positive fact to prove this possibility, at least for the same species; for it is not so with the genera. We know, in fact, that some species of the same genus can live in the fresh-water, and others in the salt-water. We are acquainted, for instance, with one species of true mussel in the Danube, and many cerites, which are equally found in fresh-water. But if the species of mollusca cannot suddenly pass from the salt-water into the fresh, may they not do so gradually? Do we not, in fact, see in certain ponds, which communicate but rarely with the sea, and the saltness of which is diminished by little and little by rain-water, mollusca living, which are truly marine, and appearing to exercise all their functions there in complete perfection? The fact is certain, and has furthermore been experimentally proved by M. Beudant. But it is not equally certain that animals habitually accustomed to live in the salt-water, and which are thus forced by natural or artificial circumstances to live in that which is almost fresh, or altogether so, can reproduce there. But though not placed out of doubt, yet the fact which has been observed of marine and fresh-water mollusca living together, in the slightly salt-waters of the Gulf of Livonia, is in favour of this opinion, and still more that which Nilson relates in his history of the mollusca of Sweden, namely, that on the coasts of Norway, in places where there is no mouth of a river, he has found uniones, anodontes, and cyclades, living mixed up with mollusca of the genera venus, cardium, and cytherea.

Adanson assures us positively, that during one half of the year the Niger has only fresh water, and that nevertheless several genera of mollusca are found there, which, during the other six months, live in the salt-waters.

The aquatic mollusca, whether marine or fresh-water, also do not live absolutely in the same circumstances: the latter, which are but comparatively few, do not, however, exhibit many differences in this respect, although some remain fixed to the surface of the ground, like oysters; such are the etheriæ, according to the discovery of M. Caillaud; others adhere to submerged bodies by means of a byssus, as the mussel of the Danube; others move about in the mud and on its surface, as the unio and anodontes; and finally, others live much deeper, and also move, as the cyclades.

The circumstances of the life of the marine mollusca are much more variable. Thus the majority live on the sea coasts, on rocks, and in places in the eddies and mouths of rivers, which are hence called littoral species. But there is a certain number of others which appear to exist only at distances more or less considerable from the shore, and at great depths, which causes them to be distinguished under the name of pelagian mollusca. The terebratulæ appear to be in this case, and it is supposed that the nautili and ammonites are still more so. In fact, the calamary, the sepiæ, and the spirulæ, to which they approximate, are animals of the high seas.

We next find, that according to their mode of locomotion, some mollusca live continually swimming or floating at the surface, or in the interior of the waters, or creeping on the rocks in the middle of the varecs which cover them; others are attached there in a fixed manner by their shell, or by a byssus, or, finally, sunk more or less deeply into sponges, in the mud, in the sand, in rocks, in madrepores, in other shells, and even in non-calcareous stones, as well as in dead or living wood.

The species which live in sponges, are in the same case as mussels, &c. which are found in the holes of rocks; but as the substance, in the excavation of which they have been accidentally placed, increases, as long as it is living, the result is, that they are finally completely enveloped in it, to that degree, no doubt, that they are as it were suffocated.

The mollusca which live in mud, in sand, or even in argillaceous earth, really operate so as to sink themselves there, in proportion as they augment in bulk; and it is evident that this is done mechanically, and by means of their foot. As for those which sojourn in hard substances, as in calcareous stones, madrepores, and shells, it has been thought that their successive sinking was owing to some corrosive juice or acid, which was capable of dissolving the calcareous stone. But, besides that this is any thing but proved, the fact observed by Olivi and Spallanzani of pholades in pieces of lava, and that of teredines in living wood, do not permit us to adopt this opinion.

The terrestrial mollusca present, as may well be supposed, much fewer variations in the circumstances of their habitat. In general it is in humid places, and more or less aquatic, that most of them are to be found; but there are also some that seem more to court dry places, and exposed to the sun, as is the case with certain species of helix.

Some persons have even believed that many species were fixed to soils of a peculiar mineralogical character; but that does not appear to be probable.

What is most certain is, that the terrestrial mollusca, in countries where the continuance of some unfavourable circumstance, such as cold or drought, forces them to suspend their vital activity, are obliged to withdraw themselves from it, and for this purpose they sink more or less into the earth, into the anfractuosities of bodies, and thus enter into a sort of torpor, analogous to that of the marmots and other animals. This is the reason why we find in some places, or rather in one and the same place, a great quantity of these animals, or of their spoils, which have probably accumulated there for a long series of years.

The scientific study of the malacozoaria is even yet so little advanced, that we know but little concerning their total number, and their distribution into the different parts of the world. It may be said, in a general manner, that no part of the earth is without marine mollusca, terrestrial, lacustral, or fluviatile; and that the proportion of the species of these divisions must be in relation with the extent of the seas, the continents, the lakes, and rivers.

We may be also assured, that almost all the families exist in the different zones of the globe, but that the genera and species of some are much more numerous in one zone than in another. Thus it would appear that there exist every where octopi, sepiæ, and loligines. It is difficult to be so certain respecting the genera of polythalamous shells; and in fact the only two of which the animals are a little known, spirula and argonauta, belong to the torrid zone. The genera of siphonobranchia are also found in all latitudes, but there are many of the subdivisions which have been established on the shells of this order, which only belong to the intertropical regions; such are the pleurotoma, tonna, lyra, vis, mitra, strombus, cones, oliva, porcelaines, and ovula, genera of which scarcely a species is known in our northern seas, and two or three only in the ocean and the Mediterranean. The number of the generic subdivisions of shells, of which species are wanting in the order of the asiphonobranchia, is not very considerable, or they are represented one by the other, so little do they mutually differ. We also possess all the genera of the families which compose the order of pulmobranchia, and they are found spread over all the earth, only in proportions a little different. Thus the species of the family of auriculacea are much more rare, and smaller, in our climates, than in the torrid zone. It is the same with the agathinæ and bulimi, dismemberments from the genus helix. The limnææ appear, on the contrary, more numerous, and even larger, in our climates, than in the warm countries, which is not the case with the genera planorbis and physis. We have no species of onchidia or veronicella which appear to represent in the hot climates the limaces of our zone, as our testacellæ replace the parmacellæ of the torrid zone. In all the other families, naked or conchyliferous, we may almost generalize the same observation, in adding that the species of the same genera are much more numerous, and especially much larger in the equatorial than in the polar regions, and more especially than in ours.

In the class of the acephalophora, we shall equally arrive at the same result. In the order of the palliobranchia, the lingulæ are only to be met with in India. We find terebratulæ, orbiculæ, and crania, in all countries. This is still more evident respecting oysters, which are every where abundantly distributed. It is not the same with the tridacne, which are as yet known only in the Indian Archipelago. The pectines and limæ are found in all seas. The umbellæ, the pinnæ, and crenatulæ, appear to belong only to the seas of warm climates. The mussels, and even the irregular auriculæ, are of all seas. It is the same with all the generic subdivisions of the family of arcacea, and of that of submytilacea. The trigoniæ, of that of camacea, have not yet been found living, except in the austral zone. Species of all the genera of conchs have been observed in all seas, but sometimes one of these genera is represented by another closely approximating. Thus our cyclades appear to be the cyrene in India, &c. It would also appear that the venus saxicava are found every where; and it is the same case with the mactræ. The myæ appear rather to belong to the northern seas, as well as the pandoræ and some solens. The oval solens belong rather to the southern climates. We find pholades every where, and probably teredines, while the fistulanæ, clavagellæ, &c. are almost constantly of the equatorial zones.

The simple or aggregate ascidiæ exist also in all zones, but nevertheless more numerous, and more developed in the equatorial than in the polar. This is still more evident with the biphoræ, which do not even commence to appear but in the seas of temperate regions.

The class of the oscabriones has species in all seas, but much more numerous and larger in those of hot countries than in others.

It is almost nearly the same thing with those of the class of the nematopods.

We likewise may say of the families, the genera, and species of the acephala, what we have said of the others, that although more numerous, and of larger dimensions in the equatorial zones, the genera are represented in all, with few exceptions.

The relations of the animals of this type with others are not in general very favourable to them; in other words, they are more often obliged to fly than to seek them; and though a certain number of them are zoophagous, there are few which attack animals of the higher classes. Perhaps there are none which do so except the brachiocephala, which feed on crustacea and fish. All the other zoophagous species attack only animals of their own class, and especially of the class of the acephala, which move with difficulty. Accordingly, we may say in a general manner, that the type of the malacozoaria exercise but a feeble action on the preceding types, while the latter, on the contrary, exercise upon it a very destructive one. In fact, a certain number of aquatic mammifera, as the cetacea, the morses, but above all, the birds which inhabit the waters, the amphibious animals, and even the fish, pursue with more or less avidity the naked or conchyliferous mollusca, break the shell of the latter, and devour them. Accordingly, this group of animals does not appear to escape destruction, except from the places which they inhabit, and the immensity of their multiplication.

The human species, in its relations with these animals, also derives many advantages from them, much counterbalancing any annoyance which they may give us.

We find, in fact, that a very considerable number of species of mollusca form a part of our nutriment, not only among demi-savage people, but even among civilized. The savage people which live on the sea-coast make great use of mollusca for their food, as we learn from Adanson, respecting the hordes which inhabit Western Africa; from Molina, regarding those of Chili; from Cock, Peron, and others, as to the New Hollanders; and Forster, regarding the people of the South Sea Islands. But even in the civilized countries of Europe, the mollusca constitute a great portion of the food of the inhabitants of our maritime coasts, especially in places where the population is generally poor, and where certain days of the week or year are consecrated by religious abstinence, as in Greece and Italy, particularly in the kingdom of Naples, and also in some parts of France.

The aliment which man derives from animals of this type is in general agreeable to the taste, wholesome, and even exciting; but it is sometimes rather hard and indigestible, especially when it is taken from the muscular parts which compose the foot, and which cannot be too much cooked.

The bivalves appear in general to be more esteemed, and of a more agreeable flavour than the univalves, because they have a smaller quantity of muscular fibre. In fact, among the first, the most in request are those whose abdominal mass is nothing, or at least very trifling, as the oysters, the mussels, the lithodomi, the pholades, and especially the teredines, according to the observation of Redi, who declares them to be much more delicate than oysters.

As the mass which composes the body of these animals, especially when they are eaten raw, contains a greater or less quantity of sea-water, which often acts as a purge, it is not astonishing that man often experiences an effect of this nature when he eats a considerable number of these animals; but it is proved that in certain circumstances, not very easily defined, and on certain individuals, the effect is much more intense, and often followed by very grievous consequences.

But it is not merely as objects of nutriment that the mollusca may be useful to man: some of them, few in number, it is true, furnish him with the materials of clothing. Such are the pinnæ marinæ, whose filaments, which constitute their byssus, have been employed from time immemorial by the inhabitants of the coasts of the Mediterranean, and especially by those of Sicily, to form tissues, equally remarkable for their beauty and the duration of their natural colour, and also for their lightness, and their property of retaining heat.

The semi-transparence presented by the valves of the genus placuna is the reason why the inhabitants of China and the Philippine Islands employ them to furnish their windows, where they answer as a substitute for glass.

The property which certain parts of univalve or bivalve shells possess of reflecting the rays of light by decomposing them, which characterizes the iridescent mother-of-pearl, has caused them to be employed as objects of dress or ornament. We also remove by art from these shells portions of greater or less thickness of the *nacre*, or mother-of-pearl; and according to their plane or curved form, their thickness or slenderness, we form the ornaments of a multitude of instruments, of tables, the pannels of furniture, and of personal ornaments for the use of women.

We have already remarked, that the human species also derives from the animals of the molluscous type many objects useful in the arts of painting and of dyeing. If it is not absolutely proved that the ink of China is formed of the matter deposited in the bladder of some species of cryptodibranchia, it is at least certain respecting the *sepia*, which has even received its name from the colouring matter which it yields, and which is so finely and so equally divided.

It is not less doubtful that the ancients extracted the fine purple colour, with which they tinted the garments almost exclusively consecrated to princes, from a species of subcephalous mollusca, of the family of the *purpuræ*, which inhabited the shores of the Mediterranean, and especially towards the coasts of Tyre, and which no doubt it would be easy to find again, or to replace by some species of our own seas, as has been proposed by many other writers. But the small quantity of this colour which was derived from each individual, and consequently the great difficulty attached to the tincture, must have led to the abandonment of this employment of the mollusca, especially when means were discovered of replacing the purple by a colour equally fine, furnished in abundance by the kermes and the cochineal.

We shall not delay long in explaining the therapeutic properties attributed by the ancient physicians to certain parts of the mollusca, because time has not respected those opinions, but has destroyed them in succession. The only one which appears to have resisted its influence is that of the calming and soothing virtues of the decoction, or rather broth, made of snails and others of the limax and helix tribe, in affections of the chest and lungs. This, however, is any thing but a specific remedy. We may add the slightly purgative quality of oysters and pectines when eaten raw, which, most probably, as we have already hinted, may be attributed to the sea-water which they contain.

After what has been said, it is evident that the principal utility of the mollusca to man consists in the article of diet; but we shall also find that these animals are less injurious to us than useful.

The octopi are perhaps the only species which, by their carnivorous instinct, may prove injurious to us, in the relation of our animal nutriment. It is indeed well known, that they cause considerable injury to fishermen employed in pursuit of the crustacea, by the great destruction which they make among these animals, because, like them, they inhabit rocky places.

Our vegetable aliments indubitably suffer more considerable damage from the voracity of the limaces and helices which inhabit our fields and gardens. But this is an inconvenience proportioned to the development of our agricultural or horticultural industry, which accumulates in a small space the substances of which these animals are fondest.

Our habitations in the open country do not appear to suffer any damage from the mollusca; but it is not the same with our constructions on the sea-shore, or with those which are destined to float upon its surface: the lithophagous venus, and especially the lithophagous mussels, and the pholades, to lodge themselves in the stones which constitute our dikes, pierce them in all directions; and although they do not do this very deeply, they nevertheless hasten the destruction of these works.

This is still more evident in the case of the teredo, the species of which select wood in which to excavate their dwelling. In those countries where the people are obliged to construct dikes supported by wooden piles to protect themselves from the invasions of the sea, as in Holland, the damage is considerable; for at the end of a few years they are so much pierced below the level of the water that it is necessary to renew them. Vessels which remain a long time in harbour or in docks are also exposed to the destructive action of these animals, more especially, as it would appear, in the seas of warm climates. In fine, even living trees, whose roots or stem happen to be submerged, are attacked by the teredo, as Adanson relates of certain trees on the banks of the Niger in Senegal.

The relations of the mollusca with the mineral kingdom, and consequently with the mass of the earth which it has contributed to form, are not without interest ; for without attempting here to resolve the physiological question whether the conchyliferous mollusca borrow from the inorganic kingdom the calcareous matter which composes their shell, or if it is formed of all materials, it is however certain that they produce no inconsiderable changes in the surface of the earth, by accumulating this matter in some places more than in others, and consequently, that they alter its physiognomy or superficial structure, the study of which constitutes geology, or as it is sometimes more properly termed *geognosy*.

The modes in which this accumulation is made are altogether different, according as the mollusca from which the shells proceed were fixed or not, lived embedded in mud or sand, or were free on the superficies of the rocks or of the soil. Thus the oysters in our climates, the pintadini or regular aviculæ in warmer regions, as well as the spondyles and many other bivalves, form by their accumulation banks more or less extended, strata more or less thick and horizontal, in which the shells are still at the present day in the position in which they originally lived, and almost without any admixture of foreign bodies. Although this is less evident respecting the cardia, the tellinæ, the lutraria, the myæ, &c., and all the genera of bivalves which live vertically sunk in sand or mud, we see nevertheless that these shells must also form sorts of strata, because the individuals newly born are deposited by their parents above themselves, so that the latter, sinking in the sand in proportion as they enlarge in bulk, depress their parents, and the individuals below them in succession, so as to remove them sufficiently from the surface of the soil to prevent their tubes receiving any water, the consequence of which is death. Then their shells, vertical while the animal was living, begin to incline by degrees, become horizontal, are filled with the substance in which they were sunk, resist the pressure of the accumulated strata, so as sometimes to remain perfectly entire, with all their asperities, or if not so, they are broken and crushed, and disposed in beds more or less free from every other shell, or even from every other foreign body. This is very perceptible in the alluvions formed at the present mouths of our great rivers, or in the creeks of our sea-coasts, where currents prevail, which justifies our presuming by analogy that at those parts of our continents where similar accumulations are to be found, there was formerly the mouth of a river, or some gorge in which the waters formed an eddy. The other mollusca, living freely at the bottom of fresh and salt water, without sinking in the sand or mud which constitutes this bottom, or sunk only in the moveable part, at their death abandon their shells. These shells, rolled, overturned, cast against the rocks and the projections of the soil by the movements of the waves, are broken, reduced to a fragmentary state, more or less fine, and are then drawn along in the habitual direction of the currents and the winds, and accumulated along the shores, especially in bays, over an extent and to a height often very considerable. The strata which result from these are thus entirely composed of fragments, more or less bulky, of shells, often rolled, which consequently have lost their asperities, and are often of very different genera, which depends a little upon the localities. It is also remarked, that in the structure of these strata the fragments are in general deposited according to the laws of specific gravity, and that they are little or not at all intermingled with mud or other foreign substances, the entire shells which have escaped the destructive action of the currents, being filled even to the bottom with detritus, or coquillaceous sand. Many specimens of this augmentation of land are to be seen in various parts of our coast. These are the depositions which, in the course of time, by the long continued action of the pressure of superior strata, as well as by the tendency of inorganic matter, thus pressed and broken, to crystallize, grow more and more solid, and are converted into calcareous rocks, which end by no longer presenting any traces of their ancient organic disposition.

Previously to turning from the consideration of this division of the animal kingdom, as a whole, to a short review of its several sections, it may be useful to advert to the general principles by which these animals have been classified. These are of the same kind which are applied to the other types of the animal kingdom. The facility of collecting and of preserving the shells or envelopes of these animals, the beauty of form and colour which frequently distinguishes them, and especially the consideration that they exist alone in the composition of certain strata of the earth, has sometimes given rise to a contrary opinion: but this was truly erroneous. Thus the principle, *par excellence*, that the entire organization should be our guide, and that it is the external organs which should represent it, and furnish distinctive characters, is equally admissible in this as in every other part of zoology. But as the *ensemble* of the organization of the mollusca is sometimes pretty rigorously translated, so to speak, by the shell, and as the latter is evidently one of the most prominent of the external parts, and of which there is the most need in the accessory application to geology, a deception has resulted in the application of the principle, and a belief that it was possible to arrive at a methodical classification of the mollusca, by a simple consideration of the shell; this unquestionably appears to be a mistake.

The consideration of the local habitat should be allowed no importance in the classification of the mollusca, still less should that of the country, because neither furnish any characters imprinted on the animal or on its shell.

That of the nature of the aliment should not be admitted as of much more importance, because, though it is possible to conceive a certain correlation of visible organs with the structure of the digestive apparatus, more or less modified for a peculiar alimentary substance, yet such is never the case with the mollusca. Accordingly we find species essentially carnivorous, as the testacellæ, close by herbivorous species, like the limaces.

The existence or absence of a protecting body is evidently of greater importance, since it is a character altogether apparent; nevertheless it is easy to see, though the number of exceptions be not very considerable, that in the same genus we may find conchyliferous species and others completely naked.

The particular form of the body, the visceral part of which constitutes a wreath more or less elevated, is yet of less importance.

The appendages, the lobes, the cirrhi which border the mantle, are of no greater import to consider, unless perhaps in the lamellibranches, in which the consideration of the tubular lobes, which prolong the mantle behind, present characters of real value.

The distinction of the head from the rest of the body, as it is complete, incomplete, or non-existent, conducts to some divisions of the first order in the type of the malacozoaria, but it is a character not always sufficiently well marked.

The number, the form, the position of the tentacular appendages which accompany the head, have perhaps something still more constant, and consequently more essential to study, for the establishment of a classification in the mollusca. We nevertheless occasionally find inexplicable anomalies; such is that of the carychia, in which the true tentacula disappear by little and little, while all the other animals of the same family possess them in a very evident manner.

The position of the eyes is also worthy of some consideration, but less perhaps than the tentacula. We find, in fact, some genera of the same family, or even some species of the same genera, which have eyes subpedunculate, and others which have them sessile.

The form, the position of the principal organ of locomotion, that is to say, of the foot and the natatory appendages, give rise to considerations of still greater value in the classification of the malacozoaria; and as the characters derived from them are evidently exterior, it is not astonishing that they have been employed so often, and with much advantage.

A better character perhaps in the relation of its importance, but unhappily more difficult to observe, which no doubt has prevented its employment, is that which may be drawn from the armature of the mouth, either at its orifice or in its interior, since it is in relation with the nature of the aliment.

Another still preferable, because it agrees with the form of the shell in general pretty well, is derived from the position, the symmetrical or non-symmetrical form, and even from the structure of the organs of respiration. But unluckily, though these organs are most frequently pretty nearly exterior, a certain degree of habit is necessary to enable us to employ this character with advantage.

In fine, the part of the organization of the mollusca which appears as yet to present the character of the greatest value, is that which constitutes the apparatus of generation, composed of the two sexes in different individuals, or united in one individual, or, in fine, formed of a single female sex. Unhappily, again, this character is entirely anatomical, and consequently of difficult application in zoology.

Lastly, the consideration of the shell alone ought not to be regarded as altogether of no value, or useless, especially when we review in succession the differences according to their degree of importance: 1. the number of pieces which enter into its univalve, sub-bivalve, or operculate, bivalve, tubivalve, and multivalve composition; 2. the position on the body of the animal, dorsal, as in all the cephala, dorsal and ventral, as in a small number of cephala and acephala, or finally bilateral, as in all the lamellibranches; 3. the indices of its relations with the respiratory apparatus, that is, the existence of an emargination or a tube at the anterior extremity of the aperture in the univalves, or of an hiatus more or less considerable of the posterior extremity in the bivalves; 4. the indices of its relation with the muscular system of the animal, simple in a great portion of the univalves, but merely visible in the patelloïdes and otides, more or less complex in the bivalves, and formed, as we have already seen, by one, two, or even many impressions of the adductor muscles, one, two, or a greater number of impressions of the retractor muscles of the foot, by the abdominal ligula, an index of the attachment of the edges of the mantle, and finally, behind, by that of the tubes of respiration; 5. the symmetrical or non-symmetrical form which involves the similitude or dissimilitude of the pieces in the bivalves; 6. the form of the aperture in the univalves, the manner in which each edge and the columella, or its vitreous depot, contribute to form or modify it; 7. the ligamentous system, or the dovetailing, so to call it, of the two pieces of a bivalve shell, that is to say, those of the hinge and of the teeth which compose it, observing that each genuine species has a peculiar system of indentation on the edges of the shell; 8. the consideration of the existence or of the absence of an operculum in the univalves, of its structure, form, &c.; 9. the total form of the shell, the proportion of the spire, and of the aperture in the univalves, the direction of the latter, and in the bivalves the proportion of the two sides of each valve, the direction of the furrows with which the superficies is worked, the system of coloration, of epidermic covering, &c.

After this rapid sketch of the degree of relative importance of the characters which the different parts of the organization of the mollusca may present, and after the observation that it is often useful to consider their spoils, or protecting bodies isolatedly, we must add, that though this last be of real importance, the true basis of classification is, as much as possible, in the general form of the body, the distinction more or less complete or non-existent, of head and body, and the organ which afterwards has most influence in modifying the shell, namely, that of respiration; at the same time observing that the same form of the shell may sometimes, though rarely, be represented in genera sufficiently different; such, for example, is the form of the haliotides, which exists in the pulmobranches, in the chismobranches, and in the otides; it is the same with the patelloïd and turriculate form, &c.

Such are the principles followed by the most recent naturalists, for the divisions and higher subdivisions of the mollusca; let us now briefly consider those which regard the distinction of species, by far the most difficult part of science in all the types of the animal series, but still more in this than in any other, in consequence of the employment which has been made of the shell alone for this distinction.

If the sex and age have an evident influence in determining the differences in the molluscous animal, and consequently in its shell, it is not less evident that circumstances not easily explained may act less profoundly no doubt, that is, little on the animal, but much upon the colour, size, and proportion of the parts of the shell. Of this we have certain proofs in species a great number of whose individuals may be seen at once, as the helix nemoralis, the limnæa stagnalis, the buccinum lapillus, the neritina of the rivers, the common patella, the eatable oyster, the eatable mussel, and many species of cardia, &c., in whom we find individuals in the same locality differing one from the other, in all those characters of the shell which we have enumerated above.

More justly may we conceive that an assemblage of circumstances to a certain point inexplicable, and which have acted for a very long period, may have influenced almost in a fixed manner a succession of individuals of the same species, and determined in shells certain differences in size, proportion, colours, system of coloration, and even in the state of the superficies, smooth or rugose, especially when they shall be compared to other individuals of the same species living for a long series of ages in different localities. These differences then really constitute, as it would appear, but simple fixed varieties, so much the more dissimilar as the localities are more remote, and which we may, if we think proper, decorate with the name of local species, but which are not real species. In fact, when we come to assemble these pretended species from a great number of different localities, we shall find that they pass one into another in a manner almost insensible. M. Defrance, who had occasion to make the same remarks on the distinction of fossil shells, inquires whether a species be genuine if that to which it approaches the nearest is to be found in the same locality? Though this rule can hardly yet be considered as very rigorous, it may assist in a subject so difficult, and one so important for geology. The minute study of living species can alone furnish analogical means to diminish the difficulty, and consequently supply geologists with data, to resolve the problems concerning the analogy of formations in the structure of secondary and tertiary strata.

Notwithstanding the great length to which this supplement on the division has already extended, we can find no better place for the insertion of a few general observations on the principles of CONCHOLOGY, inasmuch as the naked and coquillaceous do not form two distinct divisions, and that such general observations (which by the way involve all the interest in the subject to all except professed conchologists,) could with no propriety be introduced and distributed among any subsequent remarks we may have to make on the orders or genera.

Conchology is the art of arranging the shells, or rather the protecting bodies of testaceous animals, so as to enable us to recognize them promptly and certainly, without giving any attention to the animals which they have contained, or contain, or at least regarding this part as a matter of very minor importance. We have already said quite sufficient concerning the animal, and concerning the shell too, as a portion of that living animal; what we have now to say will refer merely to the envelopes themselves, which may be preserved independently of the animal, and which, in fact, may have belonged to animals of classes and even of types very different from each other; and consequently in this point of view we must follow the method of Linnæus and of a great number of other naturalists, though we cannot avoid regarding it as totally artificial.

For a long time this portion of natural history, which may almost be said to have been invented for the gratification of amateurs of the rare and brilliant, was regarded as a study nearly idle and useless by all true zoologists; and this was so far true that it was often more necessary to be acquainted with the shells in their artificial state, to which they were brought by the application of emery and other substances, and by processes which removed even one or two of their strata, than in their truly natural state, in which they were often rejected. Consequently, all those shells which either naturally, or by art, presented nothing remarkable, no singularity, were remorselessly excluded from the cabinets of collectors. The methodical zoologists would at last have ended by totally annihilating this art or study, or by degrading it into a mere pastime, if geology, from the lofty flight which she has taken in these latter days, had not found the necessity of characters extremely minute for the purposes of comparison, either between fossil shells, or of the latter with living species. It is really to this cause that conchology, properly so called, owes both the continuance of its existence, and the daily increasing efforts of enlightened naturalists, who endeavour to give it sure principles and rules, by means of which geologists may be guided in their minute researches and the very difficult problems which they propose to resolve. Conchology then, or perhaps a better word would be *ostracology*, forms among the natural sciences a branch altogether separate, which may have its proper and particular rules, and to which there would be nothing analogous, unless we should think proper, for example, to examine in detail the hairs of mammiferous animals, the feathers of birds, or the scales of fishes. It appears, however, that if we could, while we studied conchology in a manner perfectly independent, so manage that it might be contained altogether in malacology, equal utility would result both to the science of animals and to geology, or palæozoology. To this object our researches should unquestionably be directed, though at the same time we must confess the predominant weight of geology in this question.

Every art, be it what it may, has necessarily a greater or less number of terms which are proper to itself, or common terms, whose acceptations are peculiar. These we name technical, which it is of the utmost importance properly to define, so that they may be well understood, and which are conveniently employed to avoid the long circumlocutions to which we should be forced to recur by the usage of ordinary

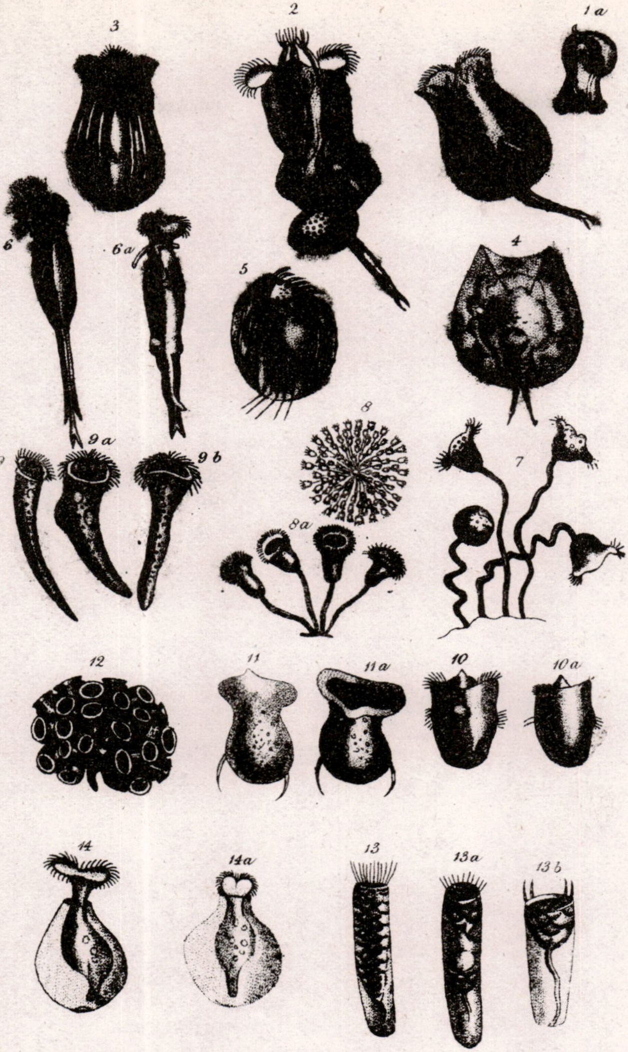

1 *Brachionus urceolareus.*	8 *V. socialis.*
2 *B. plicatilis.*	9 *V. buccina.*
3 *B. striatus.*	10 *Urciolaria appendicula.*
4 *B. bracteus.*	11 *U. cirrata.*
5 *B. patella.*	12 *U. nasuta.*
6 *Furcularia revivifiabilis.*	13 *Vaginicola inquilina.*
7 *Vorticella hemispherica.*	14 *Folliculina ampulla.*

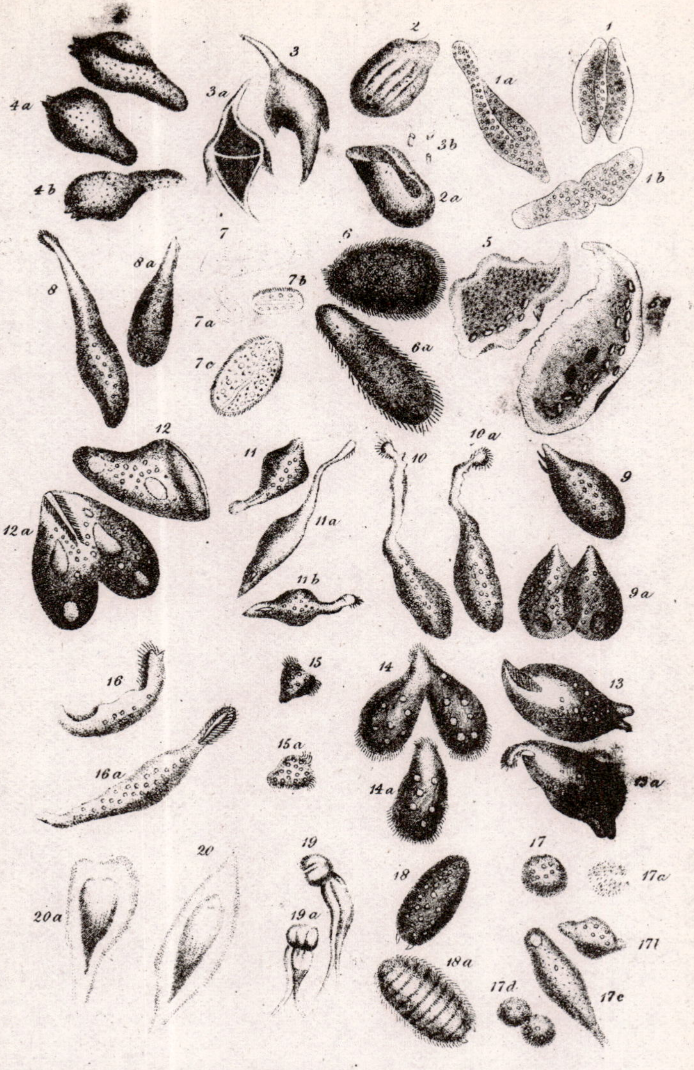

1 *Paramæcium aurelia.*
2 *Bursaria truncatella.*
3 *Bur. Hirundinella.*
4 *Kolpoda cucullus.*
5 *Kol. meleagris.*
6 *Leucophra virideseens.*
7 *Leu. nodulosa.*
8 *Trichoda anas.*
9 *Tri. ignita.*
10 *Tri. melitea.*
11 *Tri. versatilis.*
12 *Tri. aurantia.*
13 *Tri. bipes.*
14 *Tri. lepus.*
15 *Trichoda trochus.*
16 *Tri. patens.*
17 *Cercaria podura.*
18 *Cer. hirta.*
19 *Cer. catellus.*
20 *Paramæcium oceanicum.*

phraseology. Those technical terms we are about briefly to explain, and we shall conclude by a sketch of the history of this science.

We have, in reality, no other generic terms to indicate the hard, calcareous, fragile bodies which form the object of this part of natural history, than that of envelope, or better, perhaps, of protecting body or of testa; for by that of shells, or *conchæ*, we simply understand those of the molluscous animals. The Greeks had the word *ostraca*, whence *ostracodermata* and *ostracea ;* and the Latins that of *testa*, whence the denomination *testacea*, or animals covered with a testa or hard envelope. Nevertheless the vulgar appellation of *shells* (more vague perhaps in English than in some other languages) has prevailed, and the term conchology is in general use.

Quitting, however, these verbal criticisms, we understand by shells or protecting bodies, bodies of a very variable form, cretaceous, more or less thin, breaking off readily and completely, easy of preservation, and constantly in relation with the skin of an animal.

In considering at first these bodies in a general manner, and under the relation of structure, we find a first division of shells, namely, into *false* and *true*.

A false shell is that which does not belong to a molluscous animal, or rather that which is composed of a great number of small polygons applied side by side, and the *ensemble* of which forms a calcareous, hard, and frangible envelope, which is seen in the testa of the echinites.

A true shell is that which is formed of laminæ applied one within the other; the most recent and the largest being the most internal, and the most ancient and smallest the most external, whatever its form may be and the number of pieces of which it is composed.

The general study of this form then gives a division into those which are tubular and those which are not.

We call *tubular shells* those whose transverse diameter is considerably smaller than the longitudinal, and which are not enrolled or turbinated, or if so, only in a very irregular manner, and never in a spiral. These are the tubes of certain genera of *setipodes*, which have another distinctive character, namely, that the summit (*head*) is always open, which is never the case with the shells of the mollusca proper.

The *non-tubular* shells are then divided into shells of a single piece, these are the *univalves;* and into shells of many pieces, or *multivalves;* and these last into *bivalves* and *multivalves*, or *dissivalves*.

According to this, we must understand by valve (*valvula*) a calcareous piece of a very variable form, applied on or in the skin of a molluscous animal, and covering a greater or less part of it; but then we must often have recourse to the skin of the animal to judge whether a certain number of these valves belonged to one individual; as, for example, when they have no direct relations between themselves, but only indirect by means of the skin. This is the reason why for a long time one valve of the testa of the lingula has been regarded as an univalve shell.

The *multivalve* shells are of three sorts; those which are composed of many transverse imbricated pieces, as in oscabrio, those which are formed of five or many valves, symmetrically ranged on the right and left, and sometimes even placed in scales, and united together by means of the skin, as in anatifa, (these are the *dissivalves* of M. Denys de Montfort), and finally, those which are disposed in a manner almost circular, as in the balani and neighbouring genera.

The *bivalve* shells are those which, as their name indicates, are formed only of two pieces; sometimes, it is true, enclosed in a tube, or calcareous envelope, more or less developed, which some authors erroneously consider as another valve. They are always applied on the sides of the animal, and constantly in a relation more or less marked between themselves. Nevertheless, we should mention that this relation between the two pieces of a bivalve shell not being always evident, we may be sometimes led into error, and induced to regard as having belonged to an univalve, a piece or valve which belonged to a bivalve, as in the lingula, some species of camus, &c.

The *univalve* shells are, on the contrary, a testa of a form extremely variable, sometimes even almost tubular, which covers the molluscous animal more or less, and may also be entirely concealed in the interior of its skin.

The univalve shells may be considered under several different relations, which we shall just briefly mention here.

1. Under the relation of the places in which they are found, or rather of the animals to which they have belonged, it has been thought proper to distinguish them into *terrestrial*, *fluviatile*, and *marine*. But it must be owned that this distinction is often very difficult, and that its importance has been exaggerated, as far as the use that the study of fossils may derive from it.

The *terrestrial* univalve shells are generally rather thin; their external surface, most frequently smooth, presents little but the striæ of growth, and never any spines or asperities properly so called. The surfaces, both internal and external, are never nacreous. Their aperture, always entire, has very often, at least in the adult state, and only in these species, its edges thickened in the manner of a pad, or more or less thrown out externally.

The univalve fresh-water shells are also pretty generally of no great thickness; they are sometimes furnished externally with some striæ, and even with spines; and under the epidermis, which is almost always thin, smooth, and of a very deep green, we pretty frequently find that they are nacreous, or of an exceeding whiteness. Never, at least up to the present time, have any been found, the aperture of which is really emarginated, and its edges are always straight and trenchant.

As for the univalve *marine* shells, they are often very difficult to be distinguished from the preceding. In general, however, they are thicker, and much more frequently provided with pads, varices, spines, &c. Their aperture, very frequently emarginated, or elongated like a tube, more or less long anteriorly, is pretty often edged by a thick pad, which may be tuberculous, scaly, or laciniated. They are sometimes nacreous in the interior, when they are covered by an epidermis, which is scaly, pilose, and in general of a very different aspect from that of the terrestrial, and even of the fresh-water shells.

2. Under the relation of the degree of depth at which they are found, the marine mollusca have been separated into *littoral* and *pelasgian*, according as they are to be met with on the sea-shore, or at depths more or less considerable in the high seas. But it must be allowed that this division is still worse than the preceding, since no character inherent in the shell can be brought in support of it.

3. Under a relation almost anatomical, a distinction of shells has been established into *external* and *internal*. The internal shells are in general much thinner than the external, flat, or but slightly rolled, and without epidermis, of no colour but white or yellowish.

4. The size is next taken into consideration, for the separation of univalve *microscopic* shells, which are those, as may easily be conceived, which are so very small, as not to be seen but by the assistance of the microscope. But such a division can by no means be clearly exact.

5. If we consider the general form of univalve shells, without paying attention to any of their parts, denominations are

employed, which, though vague enough, are nevertheless necessary to be known.

The first distinction is that which relates to the equality or inequality of the two sides of a shell, of whatsoever form, separated by a fictitious axis, taken from the head to the base, or from one extremity to the other. A shell is named *symmetrical*, the two sides of which are perfectly equal, and *non-symmetrical* when such is not the case. Thus the bone of the sepia, the shell of the argonaut, that of the patellæ, &c. are symmetrical. The Chinese patellæ, the sigaret, and many others are not symmetrical.

The *flat* shells are those which have no cavity, as the bone of the sepia, the Chinese patella, &c.

The *tubular*, those whose diameter is considerably less than the length.

Covering or *sheathing*, those which are conical, and without any spire, properly so called, as in the patellæ.

Spiral, those which are more or less turned, and in different directions, as we shall presently explain. But it is first necessary to define some terms, which belong to the shell considered in the mass. We name,

Discoïd, those which more or less resemble a disk, and which, considering the manner in which the spire is rolled, are termed *rolled*, as in the ammonites.

Depressed, the species oval or rounded, whose form is very much flatted, and the spire very short—for example, the sigaret.

Globular, those of which all the diameters are obviously equal, in consequence of the great development of the last turbination of the spire, which is much larger than that which precedes, as in the ampullariæ, &c.

Oval, or *ovoïd*, the species whose longitudinal diameter is a little longer than the transverse, as in the porcelaines, and a considerable number of the helices.

Navicular, some shells which, when turned on the back, with the aperture upwards, have a certain resemblance to a small boat, as the argonauta.

Pyriform, when one of the extremities is thick or swelled, and rounded, and the other pointed in the form of a tail—for example, the pyrula.

Conical, when one of the extremities, being widened, is as it were cut squarely, the other being pointed and forming the summit. When it is the summit or head of the shell itself which forms the summit of the cone, the shell is named *turbinated*, as in the trochi; and it is called *conical* or *convoïd*, when, on the contrary, the summit of the cone is at the anterior part of the aperture, as in the coni properly so called.

Cylindrical, when the shell is elongated, and of a breadth or bigness pretty nearly similar, both in front and back. Such are most of the involuted shells, as the olives, &c.

Fusiform, those which, swelled in the middle, are pointed at the two extremities—for example, the fusi.

Turriculated, or turreted, those which are very much elongated; that is to say, whose longitudinal diameter is much longer than the transverse, which depends upon the manner in which the spire is formed—for example, the turritellæ.

6. The univalve shells may finally be considered in the relation of the distinction of each of their parts.

A univalve shell may be conceived always to have a summit or point where it has begun, a base which is its actual termination, and an intermediate body, with a cavity sometimes almost imperceptible, in the case where it is extremely depressed or altogether flat, and then it has really many relations with one valve of a bivalve shell. It is altogether the reverse in the tubular or tubiform shells, which much resemble the calcareous tubes of certain *setipodes*.

But before proceeding farther, it may not be superfluous to explain the position in which conchologists place the univalve shells, for the purpose of studying and naming their different parts. Linnæus, Bruguières, Da Costa, Lamarck, &c. place the shell which they study standing on the extremity opposite to the head, and the aperture in front of the observer. M. de Blainville, on the contrary, imitating Draparnaud, and many other authors, supposes it placed obliquely on the back of the animal, or what comes pretty nearly to the same thing, applied on a table, on the side of the aperture, and consequently the head, or highest point, being backwards, and upmost, while the opposite extremity is in front, and below. From this it results, that the names of *right* and *left* are applied to the same sides, according to the two different points of view; but those of *inferior* and *superior*, in the description of the aperture and its edges, are replaced by the words *anterior* for the first, and by that of *posterior* for the second.

The summit or *head* (*apex*), which is the part where the shell has begun, may be altogether flat or very projecting, straight or vertical, or inclined directly backwards, to right, or left, but it does not appear to be ever directed forwards. It may be pointed or nippled, entire or carious, and even sometimes hollow, as in the bullæ.

The base (*basis*), or the part usually opposed to the summit, is that in which the aperture, of which we shall presently speak, is invariably pierced. Under this name, however (for we follow the system of M. de Blainville), we do not understand what Linnæus and the majority of conchologists so designate. In fact, with them it is the extremity, pointed or not, opposite the head, and they name it thus, because, in their manner of designating the different parts of a shell, they placed it vertically, the head being above, and the aperture in front; but in the system which we follow, the base is all that part which rests more or less obliquely on the back of the animal. Sometimes this base is very broad and round, as in the trochi, which gives them the form of an inverted top; at other times it is small, as in the vis, &c. It may be very much elongated, as for example in the cypræadæ. It is formed entirely by the aperture in the patellæ and some others, and at other times by a portion of the last turn of the spire.

Its direction, which is usually that of the aperture, also presents some considerations which ought not to be neglected: thus, it is altogether perpendicular to the axis of the shell in the patellæ, &c.; and it is almost entirely in its direction in the cyprææ and olivi. The other shells are more or less intermediate.

The body of the shell is all that portion which is between the basis and the head. Most frequently it is hollow in the interior, and serves not only to cover the body of the animal on its upper part, but also to contain a greater or less portion of it.

Sometimes it receives the name of *discus,* as in the haliotides; but then only the last turn of the spire is comprehended under this name.

In a certain number of shells or testæ, the body is curved in no direction, neither to the right, nor to the left, nor in front, nor in rear, and is even in no wise excavated. These are named flat shells, symmetrical in the bone of the sepia and the calamary, non-symmetrical in the patella Sinensis.

Pretty often the base and summit are united by a body which is not recurved in any way, but which is more or less excavated. This constitutes the covering or sheathing shell, as in the patellæ, the emarginulæ, and especially in the dentalia.

Finally, in many cases the body of the shell is formed by its enrolment in different ways. Such it is in the true cochlidæ or *spirivalves.*

To form a just idea of this, we may conceive that every univalve shell was a cone more or less elongated, but flexible.

If it be rolled from rear to front, and from top to bottom,

absolutely in the same vertical plane, there results a discoïd shell, compressed from right to left, the summit of which cannot be visible but in the same direction, and the axis of which is altogether equally transverse. These sorts of shells the French name *enroulées*, and we *volute* (*revolutæ*). A rigorous example may be taken from the argonautæ and neighbouring genera, but not in the planorbes, which are really subvolute.

The principal differences presented by this sort of volutation, or rolling, consist in its greater or less perfection; we name *arched*, the shell which only presents an arching more or less considerable, as in certain species of belemnites.

Curved, that in which the body begins to be much more curved, as ammonoceros.

Semi-volute, that in which the turns of the spire, which we call *whorls*, do not touch, as in spirula.

Volute, when the whorls touch, as in the true ammonacea.

And finally, very *much voluted*, the species in which the whorls penetrate each other reciprocally, so as that the last whorl or circumvolution conceals all the others, and the aperture is modified by it, as in some nautili.

If, on the contrary, the rolling of the spiral cone is made transversely, or from left to right, this constitutes what are called *involute* shells.

In these species the base of the shell is almost as long as the latter, as well as its aperture, and the axis of volution is longitudinal. There are in reality scarcely ever any shells completely involuted; those which approach the most to being so are the cypræadæ and the ovulæ. Sometimes the shell does not make a complete whorl, as in the bulleæ, and then the aperture is as broad and as long as the shell itself.

Finally, the great majority of univalve shells are intermediate to both these arrangements; that is to say, that the body of the shell is the result of an oblique volutation from right to left, and from bottom to top, if we consider it from the base to the head, or exactly the reverse if we follow the growth of the shell. Those are the true spirivalve shells which we name *turbinated.*

The name of *spire* (*clavicula*) is given to all that part of a spirivalve shell which is formed by the revolution of the spiral cone.

That of *circumvolution* or *whorl* to a complete revolution of the spiral cone.

Sometimes we distinguish from the totality of the spire the last whorl, which is usually the biggest, and where the aperture is found; and it is designated under the name of the body of the shell. The face, which is found to correspond to the aperture, is the *belly*, that which is opposed to it the *back;* but Bruguières will have it that the belly is only that part of the last whorl which forms the left portion of the aperture, and on which the internal lip is attached. Be this as it may, the name of clavicle is reserved for all the rest of the spire.

The direction according to which the rolling of the spiral cone is made serves to distinguish the shells into right and left. In general the actual termination of the shell is at the right of the animal, and consequently proceeding from this point, the volutation or *torsion* seems to be made from right to left, going from base to head: such are the normal spiral shells. But it happens pretty often, that the animal being anomalous in this respect, is, as it were inverted, that is, what is usually at the right is found at the left, and vice versâ, and then the shell is equally anomalous by having its terminal edge at the left. These shells are called *left*, or *sinistræ*, or *heterostrophes*.

The consideration of the spire, properly so called, but taken as a whole, gives rise to some technical terms, which to a certain point are involved in those used to designate the general form of the shells. The spire is called

Flatted when the united whorls form a surface altogether flat, as in conus cardinalis.

Crushed when the progress in a vertical direction is but little rapid compared with that in an opposite direction. These are the shells which approach a little to those named above, *discoïd;* for example, the solaria.

Moderate when the progress in both directions is pretty nearly equal, as in the *buccina*, &c.

Elevate when the spiral cone advances more in height than in breadth.

Elongate when this disposition is still more marked.

Turreted, or *turriculate* when with a similar disposition the whorls are very exactly separated by their different divisions of thickness, as in mitra.

Decollated when in consequence of age their extremity is broken.

Coronate when the edges of each whorl are armed with projecting points, with tubercles or spires, as in a great number of coni, and in the Ethiopian voluta.

The *whorls*, or turns of the spire, also give rise to many characters, which are expressed by determinate words.

As to their number, they are reckoned either in proceeding from the summit or from the end of the spiral cone.

Their proportion one to another is expressed in ordinary terms. It often happens that the last whorl but one is bigger than the rest altogether; sometimes the last is smaller than the last but one, &c.

The whorls themselves may sometimes be quite flat, to which the name of *ribboned* has been given, as in terebra. Sometimes they are distinguished with difficulty, and may be considered as almost *confounded;* finally, they may be separated from each other by a pretty deep furrow, as in the olives: this line of separation between the whorls is named *suture* (*sutura*).

The superficies of the whorls is also to be regarded. They may be designated under the name of *carinated* when in the direction of their length they present an angle or fold more or less marked; *smooth* when they have no projections or anfractuosities; *rugose*, *tuberculous*, when their surface is charged with rugosities or tubercles; *striated* when they are striped in breadth or length; *trellised* when this takes place in both directions; *corded* when they are bordered by a projecting and knotty side; *costate* when the pad of the left lip holds on the whorls, as in lyra; *varicose* when the continuous pads of the right lip are more or less tuberculous and desected, as in murex generally.

According to the notion which we have given above of the formation of a spiral shell, we shall see that if the whorls touch not transversely, i. e. from right to left, nor from top to bottom, there will be perceptible in the middle of the shell a conical depression extending from summit to base (which is named *umbilicus* in Latin), and at the same time a vacancy more or less considerable between each whorl, as in the vermetus of Adanson, and in the true scalaria. These shells are named *disjunctæ* (disjointed.) If, in rolling, the convolutions of the cone touch each other from top to bottom, but not transversely, we have a shell strongly umbilicated, as in the scalaria; and finally, if the whorls of the spire touch in all directions, without encroaching, or more especially if they should encroach upon each other, constituting in the first case a complete spiral cone, or in the next an incomplete one, it follows that the fictitious axis is no longer free, no longer hollow, except sometimes at the base, and that it is replaced by a sort of small twisted pillar, resulting from the contact, and interfusion of the internal edge of the cone, on which it is rolled. In fact, in examining a shell of this nature from base to summit, we see in its interior a solid part more or less tortuous, to which is given the name *columella*, pillar in

English; and as this sort of pillar, when the base of the shell is very oblique, is often lengthened to its anterior extremity, it is it which in this case forms entirely the left edge of the aperture, in consequence of which that sometimes receives the name of *columellaria.*

This column or pillar is called *pointed* when it is terminated anteriorly in a point, as in lyra; *truncated* when it appears to have been cut, as in the agathini; *salient* or projecting when it forms an elongation in front of the shell, as in the terebellæ; *spiral* when the part thus extended is *twisted* like a gimblet; *folded* when a greater or less number of oblique folds are perceptible there proceeding from its torsion, as in the volutæ; and it is said to be provided with a pad when towards its extremity it presents a swelling more or less considerable, and transverse, as in some cerithia.

Externally, or to the left of the termination of the pillar, is often seen a hole, or rather a cleft, more or less deep, of a form somewhat variable, and which especially exists in young subjects. This is the umbilicus, whose formation we have explained above. From the presence or absence of this hole proceeds the distinction of shells into *umbilical* or *non-umbilical.* We call the *umbilicus, consolidated,* or *subconsolidated* when in the full-grown shell it is covered by a sort of calcareous deposition, called a *callosity,* but this likewise exists underneath. If it exhibits grains projecting in its circumference it is called *crenulated; dentated* if it is accompanied with one or many teeth, as in the turbo pica; *canaliculated* when at the interior it presents a spiral gutter, as in some turbines and many cerithia.

After having thus successively viewed the univalve shells in their ensemble, and at their external surface, we will now consider their interior and orifice.

The cavity of a shell may be entirely occupied by the animal, or the part which is occupied may be separated from that which is not by one or many partitions, which divide it into several cavities named *chambers*, *concamerations*, *lodges*, or *cells*.

The shells which have but a single cavity are called *unilocular* or *monothalamous*, as is the case with by far the greater portion of the univalves.

Those which, on the contrary, have their cavity separated into a great number of lodges are named, in opposition, *multilocular*, *polythalamous*, *chambered*, *celled*, &c.

The form of the partitions, which may be very different, has determined the names of partitions.

Even when they are simple.

Sinuous when they present, and especially on their edges at the point of junction with the shell, certain sinuosities or sections, which have been compared to those on the edges of a leaf of parsley; from the same cause they are also named by French conchologists *decoupées*, *persillées*, words for which our less flexible language furnishes no equivalent.

It is from this disposition that in fossil conchology have been formed the names of *articulated* shells, of *articulation* derived from the disposition which the pieces of foreign substance which have been moulded into these anfractuous cavities, preserve among themselves, and which are observed, after the shell itself has been destroyed. These articulations may be *compressed*, *cylindrical*, *ventruous*, &c.

These different chambers, or particular lodges, communicate more or less completely between themselves by means of a hole in the form of a canal, which traverses the partitions. This hole is named siphon, and has been studied, 1. according to the number, which is never above two, as in tne oisyphytes, but in most cases there is but one; 2. as to position, it may either be in the middle of the partition or approached to one of its extremities, from whence the names of

Medial, when it is at the middle.

Dorsal, or *external*, when it is pierced near the external edge.

Internal, *ventral*, or against the spire, when it is towards the internal edge.

3. As to its form, which may be round, oval, or triangular. In the unilocular shells the cavity is rarely divided into two only, and incompletely, by a straight lamina, more or less extended, which is named *diaphragm*, as in the septaria. At other times this lamina is more or less curved, thus forming a little tongue or horn, *ex*. crepidula, calyptreæ, &c.

The *aperture*, or mouth of the univalve shells, is the entrance of their cavity. It is really formed or circumscribed by the edges, which are only the union of the interior surface of the shell with the exterior. Linnæus gives the name of *faux*, or throat, to all that part as far as we can see into the interior of the shell, that is, pretty nearly to the last semi-whorl.

Some authors give the name *peristoma* to the entire thickness of the shell at its aperture, but most generally it is divided into two parts, designated as edges or lips, distinguished into the internal and external, or the right and left, or columellary lip.

Considered in totality, and along with a portion of the last whorl, which it terminates, the aperture is said to be falling, or inclined, when, not following the direction of the spire, it falls suddenly; inverted, when, on the contrary, it curves towards the spire.

If we consider the aperture as to its regularity or irregularity, it is *symmetrical* when it can be divided into two parts perfectly equal and similar, and *non-symmetrical* in the contrary case; thus it may be formed by the excavation, more or less considerable, of one of its edges, which ought to be taken into consideration.

As to its proportional size to the rest of the shell, it may be very large, as in the haliotides, which, in consequence of this, have received the names of *megastomata* and *macrostomata;* or it may be middling, small, &c.

The last whorl of the spire may penetrate more or less into its interior, and modify it. In that case it is said to be so modified, as in the argonauts, helices, &c. Then, according to the observation of M. de Férussac, the spiral cone is always incomplete, but in the contrary case it is complete. It is to this part that Bruguières exclusively applies the name of left lip.

But above all it may be more or less deeply *emarginated* or *entire.* This explains the term *entomastomata,* which indicates that the aperture is entire.

It may show a simple inclination to be emarginated, and then it is termed *versant.* This means, that if the shell were on its back, and filled with a fluid, the fluid would run out through a part of the circumference a little widened. Many coni are in this state.

Finally, we may mention the form which has caused it to be named *siphonostomatous* or *canaliferous,* i. e. when it is terminated anteriorly by a canal or siphon, more or less elongated; this form is in relation with a similar disposition of the animal itself.

In the relation of form, which is very variable, the aperture of univalve shells may be round, oval, transverse, having more breadth than length, angular, semicircular, narrow, or linear.

The edges of the aperture are sometimes called lips. It is divided by a fictitious line, supposed to proceed from one extremity to the other of the shell; all which is found to correspond to the right side of the animal is properly termed *labium,* and on the other *labrum.*

In fine, this aperture of univalve shells may be always open, or more or less completely closed by a piece either calcareous or corneous, flat or slightly concave, formed of concentric elements, and attached to the posterior part of the foot of the animal. This is named *operculum*, cover or lid. Its form and size are taken into consideration, but give rise to no peculiar appellations. It is not so with the manner in which it is joined to the aperture. Those opercles are *simple* which have no other relation than that of their form with the aperture of the shell; *composite*, which are articulated by means of eminences and of corresponding cavities.

The bivalve shells, which we must very briefly dispatch, may be considered pretty nearly under the same relations as the univalve, and a few peculiar to themselves. As to the relation of the places where they are found, they are divided into *fluviatile* and *marine*. No terrestrial bivalves have as yet been discovered.

The fluviatile are not very numerous, and perhaps are still more difficult to distinguish from the marine than the univalves. It may be remarked, however, that usually nacreous in the interior, they are covered with a thick epidermis of a green more or less deep, and that the summits are worn, or what is technically termed *decorticated*. None are yet known but among the species with a double muscular impression, and altogether closed or short.

The next relation under which the bivalves are considered is that of their fixity, or that of their mobility; but of this we have already spoken sufficiently in another place.

A third relation under which the bivalves may be considered is whether they are free or concealed in a tube more or less developed. In this last case the valves are altogether contained and concealed in a tube of the same nature as themselves, and open only at one of its extremities. They may be called *tubicolæ*.

Another point of view is that of the substance in which they may be found: but this has been already treated of.

In considering, for the present, a bivalve shell composed of a single piece as forming a whole, it may be termed *long, elongated, cylindrical, transverse,* &c.; but the application of these epithets entirely depends on the position in which the shell is placed for study, which, with M. de Blainville, is the same as for the univalves. The shell is considered as covering the animal, and the latter walking before the observer, head in front, though in reality few of them change place, and they are sometimes in a determinate position on the side, or even with the head under. On this depends the name given to their forms and proportions, to enter into a detail of which in this slight sketch would be quite beyond our purpose, and we should but merely repeat what has been already said.

Under the name of *multivalve* shells we do not understand those in which the two valves are covered with a tube, which we mentioned above, but only those which are completely discovered.

They constantly appertain to animals which may be said to be intermediate between the malacozoaria and the entomozoaria, or in other words, between the true mollusca and the articulated animals. They are so few in number that it has been considered nearly useless to establish peculiar terms to indicate each of their parts, or such terms enter for the most part into those which have been already indicated.

These parts may be divided, however, into three sections:

1. The *serial,* or *articulated,* so named because they are placed in a series one after the other in a symmetrical manner, in the middle and dorsal line of the animal. In a great number of cases they touch, and even overlap each other more or less. This is easy to be recognized, because their anterior edge is attenuated, and the posterior the reverse, except the first and last, which are rounded, one in front and the other behind; their exterior surface may be smooth or rugous, &c.

In a certain number of species the pieces are extremely small, and do not touch; in this case they might easily be taken for the imperfect shells of univalves, especially the first and last of the series.

2. The *lateral* when they are in a greater or less number placed symmetrical on each side of the envelope of the animal, a single one occupying the dorsal line. They may touch or exist only in a rudimentary state, but they never articulate. They may also vary considerably in form and size, and be more or less smooth or striated.

These two groups of multivalve shells have been named *dissivalves*.

3. The *coronals* or *sub-coronals*, first established by M. de Lamarck, when being disposed in a manner more or less regular round a common axis, they are solidly dovetailed together by the edges, so as to form a complete cavity, close or open inferiorly, and closed above by a small number of pieces of a form not very variable, the ensemble of which is termed *operculum*.

The form, the number of the principal pieces, as well as that of the operculum vary, but the differences which they present do not require to be designated by peculiar terms.

We shall now close this imperfect yet we fear tedious sketch, by a very brief account of the history of conchology.

Aristotle, the first in this branch of the sciences, as in so many others, presents us, if not a systematic arrangement of shells, which was not his object, at least the basis of several divisions, which have been subsequently established. Thus we find in his principal work that he has considered shells with a view to the principal relations under which we study them at present; that is to say, according to the number of the pieces of the shell, he divides them into *monothyra* or univalves, and into *dithyra* or bivalves. He then takes among the first the consideration of their turbinated or non-turbinated form; their living on land or in the water; their habits of frequenting the shores or the depths of the sea; and even their capability of motion or fixedness, according to which he names them *cinetica* or *acinetica*.

Pliny, Appian, &c. added nothing, or almost nothing, to what Aristotle had done, even in the way of simple facts, and most assuredly nothing to their classification. We must therefore pass to the writers at the period of the restoration of letters.

The first author who really occupied himself in the distribution of shells, or in establishing a true conchological system, was, as every one agrees, Daniel Major, in a sort of appendix which he placed at the end of a German edition of the treatise respecting the Purpura of F. Columna, under the title of *Ostracologia in ordinem redacta*, printed at Kiel, in 1675. This consists of synoptical tables which conduct to genera tolerably natural, but few in number, and established only on the species observed by Columna. It is to him that we are indebted for the division into univalves and multivalves, among which he places the bivalves.

In 1681, Grew, in his *Museum Regium*, or description of the collection of the Royal Society, of which he was secretary, published a systematic and synoptic table of the genera of shells, in which he includes all the testa or testaceous envelopes, and in which, without employing the terms at present received, he establishes the division of shells into simple, double, and multiple, which corresponds to our univalves, bivalves, and multivalves. Among the first he separates those which are not voluted from those which are, and among the latter those in which the whorls are apparent from those in which they are not so, as in the nautili, the cyprææ. Could we have presented this synoptic table it would have been evident that Grew had arrived at the majority of genera admitted at the present day, and that many authors have derived excellent hints from his performances.

Sibbald, in 1684, in his *Scotia Illustrata*, returned pretty nearly to the division of Aristotle, that is, he took into chief consideration the abode, from which he deduced the division of shells into terrestrial and aquatic, and these last into fluviatile and marine.

This was also done by Lister, who, living at a period when commerce had brought a great number of shells into this country, published a treatise necessarily much more complete, under the title of *Historiæ sive Synopsis Methodicæ Conchyliorum libri Quatuor*, &c., in numbers, from 1685 to 1688. We find in this work, besides some very excellent figures designed and engraved by his daughter, and characters rather more rigorously circumscribed, the introduction of the distinction of shells according to the equality or inequality of the valves. A greater degree of attention is also paid there to the hinge of the bivalve shells.

Tournefort, the celebrated French botanist, who died in 1708, also endeavoured to facilitate the study of shells, which he designated under the general name of *testacea*, and which he defined to be the envelopes of certain animals, which have the hardness of a tile or of a vessel made of baked clay; but his method was not known for the first time but by the work of Gualtieri, in 1748. This clever botanist substituted the names of *monotoma*, *ditoma*, and *polytoma* for those of univalves, bivalves, and multivalves. Among the monotomes he established the distinction of univalves, properly so called, spirivalves, and fistulivalves; and in the generic characters he paid sufficient attention to the form of the aperture. In the class of ditoma he seems to have been the first who established the divisions of bivalves close or open (*clausæ et hiantes*). He also gave some attention to the position of the hinge.

In his polytoma, or multivalves, he places at once the echini and balani.

In 1711 Rumph made known a tolerably great number of shells from the Indian seas; but he made no great addition to conchology, properly so called. He did not even separate the bivalves from the multivalves. As for the rest, in his work the univalves are simple or turbinated, as in Aristotle. It must not be forgotten, however, that he has pointed out some generic sections tolerably good, as the strombi, porcelaines, volutæ, &c.

A little later, in 1722, Langeius proposed a new conchological arrangement, but a partial one, as he treated only of marine testacea, in a work in 4to., published at Lucerne, under the title of *Methodus nova et facilis Testacea Marina pleraque, quæ huc usque nobis nota sunt, in suas debitas et distinctas classes, genera et species distribuendi, nominibusque suis propriis, structuræ potissimùm accommodatis, nuncupandi*, &c. But it is certain, notwithstanding this pompous announcement, that he has added no very new considerations to those which had been already employed by Lister, unless perhaps that which is derived from the equality or inequality of each valve, or from the relative position of the summit or head. He also pays a little more attention to the form of the aperture in the univalves, and to that of the head in the bivalves. He also establishes among these last a division of anomalous species.

It is to J. Philip Breynius, in 1730, that we owe the employment of a new character, hitherto unobserved, namely, that derived from the number of chambers in the univalve shells, from whence the names of polythalamous and monothalamous. This he performed in a work published at Dantzic.

A little before him, in 1728, J. Ernest Hebenstreit published at Leipsic a dissertation, *De Ordinibus Conchyliorum*

Methodica ratione instituendis, in which we find but few important innovations. He paid, especially in the univalves, more attention to the spire than perhaps had been given before his time; and in the bivalves, his first division is founded on the presence or absence of a hinge.

In 1742, Gualtieri, an Italian author, whose work is still often cited, from the great number of mediocre figures which it contains, published a method, in which he has employed all the combinations which his predecessors made use of, without introducing any thing new. Thus his first division equally rests on the habitation of the shells. He names *exothalassibiæ* those which are not marine, and divides them as usual into fluviatile and terrestrial: as to the marine, *thalassibiæ*, they are turbinated or not, and the latter are vascular or tubular; he admits the polythalami. He pays attention to the equality or inequality of the valves, and of their sides. Finally, he considers the presence or absence of the hinge. In general, though in this work we find a considerable number of generic sections indicated, they are not solidly established.

In the same year was published, in France, the first edition of a work, which for a long time enjoyed a degree of reputation which it little merited. It was by d'Argenville, entitled *l'Histoire Naturelle eclaircie dans deux de ses parties principales, la Lithologie, et la Conchyliologie*, in 4to. Although this work was very successful, especially in France, in consequence of the figures which it contains, its merit is of a very inferior order. In fact, the author has introduced absolutely no new consideration in the manner of observing shells, which he again divides, according to their habitation, into marine and fluviatile, but nevertheless places the helices among the latter. Each section or subdivision is divided, according to the number of pieces, into univalves, bivalves, and multivalves, for the first, and into univalves and bivalves only for the second. We should also observe that the class of multivalves is still worse handled than in any other system. As to the genera, those of the univalves, though very few in number, are pretty well characterized by the form of the aperture; but it is not so with those of the bivalves, in which there is no reference made to the hinge. In fact, d'Argenville has almost always followed Lister; and when he has not, he has done mischief. He has often strongly criticised him, and always erroneously.

Immediately after d'Argenville comes another entirely systematic writer, who has not the advantage of having given good figures. This is Klein, who has almost invariably attached himself to changing every thing which Linnæus attempted to establish. He published, in 1753, a new system of conchology. He comprehends them all in the testæ, which he divides into *cochlides, conchæ, niduli testacei, echniodermata,* and finally into *tubuli,* or *marine tubes.* Under the name of *cochlides* he understands the turbinated shells, which he divides into two sections: simple cochlides, which he defines to be a spiral canal, resulting from a single circumvolution of the shell, and composite cochlides, which are those in which the circumvolutions of the shell appear double; so that the testa seems to be composed of two cochlides. Although his definitions are bad enough, we can see that the first section comprehends the spirivalve shells which have not their aperture terminated by a siphon, or rather the last whorl of which is not terminated in a point like the spire; and that he understands, on the contrary, by his composite cochlides, those which are pointed in front as well as behind. Although this consideration is evidently new, it is clear that it conduced but little to a good division. Another innovation of Klein is his having separated, one cannot well tell why, the conchs, *conchæ,* into *monoconchs,* which are the patellæ and the neighbouring genera; and into diconchs, *diconchæ,* which are the ordinary bivalves, an innovation which, to a certain point, has been adopted even by later writers. Not admitting multivalves, he places the anatifæ among the conchs, under the name of *polyconchæ*, while the balani form a division under the name of *niduli testacei*. The bivalves are then divided according to the consideration of the resemblance or dissimilarity of the valves, and their more or less complete closure. He has, besides, proposed, rather than established, a great number of genera, which have since been adopted; but the characters which he assigns to them are so vague and so ill defined, that it is not wonderful that this writer should have remained in partial oblivion.

We shall yet place before Linnæus, although the first editions of the *Systema Naturæ* had already appeared, the celebrated French traveller Adanson, because it appears almost indubitable, that it was from the latter's "Voyage to Senegal," published in 1757, that Linnæus has taken the most considerable part of his fixed general principles of conchology. Adanson, as we have already noticed in treating of the mollusca, took into consideration at once both the animal and the shell; he has nevertheless carried some innovations into conchology, properly so called. Thus, beside a profound study of each of the parts of shells, and an exposition of the characters which may be drawn from them, he has, as it were, established upon each of them a particular system. He has, among other matters, divided the bivalve shells according to the number of the muscles, or of their attachments; and above all, he has introduced the consideration of the opercula, which before his time had been almost entirely neglected, or merely separately viewed under the name of marine *umbilici*, without any reference to the shells to which they had belonged. It was according to this character that he established, in the family of the helices, two sections, the first the univalve helices, and the second the operculated helices, which he considers as forming the passage to the conchs or bivalves, but erroneously. We should also observe, that he appears to have been the first who arranged the oscabriones with the patellæ, the section of his multivalve conchs containing only the pholades and teredo. Linnæus, who in the first edition of his *Systema Naturæ* had not proved that he was really *au fait* at this part of the natural sciences, showed in that which followed the publication of Adanson's work, that the same principles might be applied to it, which he had so happily imagined and successfully employed in botany. He nevertheless created no very new characteristic distinction in the primary sections, nor even in the secondary, since he divides the testæ into multivalves, with which he commences, and in which he ranges the oscabriones into bivalves and into univalves, which he subsequently divides into turbinated and non-turbinated. But he has introduced in the exposition of the characters, in their circumscription, and in the creation of conchological language, that precision and that clearness, which must ever cause him to be regarded as the model and the master of all systematic naturalists.

Pretty nearly about this period, viz. in 1769, began to be published the great work of Martini, continued and terminated by Chemnitz in 1788. As we must regard it rather as a collection of figures of shells, than a true system of conchology, we shall content ourselves with mentioning that the order which has been adopted by the last, partakes at once of that of Gesner and Lister, the primary divisions being still derived from the habitation of the animals. In other points he follows Linnæus closely; and it may be said that his sections are tolerably simple, and tolerably consonant with natural relations.

In 1776, Da Costa published, in our own language, some true elements of conchology, in a work entitled "Elements of Conchology." His system evidently differs but little from that of Linnæus; nevertheless, he appears to have insisted more on the predominance of the characters drawn from the form of the aperture in turbinated univalve shells, and of the hinge in the bivalves. He was the first too, as it would appear, who proposed to change the terms, which in reality are somewhat obscene, especially when translated into any modern language, which were imagined by Linnæus to designate certain parts of the bivalve shells. He has, besides, sufficiently augmented the number of the genera of the Swedish naturalist, and has constantly united a very passable figure of one species of each. In general, his work is very instructive, although he has not introduced any very new consideration into the science.

We shall pass over in silence a tolerable number of authors, such as Müller, de Born, &c. who have added scarcely any thing to conchology, except some new species, and proceed to speak of some French naturalists, who have done more for this art than any of their predecessors: we particularly mean de Bruguières and de Lamarck.

Bruguières, in 1792, almost entirely followed Linnæus; but we must do him the justice to say, that he has much more clearly circumscribed and characterized the genera, which obliged him considerably to augment their number. The descriptions of the species, in the small number of genera which he was able to treat of in detail (death having cut him off long before he could terminate his work), are well done, and complete. In a word, he should be regarded as the conchologist who first began to introduce into the science that exactitude and those details which have enabled us to employ it in palæozoology, or in the comparison of fossil remains. We must nevertheless observe, that he has introduced no new consideration.

M. de Lamarck brought to still greater perfection the method or mode of view of Bruguières, his friend, not only in not confining himself to the mere consideration of the shell, and in viewing it as a part of the animal, but also in conchology properly so called, by a great number of new generic sections, by the employment of a terminology still more rigorous; and, finally, by the introduction, as the basis of a principal division of the bivalve shells, of the number of the muscular impressions, in 1807, which was adopted in 1810 by M. Oken. He was of opinion, however, that the oscabriones should be placed with the patellæ, contrary to the felicitous conjecture of Linnæus. In general, as one might easily be convinced, by a complete exposition of his new system (for which, however, we have no room here), he has entirely abandoned the division of shells, established by the majority of the conchologists his predecessors, on the number of pieces of which the testa is composed, and has rather chosen the general form of the shells on which to establish his first four divisions into subspiral, cardiniferous, subcoronal, and vermicular; in fact, he could no longer admit the division of univalves, bivalves, and multivalves, since he places the oscabriones among the subspirals, which certainly no one would do, who proposed merely to make an arrangement of shells. In general, it would appear that M. de Lamarck, in this systematic arrangement of shells, was too anxious to place it in a direct relation with that of the animals, which unquestionably must render it more difficult, but at the same time, perhaps, also more interesting as regards true science.

Since, and during the publication of the method successively brought to perfection by M. de Lamarck, other conchologists adhered almost rigorously to the system of Linnæus, extended by Bruguières, such as M. Bosc, Montagu, &c., or carried to excess the generic sections or subdivisions, as did M. Denys de Montfort, in his Systematic Conchology, printed in 1808, but which contains only the univalve shells. This author, absolutely paying no attention to any thing but the testa, has necessarily considerably multiplied the genera, by desiring too much to specialize or render rigorous their characters. But we must not deny that a considerable number of them ought to be, and have been, already adopted, and that he was the first to call the attention of conchologists to those extremely minute shells termed microscopic; and that, although this part of his labour ought certainly to be considerably modified, conchology is not the less indebted to him for a real service. He has also separated from the multivalves the shells or testæ of the anatifæ, under the name of fissivalves.

A few years afterwards, M. Megerle proposed a new distribution of shells, but none of it we believe has been published, except the part which treats of the bivalves, in the Berlin Magazine for 1811; and although he has entitled it a new system of conchology, it is evident that he follows Linnæus almost scrupulously, with this difference, that he has established a tolerably great number of new genera, which have since been proposed and admitted by others.

Finally, M. de Blainville (to whom, as we have already said, we are indebted for the principal part of this essay), in the Memoire read at the Philomathic Society, and inserted in its bulletin, although his classification referred essentially to the animals, and not simply to their envelopes, has introduced some new considerations into conchology, by shewing that the shell, especially in the univalves, is essentially the protecting body of the organs of respiration, of which it follows, to a certain point, the general form and position.

He has also called attention to the employment of a new character, drawn from the symmetry or non-symmetry of the univalve shells, in relation with the organs of respiration, and has replaced the oscabriones among the multivalves.

We might very considerably lengthen this critical analysis of the works of authors who have written on conchology properly so called. But it would be superfluous to speak of those who have added almost nothing to the art of classifying shells, although they have frequently proved more useful to real science, by making known a great number of new species. Those zoologists who have considered the shells only as making a part of the animal, and who in general have diminished rather than augmented the number of genera of shells, we of course pass over in silence.

As we have inserted four plates in explanation of the terms made use of in conchology, it becomes necessary to advert here to such of the references as could not be engraved on the coppers.

The figures of *univalve spiral shells* include,

1. An involute shell, with the spiral horns separate, and the siphon ventral. *Spirula Australis* (*Peronii*, Lam.,) *Nautilis spirula*, Gm.

2. A multilocular shell, the spire invisible, siphon medial.

3. A monolocular shell, subinvolute, columella plaited. *Voluta musicæ.*

4. An involute shell, emarginated or notched, narrow mouthed. *Oliva littorata*, Lam.

4. a. Shell without columella. *Id.*

4. b. Shell with sulcated suture. *Id.*

5. A globular shell, aperture semicircular, left edge septiform. *Nerita peloronta.*

6. A turriculated siphonostomatous shell; a. anterior extremity (inf. Lin.)—b. the canal—c. right or external edge—d. emargination, or notch—e. posterior sinus—f. umbilical cleft—g. left or inner edge. *Pleurotoma* (*Murex*, Lin.) *Babylonia.*

7. A turriculated shell, columella smooth and truncated. *Helix* (*Bulimus*) *zebra.*

8. A monolocular ampullaceous entomostomatous shell; a. emargination, or notch. *Buccinum dolium*, Lin. (*Dolium maculatum*, Lam.)

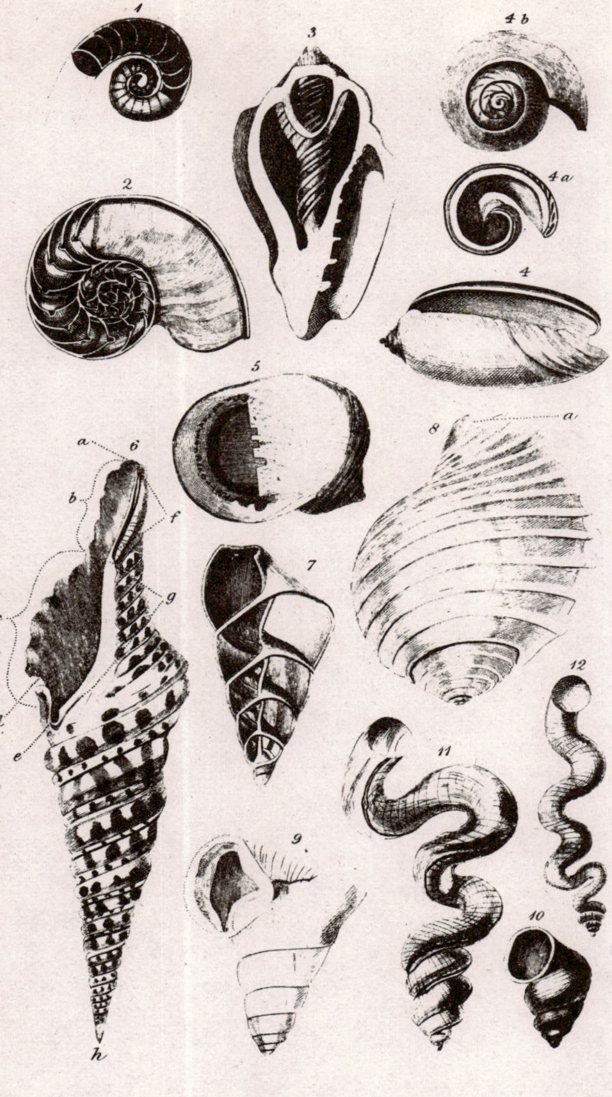

Univalve spiral shells.

Univalve non-spiral shells and Opercula.

9. An heterostomatous shell, lip subcontinuous and folded back. Gen. *Pupa*, Lam.

10. A cricostomatous shell, lip continuous. *Cyclostoma elegans*.

11. A tubular cricostomatous shell, fissure subspiral. *Siliquaria anguina ; Serpula cnquina*, Gm.

12. A tubular shell, cricostomous and subspiral. *Vermetus* (*Serpula*) *Adansonii*.

The figures of univalve non-spiral shells and opercula include,

1. A symmetrical tubular non-spiral shell. *Dentalium entalis*.

2. A symmetrical shell, with the beak or apex posterior; a. posterior—b. anterior.

3. A symmetrical shell, with the beak or apex anterior. *Patella cymbalaria*, Lam.

4. A non-symmetrical shell. *Siphonaria radiata*.

5. A non-symmetrical shell, with the apex or beak posterior. *Hipponix*.

5. a. It's foot.

6. A subspiral shell with a perforated disk. *Haliotis*.

The following are opercula:

7. Multispiral operculum. Genus *Trochus*.

8. Paucispiral op. Gen. *Turbo*.

9. Unispiral op. Gen. *Nerita*.

10. Subspiral op. Gen. *Phasianella*.

11. Unguiculated op. Gen. *Murex*.

12. Unispiral inarticulate op. Gen. *Natica*.

13. Subunguiculated op. Gen. *Purpura*.

14. Squamous op. Gen. *Helicina*.

15. Lamellateous op. Gen. *Buccinum*.

16. Radiated op. Gen. *Navicella*.

The figures of bivalve shells include,

1. *Cytherea lutea* (*Venus chione*, Lin.) in a normal or perpendicular position.

2. Back view of the same in the same position.

3. Interior view of the right valve of the same, in the same position.

4. Exterior view of the left valve of the same, in the position of Linnæus and Lamarck.

The letters A and P in each figure indicate the anterior and posterior parts of the shell, L and R the left and right.

The figures of multivalve shells include,

1. Interior of the right valve of *Pholas dactylus*.

2. Left view of *Pholadidoides Anglicanus*.

3. *Teredo nigra*, De Blain., with a portion of the animal.

4. Posterior extremity of *Teredo navalis*.

5. Articulated palette of a *Teredo*.

6. *Balanus* (*Lepas*, Lin.) *tintinnabulum*. U. S. V. upper side valve, L. S. V. lower side valve.

7. Operculum of *Balanus squamosus*.

8. Left view of *Polylepe vulgaris*, Blainv. D. back, or dorsal valve; V. belly, or ventral valve.

9. Upper view of *Lepas diadema*.

10. Valves of *Chiton squamosus*.

11. Valves of *Chiton raripilosus*.

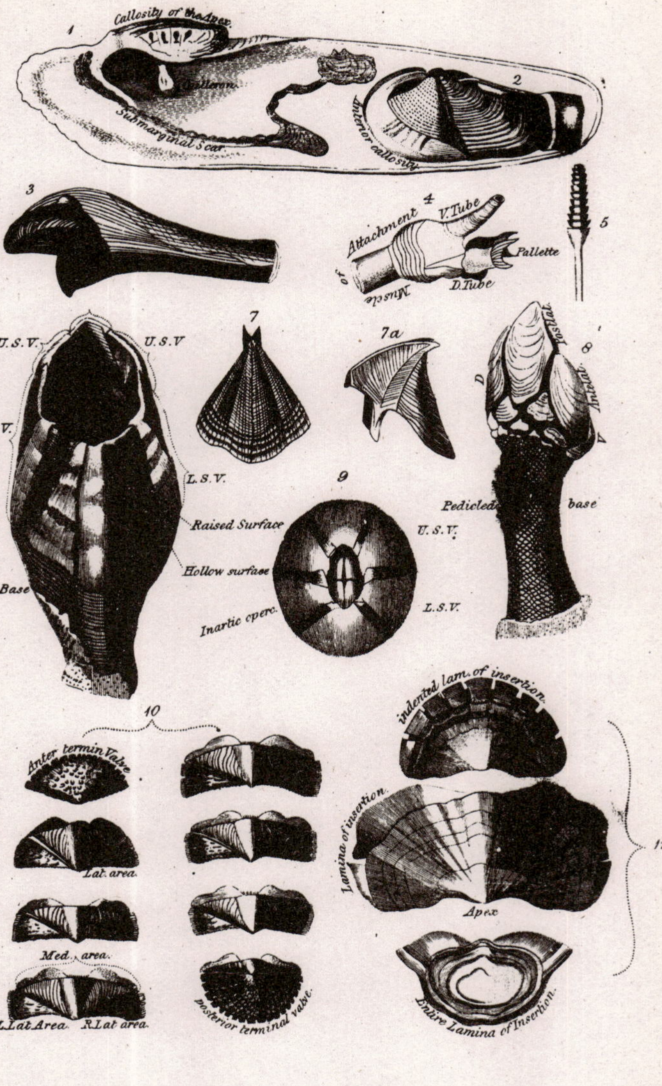

Multivalve Shells.

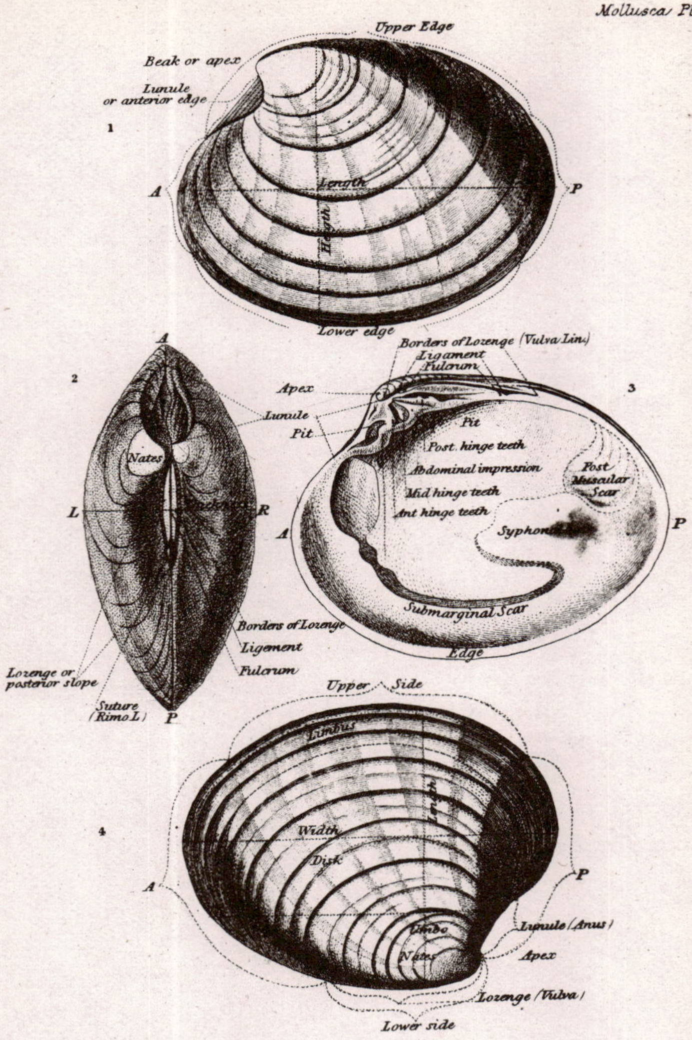

Bivalve Shells.

SUPPLEMENT

ON THE

CLASS CEPHALOPODA.

THE division of molluscous animals, however numerous in species, is so concisely arranged by the Baron, that it has been deemed inexpedient to separate his text by inserting our own supplementary observations at the end of each class. Having translated this portion of the *Règne Animal* uninterruptedly, we shall proceed in like manner with the supplements.

It would be repetition to speak of the class cephalopoda in general after our long preliminary article on the division. We begin with the genus SEPIA.

We are chiefly indebted to the ancient writers for all we know respecting the natural history of those remarkable mollusca. Although the species of this genus appear to have been rather neglected by zoologists, it would seem that sepiæ are to be found in all parts of the world, and alike in the different zones, frigid, temperate, and torrid.

It is only in the sea that the sepiæ exist, but at some distance from the coasts, and probably at all depths. They are, in fact, along with the calamary, or *loligo*, the only animals of the molluscous type which may be regarded as not littoral.

We may be assured that they never come out of the water, which cannot be averred with equal certainty respecting the octopi.

Their movements are rapid and in all directions, pretty nearly like those of fish, and performed by the aid of the branchial sac, and of the circular fin which surrounds the body, the tentacular appendages being pressed close one against the other into a packet pointed in front, and the branchial appendages retracted within their cavity. The first do not separate except when the animal attempts to seize any prey within reach, and they especially serve to retain it, and submit it to the action of the powerful teeth with which the mouth is armed. As to the branchial appendages, it is probable that the sepia can put them forth with rapidity from their cavity, and, as it were, shoot them upon an animal at some distance from itself, to bring it back within reach of the tentacular appendages. We may also equally conceive that they answer the purpose of hooking the animal to the rocks at the bottom of the sea, and thus sheltering it from the storms and tempests with which that element is frequently agitated; but this, though probable, can hardly be considered as more than conjecture.

The sepiæ are evidently carnivorous. They probably feed upon fishes, and especially on the swimming crustacea, which live at some distance from the coasts, and which they overtake and seize after a pursuit of longer or shorter continuance, like the loligo, and not by placing themselves in ambuscade, like the octopi. We must nevertheless add, that Aristotle regards the sepia as a very cunning animal: he says that it not only casts forth its ink when it is afraid, as do the octopi and loligines, but that it also makes use of this liquor, which it is true is more abundant and more coloured than in the former, to create around it a thick cloud, in which it envelopes itself, either to escape the pursuit of the fishes, or to attract the fish by rendering itself invisible. He adds that it will then seize very large fish, and even mullets. Unfortunately, this observation, although many authors have adopted it, which has rendered the sepia yet more celebrated than its congeners, has no other foundation than the authority of the Greek philosopher, and has not been confirmed by any modern naturalist.

The animals of this genus do not probably live in troops, and still less in a social state. But it appears that this is not the case as regards individuals of different sexes; and Aristotle attributes to the male a sentiment of friendship for its female sufficiently strong to induce it to come with courage to her assistance if she has had the misfortune to be harpooned. But as this sentiment is not reciprocal, it is probable that Aristotle alludes to the means of procuring a number of male sepiæ, still employed in the Mediterranean, and of which we shall presently speak.

The internal differences which characterize the sexes we have explained in our general article, and many of them had been already noticed by Aristotle. We must add here, that the males are more brightly coloured, and the spots, or zebra-like stripes of the back, are much blacker and more numerous than in the females.

It would appear that it is at the end of the spring or the commencement of summer that the season of love takes place with these animals, and that they fecundate, if we may judge at least by the period in which eggs are found containing young ones; this most usually happens in August or September. Aristotle, however, in one place says that the sepiæ reproduce at all seasons of the year, though he adds a little further on, that they are full in the spring.

We are ignorant of the generative relations of these animals, or even if there be any true sexual intercourse previously to the female laying her eggs. No modern observer appears to have afforded any information on this subject; and it would seem that the text of Aristotle which speaks of this particularity in the natural history of the sepiæ has been interpreted in a different manner by translators and commentators. Most of them, however, are inclined to believe that there is no intercourse, since they say that the male bedews the eggs with the seminal fluid when they have been deposited by the female, and that it is this fluid which, being viscous, attaches them one to another, and forms them into a cluster. This appears to be more than doubtful. Be that, however, as it may, the following is what Aristotle says respecting the female. " Fifteen days after she has been full she throws out her eggs, near land, among the algæ, the reeds, and other bodies which may be found upon the shore, in its anfractuosities, and even around the sticks which fishermen have placed there for the very purpose. She does not lay them all at once, but at several attempts, as if she were in a state of suffering. This operation lasts fifteen days. After the laying, the female herself sheds her ink upon her eggs, which turns them from white to black, and causes them to increase in bulk. It is then that the male bedews them with the seminal liquor; a fact," he adds, " which, though not having been observed but with regard to the sepia, ought, however, very probably be extended to the loligo and octopus."

We have just said, that according to the same author, it is this fluid which unites the eggs, and gives to them the appearance of a cluster of grapes. In fact, on our coasts these heaps of the sepia's eggs, more or less considerable, are designated by the name of sea-grape, in consequence of their form and colour, which is most usually black; some, however, have been observed which were altogether white, and which contained young sepiæ as much advanced as those which were in eggs of the finest black. Accordingly we cannot be assured to what the coloration of these eggs is attributable; but it is not probable that the cause which Aristotle assigns for it should be the true one. Neither do we think that their agglutination, which is extremely irregular, and which takes place by the pedicle that terminates them, is owing to the viscous quality of the semen of the male, which would be shed on their upper part, but rather to the viscosity of the adventitious membrane of each egg; and it really seems probable that an actual intercourse takes place between these animals in the same manner as in all the cephalous malacozoaria. Aristotle also tells us that the female sepia, after having totally got rid of her eggs, hatches them in the place where she has deposited them. She is often to be seen, he avers, with the body resting against the ground and over her eggs.

The eggs of the sepia have an oval form, attenuated at the two extremities, one of which is free, and the other prolonged into a pedicle of greater or less length, and twisted round some foreign body, or even joined to a greater or less number of other eggs, in a manner completely inextricable. It is thus that the clusters are produced. Their bulk and the number of the eggs which compose them are extremely variable; though most frequently, as we have observed, of a black colour, they are sometimes of a yellowish white, a little transparent. In studying their organization we easily perceive that the envelope of the egg is nothing but a gelatinous matter, more or less thick, which may be divided into an indefinite number of laminæ, but which is not really organized. In the interior is the egg, properly so called, composed of the germ and of the vitelline mass, in an inverse ratio of development, according to the period at which it has been laid. The vitelline mass is almost of a white colour. Contained in its egg, the little animal has the head and eyes much bigger than they are subsequently to be, as Aristotle has most justly observed. In proportion as it is nourished by the vitellus it acquires more and more resemblance to the mother, and, while it resembles her completely, it continues for some time to swim in the interior of the fluid which fills the egg, and which has taken place of the vitellus. Accordingly, when accidentally at this period the envelope of the egg is broken, the little animal issues forth, and swims with as much facility and elegance as the mother. Its eyes perceive obstacles in the same manner; in a word, it is in full possession of all its faculties. In the ordinary state it is probable that the parietes of the envelope, distended by the fluid, which is doubtless introduced there by transudation from the exterior to the interior, conclude by bursting, which gives issue to the young sepia. We are ignorant how much time is necessary for its complete development from the period when the egg is laid, but it is probable that it cannot be very long, since the eggs of the sepia are often found on the coast at the end of July, the little ones in which, extracted artificially, are altogether complete.

This little sepia is then scarcely five or six lines in length, though destined to arrive at nearly a foot long, with a proportional breadth. We are altogether ignorant of the time which is necessary for this growth; nevertheless, if we give credit to what Aristotle again affirms, that the sepiæ, like the loligines, live but a short time, and rarely attain to their second year, we must admit that their growth must be extremely rapid, which appears by no means probable.

The sepiæ exert no great influence on the external world, if it be not upon the animal kingdom, in consequence of the species of fish and crustacea on which they feed, and because they themselves are food for a great number of marine animals, for fish, cetacea, &c.

Even man employs them sometimes as food, as well as the calamary (*loligo*), at least on the coasts of the Mediterranean and Atlantic, where they constitute a portion of the aliment of the poor. The ancients likewise ate of them, according to Athenæus.

Nevertheless these animals are but little sought after now by fishers, but for their bone, which has been called *sepiostarium*, and their ink.

The first, or the bone of the sepia, is employed to be put in the cages of those little granivorous birds which are reared in a domestic state, probably to replace the little grains of silex, or lime, which they are in the habit of swallowing when at liberty, or to wear the extremity of their bill, which without that might grow to an injurious length. But more frequently it enters into the composition of those decorating powders under the name of coral powder, which is used for the purpose of cleaning teeth, and removing the calcareo-animal substance called tartar, which is deposited on their surface during the night.

As to the ink of this animal, it composes almost by itself alone that colour so agreeable from the equality of its tone, its warmth, and even its tint, called *sepia*. But it is more than probable, notwithstanding all that has been said on the subject, that it does not enter into the composition of the China-ink, or as we more generally call it, Indian-ink, which, as it is now understood, is formed of lamp-black, extremely divided, and mixed with a certain quantity of gum, and aromatized with some substance with which we are not precisely acquainted.

The genus OCTOPUS closely approximates to sepia, and, indeed, was considered but as one of its species by Linnæus; the ancients gave it the name of *polypus*, a name now exclusively consecrated to a tribe of the radiata.

From the organization of the octopi it is easy to perceive that they are animals whose sensations must be pretty nearly similar to those of others of this class, but their mode of locomotion differs from that of the majority; in fact, they do not swim with the swiftness and elegance of the sepia and the calamary; they perform this movement rather by whirling in an irregular manner enough, the head usually downwards, and by rowing with the assistance of their long tentacular appendages. On the other hand, they can walk, or to speak more correctly, draw themselves along over a resisting ground at the bottom of the water, or even on the shores, in the anfractuosities of the rocks. For this purpose they attach one of their arms, previously considerably extended, to a solid body, and then make use of it to draw towards this point the rest of their body. It has also been supposed that these animals can walk with the head downwards, and by means of their eight tentacula; but this is less probable. Aristotle, nevertheless, declares positively enough that this is the only molluscous animal which makes use of its arms for walking. According to this notion he admits, as do Pliny and all the ancient writers, that this animal issues from the water, and that it sometimes visits the land, avoiding however all the smooth places. Ælian and Athenæus add that it can also mount on trees, which is much more doubtful; for, in truth, what should it go in search of there? They suppose that it must be fruits!

Most generally the long arms of the octopi serve the purpose of enlacing their prey, and of attaching themselves to it by means of the numerous suckers with which they are armed, and the action of which it is easy to conceive. In fact, beside the trifling adherence which may be owing to the viscosity produced by these organs, each nipple acts absolutely like a cupping-glass, its edge being fixed, and a capability given of producing a vacuum by the contraction of the longitudinal fibres of its bottom. Now, as the number of these cuppers, or suckers, may amount to several hundreds, we may easily conceive that the adherence of the octopi to a body is so strong that it is almost impossible to tear them from it otherwise than by cutting off the arms, and that they will even adhere to it for some time after death. This adherence will produce a pretty strong redness in the part affected by it, but it does not seem probable that this ever amounts to inflammation.

The octopi are animals eminently carnivorous, and particularly live in the anfractuosities of rocks, where they place themselves in ambuscade, concealing their body, properly so called, in the cavern which they inhabit, and letting nothing come forth but their arms, which they make use of to reach, enlace, and draw in their prey. They sometimes, however, do this more openly; in fact, Belon informs us that he saw an octopus fighting for more than an hour with a crab, in the port of Corcyra. Aristotle says that this animal has the faculty of changing colour, and assuming that of the bodies which surround it, and this for the purpose of more easily catching fish: he adds, that it does so likewise when under the influence of fear, and at the same time it darts its ink, the colour of which is rather red than black.

It appears that the octopi make their principal food of crustacea, as Aristotle observed long ago. M. de Blainville informs us that he has many times heard the fishermen complain of the injury done them by these voracious animals, not only from the quantity of crustacea which they destroy, but chiefly by frightening those of which they are unable to gain possession, and forcing them to quit the localities which they had previously inhabited. The octopi also feed on conchyliferous mollusca; and Pliny relates concerning them the trick which has also been attributed to apes, of placing a little stone between the two valves of oysters, of which they are extremely fond, so as to prevent them from closing, and that then they extract the flesh: on this Pliny cries out, "Such is the wonderful intelligence of animals, even the most stupid!" But how could an octopus take up a little stone and place it so adroitly, even supposing that the semi-hiatus of the oyster, continually filled by the tentacular cirrhi of the edges of its mantle, would permit it?

The habitation of an octopus is discernible by the debris of shells and of fish, the flesh of which it has eaten. It has also been asserted, absurdly enough, that when pushed by hunger it will gnaw its arms, which possess the singular property of shooting forth again. But Aristotle and Pliny, more justly, attribute the fact of octopi being often found which have some few appendages less than usual, to their having been eaten by the conger eels. Belon, in fact, informs us that he saw, near Epidaurus, some muræenæ whose stomach contained arms, or tentacula of the octopus. These fish, indeed, inhabit rocky places, as well as the octopi. Rondelet tells us that the octopi are very fond of the branches of the olive, and also of the fig-tree, and that these substances may be made use of as a bait to catch them. Whether this observation be peculiar to Rondelet or no, it appears to be very doubtful.

The moderns do not appear to have completely observed the mode of sexual intercourse among the octopi; nevertheless Rondelet assures us that it takes place, as among the sepiæ, mouth to mouth, and interlacing with their arms. Aristotle says that the male is distinguished from the female by the form of one of the arms, where fishermen aver that the male organ is situated. But nothing appears to confirm this popular report, and indubitably there is no sort of excitatory organ in the anus of the octopus. He adds, that their intercourse takes place like that of the other mollusca, by which we essentially understand the sepia and calamary; while Pliny affirms that they unite by means of their pointed and forked tail, an organ with which undoubtedly no species of octopus is provided. He probably means the foot, which is more pointed in these than in others of the class, whitish, and cleft at its extremity. This, according to Aristotle, is placed upon the spine, a part which has no more pretensions to existence than the aforesaid tail. It has been said by observers that the males are much more numerous than the females, and that in intercourse there is a very strong adherence between the two individuals; in fact, the fishermen of the French coast have a peculiar mode of fishing for the octopi, as for the sepiæ, by attaching a living female individual to the end of a cord and letting it go, the male then approaches, and on drawing the cord two individuals are brought up, interlaced together. It is only necessary to repeat this operation to get possession of all the male individuals in the district. This fishery takes place towards the middle of spring, and nevertheless, according to Aristotle, it is in winter that the intercourse of the sexes takes place, so that the female deposits her eggs in spring. They form a mass, more or less considerable, according to the age of the individual, which mass the Greek philosopher compares to the fruits of the alder, or of the wild vine. Their number is considerable, and the united mass which they form is much greater than the part of the body from which it issued. This shows that these eggs are like those of many other aquatic animals, which swell considerably after they have been ejected. It is always in some hole or cleft of the rocks that the octopus deposits its eggs: Aristotle had already made this observation when he said that the octopus seeks a convenient place for the deposition of her eggs, such as the interior of a shell, the bottom of a vessel, or of some other cavity, to the sides of which it suspends them. He adds, that this animal hatches its eggs, that is to say, that it sometimes places itself upon them, and that sometimes it fixes itself at the entrance of the hole where it has placed them, disposing its arms in such a way as to cover them more securely. During this time the animal grows thin, for it eats nothing. He says also that fifty days must elapse before the little octopi come forth from their eggs. It is probable that, as these eggs are conformed altogether like those of the sepiæ, the young ones are in possession of all their faculties on issuing forth from the egg.

We are not precisely acquainted with the duration of the life of the octopi. Aristotle says in a general manner that it is not long, and that the major part of them only live two years; that then they grow soft, become decomposed, and are, as it were, annihilated. But it seems by no means probable that an octopus, whose egg is of such small dimensions, can, in two years, attain to the size at which it eventually arrives. As for the reason which Ælian assigns for this shortness of life, namely, the exhaustion occasioned in the male by sexual intercourse, and the number of births in the female, it lends but little support to the pretended fact.

It appears nearly certain that the octopi remain concealed during the winter; the fishermen never meet with them during this season. The ancients tell us that this hybernation lasts two months; how then is it possible that the winter can be the period of their sexual intercourse?

We are ignorant respecting the full size to which the octopi may attain. We find in the recitals of certain travellers, and even of some naturalists, that a species exists to which the name of *kraken* has been given, which arrives at an immeasurable bulk, so as to resemble an island when it approaches to the surface of the water, and to overset the largest vessel under full sail when it becomes entangled in their cordage. But we may be assured, without any fear of deceiving ourselves, that this is but an exaggeration of what has been said by the ancients, especially Pliny, concerning an octopus, which, according to Trebius, had a head of the size of a barrel containing fifteen amphoræ, and whose tentacular appendages, which, as well as the head, were presented to Lucullus, were thirty feet in length, knotted like clubs, and so thick that a man could scarcely embrace them round; the suckers resembled basins, and the teeth were proportional. All that was preserved of the body weighed seven hundred pounds. There are other traits still more curious in the history of this most marvellous octopus. It was observed at Castera, in Bœtica, in Spain, and was accustomed to come forth from the sea into the reservoirs or depots for salted fish, &c. and devour those provisions. The pertinacity of its robberies at length aroused the indignation of the keepers; they built very lofty palisades, but all in vain, this persevering polypus succeeded in getting over them by taking advantage of a neighbouring tree, so that it could not be taken but by the sagacity of the dogs, which, having marked it one night as it was returning to the sea, intimated the affair to the keepers, who were struck with terror and astonishment at the novelty of this tremendous spectacle: in truth, the animal was of an immeasurable bulk; its colour was changed by the action of the brine, and it exhaled a most intolerable odour. Nevertheless, after a desperate combat with the dogs, which Pliny depicts with all the vigour of his poetical style, and by the efforts of men armed with tridents, it was at last killed, and the head was brought to Lucullus.

Ælian also tells us that in the course of time these animals, which he, in common with all the ancients, calls polypi, arrive at a most extraordinary bulk, so as to equal in size the largest cetacea. On this subject he favours us with a story pretty nearly similar to that of Trebius, of an octopus which, having devastated the magazines of the Iberian merchants, was *besieged* by a great number of persons, and cut in pieces with hatchets, just in the same style that wood-men cut down the thick branches of trees.

Aristotle, indeed, tells us that there are polypi whose arms are as much as five cubits in length, which would make about six feet. But this is a long way behind the narrations of Trebius and Ælian, and falls still shorter of the wonders of the northern romancers concerning their *kraken.* Denys de Montfort, in his history of the mollusca, has thought proper to collect all that was written before his time upon this subject, but he has not limited himself there. Thanks to the usual fertility of his imagination, he has arrived to such a point of exaggeration, that although he boasted of having made naturalists believe whatever he pleased, he has failed in persuading any body to adopt the history of the kraken. It was about the same time when this author wrote, that they began to speak in the United States with so much assurance concerning the colossal sea-serpent, which turned out after all to be nothing but a tunny of ten or twelve feet long.

The ancients tell us that the octopi are the enemies of the locusts and of the lobsters, which dread them, while they are themselves pursued by the murænæ, which devour their arms. They likewise inform us that their bite is stronger than that of the sepiæ, but not so venomous. Ælian adds, that it is said by fishermen that the octopi are attracted to the land by the fruit of the olive-tree.

The octopi do not seem to be very hurtful to the human species, except by the destruction of such crustacea as serve for our nutriment. The mode in which these animals twist themselves, with the assistance of their arms, provided with suckers, and the force with which they do this, is the cause of the horror which a man experiences, who, at the moment in which he is swimming in the sea, finds himself thus enlaced and in peril of drowning.

In many countries some species of octopi are eaten. The ancients appear to have held them in request; and even at the present day, in the Mediterranean, and particularly the isles of Greece, sailors eat a considerable quantity of them. But it appears that their flesh is always much harder than that of the calamary or loligo, and that it even requires to be beaten with sticks to render it digestible. This is done by the Greek sailors for more than half an hour before they cook the octopus.

Octopi are found in all parts of the globe, but perhaps more particularly in the seas of warm climates. Peron and Lesueur have observed them in the seas of New Holland. The common octopus seems to extend even as far as Greenland, though it is but rarely found there.

The distinction of species is not easy, and has not perhaps hitherto sufficiently occupied the attention of naturalists. The ancients, and Aristotle in particular, had, however, distinguished four at least, but Linnæus confounded them all under the name of *sepia octopus*. M. de Lamarck very properly separated octopus from sepia, and has characterized three species.

Some *octopi* form a second division, called OCYTHOE by M. Rafinesque, and which are placed with the argonauts in the text. Among these is the polypus of the ancients, *ocythoë antiquorum*.

This species, common in the Mediterranean, has never been observed but in a shell of the argonaut genus, so that a great number of authors have admitted that this shell belonged to and was formed by the octopus; while others, on the contrary, think that it is only parasitical. The Baron, as may be seen by the text, leans to the former opinion; and though the latter is discussed and defended by M. de Blainville at great length, we shall not follow him into the controversy.

One of the species, however, which compose this division being, as it would appear, pretty common in the Mediterranean, was observed long ago by the ancients, who have related how these animals navigate on the surface of the sea, in the shell, which serves them as a bark, and by the assistance of the oars and sails, which their simple and palmated arms conduce to form. They are all pretty nearly agreed in affirming that it is only at the moments of a calm that these octopi or argonauts navigate thus, and that as soon as any danger menaces them they sink their skiff, overset it, and fall to the bottom of the water. But they do not agree respecting the manner in which their organization allows of this kind of navigation, or even the mode of it, so that in fact they appear to be in the same predicament as most of the moderns, who never themselves have had the good fortune to behold this elegant method of sailing on the surface of the waters. In fact, though M. L'Abbé Ranzani has proposed a rectification of the text of Aristotle, a little after the manner of the ancient commentators, who, by adding, retrenching, or modifying the words of which they have need, necessarily arrive at the sense which they desire, it is evident that this celebrated philosopher speaks of a fine membrane, like a spider's web, which is found between the arms, like the membrane which unites the toes of ducks. This is exactly what is found in the common octopi. Pliny also mentions the thing too clearly to admit of a doubt: he says, however, that the membrane exists between two of the arms only. He adds a tail, so that the animal is completely equipped with sails, oars, and helm. Ælian, Athenæus, and Oppian, though they have a little modified what Aristotle has said respecting this navigating octopus, rather follow him than Pliny, since they say nothing about the tail. Belon, Rondelet, Gesner, Aldrovandus, and his abridger Jonston, although most of them lived on the coast of the Mediterranean, did never themselves observe these navigating animals; accordingly, all that they say concerning them is drawn from Aristotle and other ancient writers. They have done the same as to the mode of navigation. As to the figure which they have given, it is exact as far as the shell goes; but that of the animal which inhabits it, is either solely the product of their own imagination or taken from an animal altogether different from the one that is usually found there.

The Dutch naturalist, Rumph, appears to be the only one among modern authors who has really given us the description of this argonaut, or navigating octopus. "One individual," says he, "had the body soft, fleshy, furnished with eight feet, six of which, shorter than the remaining two, were provided with suckers, as in the other sepiæ. The two longer, or the hinder ones, double the length of the others, were smooth, rounded, and furnished with suckers likewise, but they were widened towards the end, in the form of oars. Between these tentacula there was no membrane, such as is described in the octopi of the Mediterranean. In another individual, taken in 1693, in a shell of seven inches long and six in height, the six ordinary arms, from twelve to fourteen inches in length, were very slender, and attenuated at the end, while the upper two were much stronger and thicker, their bulk equalling that of the finger. They were furnished at their anterior extremity with a thin and broad membrane, more narrow behind than in front. This molluscous animal," he adds, "on the skin of which there were spots of a reddish brown, similar to those remarked in the octopi, and like them varying in shade, is free in its shell, not being attached there by any thread, as is the case with the chambered nautilus; accordingly it issues forth from it with facility, and comes to float upon the surface of the waters." Notwithstanding this, the Dutch observer says that it is very uncertain if it can live without its shell, because some individuals which he had by him, recently taken from the sea, died immediately, though they were put into water. At the bottom of the sea this animal walks by the assistance of its arms, the keel of the shell being uppermost. It is also in this position that it re-ascends, but the instant it is arrived at the surface, it throws out the water which its shell contained, sets it floating, and spreads its arms around; sometimes it fastens itself by means of its arms to large leaves of trees, or pieces of floating wood, transported by the waters, and lets itself be carried along. It is frequently seen, as the same author tells us, in this position. But he also admits that it can swim, and in this case its shorter feet spread out like a rose; the hinder feet, palmated, and much longer, issue forth from the shell, the animal suffers them to drag along in the water, and thus, by their means, directs its fragile bark. Finally, he also describes its navigation by the aid of the wind, but differently from his predecessors, since he says, that in this case it derives the principal assistance from the raised edges of its vessel, which it presents to the gale; then it withdraws its body very much back into the shell, and steers its bark with two arms, which serve to direct it. He adds, however, that the palmations of the two long arms might be of as much use to it in sailing as in rowing, though his own opinion is, that this molluscum effects its navigation without sail, and with the raised edge of its shell.

It is never met with but at sea, and always solitary, shooting the water with force through its excretory conduit. Its ink is of a bluish brown. Rumph says that he found in the belly of some individuals some round white eggs, united in a mass, and marked at the upper edge with a black point, while, in the volutation of the shell itself, there was at the same time another small mass of eggs or spawn, resembling in form and colour the spawn of fishes, contained in a common envelope, extremely fine. He adds, that even when the shell is not bigger than one's finger, an ovary is nevertheless found there, which reposes on the shell in the form of a cushion. All this, however, is by no means clear, and more especially, as he says a little farther on, that new observations have proved to him that the eggs are found out of the body, in the hollow of the keel, but attached to the animal.

Notwithstanding all these observations, which prove that Rumph really saw an octopus with dilated tentacula, he has left no figure of it but one done by his son; nor has the subject been renewed by any author, not even by Halma, the editor of his great work.

With respect to the argonauts it is unnecessary to add any thing further, as all that we have now said entirely refers to them.

We now come to speak of the CALAMARY, which was called τεῦθος by the Greeks, and LOLIGO by the Latins. The word calamary itself is derived from that of *calamarium*, which in low Latinity signifies a portable writing-desk, or escritoire, with ink, pens, and a pen-knife. It has been given to these animals because their body has a little of the cylindrical form of these sorts of escritoires, and contains a sort of pen in the back and ink in the interior.

The organization of the calamary is very like that of the sepia, but the body is usually more elongated, and almost cylindrical. The product of the female is a very considerable mass of oval eggs, disposed in series round an axis in the form of a cord; this cylindrical mass is about three feet in length and two in diameter. Bohadsch, who observed one of these dimensions, having counted the number of series, and that of the eggs in each, found that it contained 39,760 eggs. They are at first of a yellowish colour, but afterwards they become limpid, and finally blue.

The calamaries appear to possess both general and particular sensibility, still more developed than in the sepiæ and octopi. Their sight especially seems to be very fine; their muscular activity is not less great: they move with the greatest rapidity in the deep sea, which they never quit unless driven from it by some violent impulse, as is the case with the flying-fish; for this they employ the fins with which their sac is furnished, or rather the contractions of the sac itself, in expelling the water which it contains. In their general movements from place to place they hold their tentacular appendages motionless, and crowded into a point one against the other, in front of the head; nor does it seem that the branchial appendages are developed. It appears that they do not separate or move the first but to retain their prey, and the second only when they await it at a distance, or employ them to attach themselves to marine bodies in tempests and violent currents. They more especially inhabit the high sea, and die in a few minutes after they have been taken out of the water, in a sort of convulsion. They cannot even be preserved alive in a vessel filled with sea-water, except when it is very large, and the water very frequently renewed.

They pursue and destroy their prey by main force; it consists principally of crustacea and fish. These they seize with their tentacula, retain by the assistance of their suckers, which are often armed with hooks, and break and triturate to a certain point with their jaws.

We are entirely ignorant of the duration of life in these animals, and whether their growth be rapid or no; nor have we much more certain notions respecting the mode of intercourse between the sexes. The two individuals differ in size, the female being a little smaller than the male, accordingly the dorsal cartilage is always more narrow in one than in the other.

We do not know whether any actual coupling takes place. Belon asserts the affirmative, which does not however seem to be probable. It is more likely that the mode of generation is the same as in fishes. The singular mechanism of the tubes containing the seminal fluid is perhaps destined for the purpose of acting upon the eggs. The fœtus enclosed in the egg undergoes its development precisely like that of the sepia; at first it is imperceptible in the fluid by which the egg is filled, afterwards a sort of vitelline mass may be observed, then the young animal, which makes its appearance in one point, grows by little and little, appearing to embrace this mass with its long tentacula. These, in fact, exist some time before the ordinary tentacula make their appearance. Finally, when on the point of issuing forth, the young calamary differs very little from what it is afterwards to be. Its back is already spotted with red.

The calamary is very generally used as food, and especially in Greece; it is rather an insipid sort of aliment. Fishermen employ these animals as baits, cutting them into strips.

The calamaries are found in all seas, and even in great abundance at rather small distances from the shore. Many species, however, forming distinct groups, may appertain only to certain countries. But the study of the species is not as much advanced as might be wished. Molina, in his natural history of Chili, defines after his fashion those which exist on the shores of this country, in the South Sea. Though it is probable that they are new species, it is impossible to be assured of it, and particularly to insert them with any certainty in the catalogue. The first is his *sepia unguiculata ;* query if it be a calamary? the second is *sepia hexapus*, which would be very singular if it were true that the body, which he says is of the size of the index finger, is fractured into four or five articulations, which decrease in bulk towards the tail, and that, on touching it with the hand, one experiences a sort of electric commotion; his third species is called *tunicata*, and has the whole body over the ordinary skin enveloped in a black and pellucid covering. It is very large, for he adds that some were brought to him weighing five hundred pounds.

NAUTILUS is a denomination employed by the ancient naturalists, both Greeks and Romans, and among others by Aristotle and Pliny, to designate two animals which can make use of their shell as a little boat, to float upon the surface of the water. This observation, which perhaps was derived from the form of the shell, was afterwards confirmed, at least in appearance, by circumstantial details, added by Pliny, Ælian, Oppian, and Philo, on the mode of navigation of these animals. Be this however as it may, Aristotle says, upon the subject, after having previously spoken of the polypi, or naked octopi, "there are yet two genera of polypi, but they inhabit shells; the first is named nautilus by some, and nauticus by others. The animal is similar to a polypus (*octopus*), and its shell has a concave cochlea, but the animal is not attached to it. This animal, which is small, usually seeks its food along the coasts; sometimes the waves cast it ashore, and the shell falling, it is surprized and dies on land. The second, which has a shell, is like the snail; it does not come forth, but remains there, like the helix, though it sometimes stretches forth its arms out of the shell." A little further he says, "the polypus nautilus is of the nature of those animals which may pass for extraordinary, for it can float upon the sea; it rises from the bottom of the water, the shell being reversed, so that it may do this the more easily, and that the shell may be empty. But when arrived at the surface it turns it round again. It has between the arms a kind of tissue, similar to that which unites the fingers of palmipede birds, and which only differs in being much more slender, and of the consistence of a spider's web. The animal makes use of this tissue where there is a little wind, at the same time letting fall, by way of helm, the arms on each side. On the least appearance of danger it dives into the sea, by filling its shell with water. As to the origin and growth of this shell there is nothing certain. It does not appear to be engendered by sexual intercourse, but to be produced like other shells; but yet even this is not evident, any more than whether the animal can live without it."

Pliny, at the end of his 29th chapter, book the ninth, devoted to aquatic animals, speaks only of Aristotle's first species of nautilus, which he says was called *pompylos* by some persons. That Pliny speaks of the same animal, rests, however, on the supposition of Gaza, that we must read in Aristotle as Pliny gives none of the details which are furnished by Aristotle; he confines himself solely to mentioning the mode in which it navigates, by raising its first two arms, between which it extends a membrane of extreme tenuity, rowing with the others, and steering with its *medial* tail, an organ which never existed in any species of octopus.

Oppian, in his *Halieticon*, also speaks only of the first species of Aristotle's nautilus, but, like a poet, he considers only its mode of navigation, which he believes to have served as a model to the man or *the god* who first dared to open for himself a path upon the waters.

"Concealing itself," continues Oppian, "in a concave shell, it can come to land, but it can also raise itself to the surface of the waters, the back of its shell being upwards, lest it should fill with water: as soon as it has arrived at the surface it turns its shell, and navigates like the most skilful sailor. For this purpose it extends like sail-yards two of its feet, between which there is a thin membrane, stretched like a sail by the wind, while the other two, which touch the water, direct, as with a helm, the house, the vessel, and the fish. If it perceives any danger it furls its sail, draws in the helms, and plunges into the deep, being rendered more heavy by the water which it has caused to enter into its shell."

Ælian, under the denomination of polypus nautilus, likewise speaks only of the first species of Aristotle, without even adding the second name given by Pliny. He also tells us of the mode of navigation, with some little variety: to come to the surface the animal ascends, the back of the shell being upwards, to prevent it from filling with water; but once arrived there it turns it. If the sea be calm it then navigates as in a small boat, rowing with two of its arms, which it has extended one on one side and the other on the other side of its *innate* or *connate* shell; but if there should be a little wind it proceeds by converting its oars into a strong helm, at first extended in front, and it raises the other limbs, which are united by a very slender membrane, which it half opens, and it thus navigates as long as it perceives no danger: on any alarm, however, it sinks to the bottom, by its own proper weight, and by that of the shell, which it has filled with water, conceals itself, and escapes its enemy. As soon as the danger is past it reascends, and navigates afresh. It is from this operation that it has received its name. Thus Ælian has imitated Pliny in mentioning nothing concerning what was certain in Aristotle, and he cuts the difficulty by adding the epithet of *innate* to the shell.

Philo, a Greek writer, much more modern than those now cited, in his treatise on the nature of animals, and which is only a pure compilation, or rather an abridgment of Aristotle, has added nothing to what the latter has given us concerning the nautili.

Belon, the most ancient of the naturalists, who wrote at the æra of the revival of learning, appears to have been the first who considered the handsome partitioned shell which we receive from India as another species of nautilus, and we shall presently see that it is the only one to which the name should remain. Nevertheless he has not referred it to the second species mentioned by Aristotle, but to a shell which is evidently but a large species of *dolium*, and which, presenting the character of being very slender and light, might in truth belong to a swimming animal. It is unfortunate, however, that this is a true gasteropod, and that Aristotle has positively declared that the animal of his second species is a polypus or octopus. Belon adds nothing to the description of Aristotle's first species but a bad figure, evidently designed from imagination.

Rondelet manifestly copies Aristotle, and likewise gives a figure of the shell, which is supposed to be that of the first species; but the animal which he represents as belonging to it is certainly an octopus, with simple arms and a single range of suckers. He speaks of the shell which now constitutes the genus nautilus only as a cochleïd, vulgarly called *margaritiferous*, and blames Belon strongly, and in a very unsuitable manner, for having supposed that this could be a species of nautilus.

Gesner, in his article *nautilus*, compares, according to his custom, with much care, all that has been said by the ancient Greek and Latin authors, adding what he found in Belon and Rondelet, as well as their figures. But he also mentions, in his corollary, that he had received from an English physician named Falconer, the figure of a molluscous animal, the shell of which was evidently that of Belon's second species.

Aldrovandus copies Gesner very closely, and his abridger, Jonston, of course does the same.

Bonnani (*Recreatio mentis*, p. 88.) adopts the opinion of Belon with less hesitation, and gives two good figures, the one of Aristotle's first species, and the other of that which Belon thinks to be the second. The first, he tells us, has the name of *polpo moscardine* or *moscarolo* in Italy, and that it is found in the Adriatic sea, on the Italian coast, where the fishermen catch it along with the sepiæ, near the rocks upon the shore, where it frequently seeks its food.

Rumphius, the only author, except the physician cited by Gesner, who has seen the animals of both shells regarded hitherto as nautili, again confounds under this name the species without partitions or chambers and those which have them. But his denominations were not derived from this character, which was never properly marked until after the dissertation of Breynius, *de Polythalamis*, and he distinguishes three or four species, in only one of which, belonging to his second section, does he describe the animal.

The majority of authors on conchology at this period did the same as the foregoing, until Gualtieri, enlightened doubtless by the labours of Breynius, divided the nautili into two genera, and with a singular sort of caprice reserved this name of nautilus to the species with which Aristotle was certainly not acquainted, although Belon had considered it his second species, and gave the name of *cymbium* to that which this celebrated philosopher had regarded as the type of the nautili.

Nevertheless this mode of consideration was not adopted by every body. Thus D'Argenville always comprized under the same name the nautili without chambers and those with them, indicating many species in each group. Davila did so likewise, and many other authors. But at length Linnæus, having admitted the division proposed by Gualtieri into two genera, restricted the appellation of nautilus to the polythalamous species, and that of cymbium, which the Italian conchologist had proposed for the monothalamous, was changed into the denomination of argonaut.

From this it is evident, that under the name of nautilus the modern zoologists can by no means employ the observations which the ancients have left us on the octopi which navigate in a shell, but merely what Rumphius has given us in his *Curiosities of Amboyna.* This is the nautilus of the text, to which we refer for characters. The animal, though very common in the Indian seas, is known to us only by the unsatisfactory figure and incomplete description of Rumphius; for it appears evident that the details added by Denys de Montfort are the product of his own imagination, or mere conjectures founded on the description of the Dutch writer. The description, of course, we will not follow here. Rumphius tells us that this animal, when it is desirous of sailing, is always alone. It puts forth its head, all its tentacular appendages, and extends them with the membrane it has behind. Often it draws itself along with the body upwards, the head and tentacula under, and more frequently still it is at land, or in some muddy cavity where fish occasionally retire; in fine weather it remains but little upon the water.

He concludes that this animal is but slightly attached to its shell, from the quantity of empty testæ which are found either floating on the surface of the water or cast upon the shore. But this circumstance is more probably attributable to another cause; namely, that being without cover, or defence, the nautilus is an easy prey to the voracious crustacea. This appears to be fully proved by the empty shells being damaged on the edges.

This animal is in no great request for the table, because its flesh is extremely hard; but its shell serves for the purpose of making drinking-vessels. According to the Dutch author just quoted the savages make spoons of it, which they call *papeda* in the eastern islands.

From his time no person has described from observation this singular molluscum, which in all probability exhibits numerous relations with the octopi, but which likewise must differ considerably from them. The form and number of the tentacula are very different. It appears even that these animals are not provided with those singular suckers which are found in all the species of cryptobranchia that are known, unless we believe that the digitations about the size of a straw supply their place. The disposition of the mantle is also altogether different; it forms, in fact, a large advancing piece, which lines the whole cavity of the shell, and which, very probably, even out-edges it, or contains a portion of it in its interior. But more especially, this animal is provided with a shell which resembles nothing that we are at present acquainted with in the living state, if it be not that of the spirula. It is tolerably thick, particularly if we compare it with that of the first nautilus of Aristotle, to which the name of papyraceous has been given, in consequence of its tenuity. Its mode of growth is altogether similar to that of other shells, but the manner in which it is rolled, from front to rear in one and the same vertical plane, so that it may be cut into two completely similar parts, is a character without any parallel in the living state, except in the spirula. It is the same with its concameration, or the division of the posterior part of its cavity into a great number of lodges, formed by simple or smooth partitions, convex behind, concave in front, and at the middle of each is an orifice, which is prolonged into the internal of the following partition. These partitions, and the chambers which they form, are evidently produced by the advance of the animal, determined by its growth, during the whole active duration of its life: something of the same sort takes place in certain spiral shells. The siphon is evidently the point of attachment of the terminal muscle of the body, analogous to the columellary muscle of the same spiral shells, and which necessarily follows the progress of the entire animal. The animal itself is only enclosed in the first lodge, whose size and even form have no resemblance with those of the other abandoned lodges. From all this it may be conceived that the number of the chambers and their proportional extent vary in each individual. In regarding the composition of this shell we find it to be of two parts, one exterior, of but little thickness, of the usual nature, coloured by irregular bands, transverse, reddish or ochreous on a white ground, which becomes more and more predominant as we approach the aperture, and of an internal stratum, thicker, and nacreous. This stratum spreads even over the back, from the return of the spire to the interior, and more or less externally, so as to close more or less completely the two lateral umbilici situated at the two extremities of the transverse pillar. It is upon this nacreous part that may be easily seen, both to right and left, and pretty deeply, two broad muscular impressions, rounded, and which tend to unite together.

We have said above that this shell has evidently many relations with the spirula, with these differences, however, that the latter must be still more interior or more concealed by the mantle or its lobes, since it is constantly colourless, that the cavity occupied by the posterior part of the animal is much less large, and particularly that its circumvolutions are much less close. The partitions are of the same form, the siphon being only almost marginal. Nevertheless, if we should trust to the notes taken on the animal of the spirula by MM. Lamarck and de Roissy, who examined it when in the possession of Peron and Lesueur, and even the figure given by the latter, it would differ considerably from what we find in Rumphius on the animal of the nautilus.

We find a similar approximation between the partitioned or chambered nautilus and a great number of those fossil shells which are called ammonites. These shells are likewise partitioned, and rolled regularly in the same vertical plane from front to rear. The cavity which contained the hinder part of the animal appears proportionally less large than in the nautilus, and its orifice dilates a little into a trumpet form. The siphon, besides, is much more frequently dorsal; but more especially the circumvolutions are less close, holding a sort of intermediate relation between those of the spirula and those of the nautili, so that all the turns of the spire are visible, and the aperture is not modified by the last but one. Finally, the partitions are more or less sinuous, which doubtless is occasioned by the form of the posterior part of the body of the animal. As to the extreme thinness of the shell, that appears to have been somewhat exaggerated, and judged of only by a few remaining laminæ of the nacreous part, which is never exterior in any of those shells, nor even in that of the nautilus.

We have much pleasure in referring to a most excellent memoir on *Naut. Pompilius* of Lin., by Mr. Owen, with elaborate figures of the animal, its shell, and various parts, published by direction of the council of the College of Surgeons. The reader will find the most satisfactory information on the subject, and the scientific public will earnestly hope that the present memoir will be the first of a similar series.

The shell of this chambered nautilus (*nautilus pompilius*, L.), which is nearly eight inches in its highest part, is very common in the East Indian seas, and especially towards the Molucca islands. This is the one which was the subject of the observations of Rumphius, which we have cited a little farther back. It is usually brought into Europe in consequence of its fine mother-of-pearl, much in request with cabinet-makers and jewellers. The smallest and most excavated partitions are used to make pendants for the ear. The orientals, by removing the stratum of this shell, which is not nacreous, form drinking-vessels of great brilliancy, on which they engrave divers figures. Formerly the same use was made of them in Europe, and such vessels were found only in the houses of great men and princes. At present they are chiefly confined to the cabinets of the curious.

This SPIRULA is a true cephalopod, provided with a sac which envelopes the posterior part of the body. The anterior is external, and the head, which terminates it, supports ten arms, arranged like a crown around the mouth, two of which are longer than the others. M. de Lamarck adds, that at the posterior extremity of the sac is seen a shell encased, presenting externally only a single discovered portion of its last circumvolution. It was even this resemblance of the animal of the spirula with the sepiæ that first induced M. de Roissy, in his "General History of the Mollusca," and subsequently M. de Lamarck, to conclude, in a more rigorous manner than had been done before, that all the polythalamous shells had belonged to cephalopods. Unfortunately, the only individual which had served for observation to the zoologists now cited, and which had been found by Peron and Lesueur dead, and floating on the surface of the sea, in their passage from the Moluccas to the Isle of France, was lost or mislaid in the collections of the museum in the Jardin du Roi, so that it was impossible to confirm by an attentive observation that which probably had only been the result of a rapid and superficial examination. What seems to prove this is, that the figure given in the French Encyclopedia corresponds very incompletely to that given by Lesueur in his "Voyage in the Austral Ocean," and in which (singular enough) the animal, which had only been seen dead, is yet coloured of a lively carnation red. It would not, therefore, be at all surprising if the similitude of the animal of the spirula to the sepiæ were exaggerated. We find, indeed, that M. de Freminville, a judicious observer, has declared that this animal, which he saw in the living state, is totally different from what we would be led to believe from the descriptions of others. So that it appears that this subject yet remains in considerable doubt; still it seems probable, even from the incomplete figure given by Rumphius of the *nautilus Pompilius,* that the resemblance is not so great as the French naturalists and travellers just mentioned would lead us to believe. As for the shell, it is equally likely that it is altogether interior, which may be presumed from its slenderness, its fragility, and the total absence of coloration.

As yet but one species of spirula is known, *spirula Australis,* or *Peronii ;* it is the *nautilus spirula* of Linnæus. This is a handsome and very slender shell, altogether white, and nacreous in the interior, and appears to be very common in the Atlantic Ocean, in its intertropical parts, but probably it is chiefly to be found in the high seas. In St. Domingo, however, and in the other islands of the American archipelago, the shells of the spirula are so common on certain coasts that the soil over which one walks is completely strewed with them; we may therefore indulge a hope that before long zoologists will be in possession of more certain data respecting the true characters of the animal which produces this handsome shell.

SUPPLEMENT

ON THE

SECOND CLASS OF THE MOLLUSCA.

THE PTEROPODA.

To the text on this small class we must add proportionally little.

The first genus (CLIO) was established in 1774, by Pallas. Brown, some years previously, had given the name to some other little animals, tolerably approximating, but which are contained in a gelatinous case. Pallas gave to this genus the name of *Clione*. But as Linnæus and others had united all the pteropods under the general name of clio, MM. Peron and Lesueur thought fit to separate this genus anew, in their labours on the family, and to give the name of Cleodora to the true clio of Brown, and leave that of clio to the animal so called by Pallas. The characters of the genus clio thus circumscribed, are a free and naked body, more or less elongated, a little depressed, pointed behind, and without any other fins than the lateral appendages, considered as being branchiferous. The head is very distinct, provided with six tentacula, conical, retractile, separated into two groups of three each, capable of being entirely concealed in a sort of prepuce, which itself has a kind of small tentaculum at its external side. The mouth is altogether terminal and vertical; the eyes almost superior; and there is a sort of sucker under the neck. These characters are established upon the most common species of the genus, namely, the *clio borealis* of Linnæus, a very small animal, almost entirely gelatinous, which is found in great abundance in the northern seas, where it is said to be known under the denomination of food for the whale, because it is allowed to constitute a considerable portion of the nourishment of that vast animal.

With regard to the habits and manners of the next two genera of pteropods (*Cymbulia* and *pneumodermon*), we know nothing whatsoever. The same may be said of the limacinæ, except that the species which is known (*clio helicina*, Gm.) is scarcely less abundant than the clio borealis in the northern seas, and is also considered to constitute one of the principal aliments of the whale.

There are certain peculiarities in the organization of the HYALÆA, which, as they have been so briefly touched on in the text, it may be necessary concisely to notice here.

M. de Lamarck was the first who established under this denomination a very distinct genus of the mollusca, though little was then known concerning it, except the shell, which Forskahl, and subsequently Gmelin ranged in the section of the *terebratulæ*. The first of these authors, indeed, has mentioned something concerning the animal, which he observed in the living state. But he has spoken of it in a manner so obscure as to place it, as we have seen, among the bivalves, in which he was imitated by MM. Cuvier and de Lamarck, in their earlier works. Lamartiniere, who was the naturalist on the expedition of La Peyrouse, as well as Forster, who belonged to that of Captain Cook, was disposed to place it with clio. After M. Cuvier had published the comparative anatomy of this animal, no other variations took place, and all subsequent zoologists have placed it where it should be, namely, in the neighbourhood of the Clio, as it stands in the text.

The body of the most common hyalæa, of the size of a tolerably middling nut, at least when it is contracted, is formed of two parts, separated by a very well marked contraction. One anterior, containing the head and a sort of thorax; and the other posterior, which might be named abdomen. The latter is always covered by a sort of shell of a rather singular form, and which might, and has been taken for a bivalve, the valves of which were united closely, or continued to the place of the hinge. The fact is, that it is rather a kind of sheath, very symmetrical, very much depressed, and in which the anterior aperture is prolonged into a very narrow emargination on each side. This sheath, very thin, although hard, of the colour of horn, and translucent, is pretty nearly square. The posterior edge, on which the two laminæ are confounded, is divided more or less deeply into three points, the middle one of which, always longer, is pierced at its extremity. The lateral edges are straight and cleft, for a greater or less part of their length. As for the anterior edge, it presents the aperture of the sheath or shell, which is rather narrow and transverse. The upper part advances much more than the lower, because the upper lamina, which is almost plane, with four sorts of keels, radiating from the middle point, is prolonged, forming in front, a sort of blunt apophysis. The under plate, on the contrary, is much more hollow, considerably more gibbous, and as it were hemispherical, and its anterior edge is rounded.

This shell is completely uncovered, and is attached to the animal only by its middle or pierced point, to which the dorsal muscles, or those of the pillar are attached, and by the edges of its aperture, to which those of the mantle adhere.

Although it really appertains only to the abdomen, it appears that the animal can shelter the anterior part of its body under the advancement of the upper lamina over the lower, but it can never draw back into the shell itself. Perhaps even in a state of repose, the animal envelopes its shell with the lobes of its mantle, as is the case with some other mollusca.

When this shell has been removed, (which cannot be done but by breaking it, if we wish to preserve the animal, because the entrance of the shell is much more narrow than its cavity,) we see more evidently the separation of the body into two parts. Both one and the other are perfectly symmetrical. The posterior or abdomen presents the very form of the sheath or shell, which it fills exactly. Accordingly, it is much more plane above, and, on the contrary, considerably more curved underneath. It is entirely enveloped by a mantle very thin in the middle parts, where it is adherent, and, on the contrary, more thick in the whole circumference which borders the cleft of the shell, and which is more or less free. Above it is prolonged like the upper plate of the shell; underneath it equally edges the inferior lamina, but there it is thicker. It becomes more especially so on the sides, where it edges the cleft of the shell, and where it is divided into two lips, but which, nevertheless, are not cleft in their length. At the posterior extremity of their union, there seems to exist a sort of strip, which is only their prolongation, and which may sometimes be much longer than the shell.

In the space which separates the two united lips from this lateral part of the mantle, is formed a sort of fold, or projecting plate, equally muscular.

From what Forskal says of the edges of the mantle, it appears that in a living state they have the capacity of extending very considerably, and become very thin, and even translucent.

From this description of the mantle, it follows, that it is really open only in all its anterior part, especially above and on each side, and by no means on its lateral parts, which correspond to the lateral parts of the sheath, neither does it present any trace of the arrangement of gills. Those which Cuvier has described and figured as such, are really only the muscular fibres of the lateral prolongations of the mantle. As in its central parts, this envelope is very thin, one may perceive, transversely, above, the true gill on the right side, and the ovary forming a much more considerable mass to the left. All the lower part is occupied by the liver, and the second portion of the oviductus.

The anterior or cephalo-thoracic part of the hyalæa is much more complicated. It is separated from the posterior by a very sensible contraction, which has been named erroneously a neck, from whence it has followed that the divers organs which are found there have been mistaken. The fact is that we must distinguish there the trunk properly so called, terminated anteriorly by the head, which is not separated from the foot, or from the locomotive appendages, which are enormously developed on its sides.

The trunk consists in a very narrow band, but little distinct, and especially so underneath; for above it forms perceptibly enough, an oval, elongated, plane projection, between the locomotive appendages, and is terminated by the head. The extremity of the male apparatus of generation, is even seen a little behind, through the skin, which is very thin. On this part, and at its anterior extremity, are the tentacula. They are rather small, but very visible, cylindrical, and formed by a sort of sheath, in which are contained the true tentacula, a little swelled at their summit.

The existence of eyes in this animal has not been precisely ascertained, but it appears very probable.

More in front, and a little to the right, at the root of the tentaculum on that side, is an orifice a little infundibuliform.

It is that of the male apparatus of generation. We cannot say that it is really central.

Underneath, the middle projection of the trunk, which is always of a very deep colour, and which can only be well seen in front, is terminated by a small inferior cleft, disposed in the direction of the axis of the body. This is the mouth. It is placed in the summit of the angle formed by two small labial decurrent bands, and which, separating considerably, are lost under the locomotive appendages.

These appendages, which remain to be described, are nothing but what is called the foot, in the gasteropods, and which assumes its development, not in the middle and inferior part, but especially on the sides, and in front, so as to pass considerably beyond the head; and as these two lateral parts to proceed thus forward, must approximate to each other, a sort of deep emargination has been the result on the lateral parts of the head. From this character is derived the name of the class *pteropods*. Behind, however, there is no trace of separation, and there the external edge, after having formed a fold in front of the lateral aperture of the mantle, is continued without interruption, and is curved in front of its inferior edge, forming a sort of bulky transverse fold, from which it results that this foot, much thicker at the root, and at the middle of its dilatation, and rather thin in its circumference, may not only serve for swimming, by acting like sorts of wings, but that it can very probably creep a little on the surface of submarine bodies, and particularly form a sort of sucker when the animal is in a state of repose.

To complete the external description of the hyalæa, we have only to speak of the orifice which terminates the digestive apparatus, and of that of the apparatus of generation. The one is formed at the posterior extremity of the furrow which separates the two lateral lips of the mantle, on the right side, and the other exists also at the right, but is situated much more forward, and in the emargination which separates the two parts of the body, in front of the orifice of the branchial cavity of this side.

We have already viewed the exterior envelope of the hyalæa, in the relation of its disposition and form, let us now consider the mantle in its uses.

It would appear that it is not of a very mucous texture, for the shell is very thin; but it is evidently very contractile, and probably very sensible in its exsertile parts. The local, or special sensibility is carried on by the tentacula, of which we have spoken above, and which are but little developed, and probably by the eyes.

The exsertile parts of the mantle, are, moreover, capable of great extension. They have particular muscles which draw them inwards, and they may easily be seen on the back, and under the belly. These white fibres, which are evidently muscular, proceed transversely, and in directions almost parallel to each lip of the lateral lobes. These our author considers as branchial vessels.

The foot of the hyalæa being the principal, if not the only organ of locomotion, the muscular arrangement there becomes more complicated. These animals belong to the ocean, and are seldom found near the shore. Their manners and habits are little known with the exception, that they swim with great swiftness. On the rest of this class we have nothing to add.

SUPPLEMENT

ON THE

THIRD CLASS OF THE MOLLUSCA.

THE GASTEROPODA.

THIS class, though numerous, and divided into eight orders, we must embrace in a single supplement, as our object is now to confine ourselves as much as possible to popular matter.

The first order is that of the PULMONARIÆ, the first genus of which, LIMAX, we shall consider in a general manner, without attending to the subdivisions.

The limaces in general, to which we give the name of *slugs* vulgarly, are naked mollusca, or at all events have no apparent shell. Their organization exhibits a very great analogy with that of the *helices* or snails. The dermo-muscular envelope, very thick, and especially in the under part, forms a long and simple cavity, in which the viscera are contained. The dermis, which cannot be separated from the contractile stratum underneath, has a greater or less number of tubercles at its superficies. The colouring pigment at its surface is often very thick, whereas the epidermis is very thin. A great number of pores are there distinguishable, which send forth a great quantity of mucosity to its surface. This appears particularly to issue in greater abundance from a sort of white sinus of no great depth, surrounded with tubercles, and which exists at the posterior part of the back of the red limaces. Into the thickness of this skin, it appears from desiccation, that there enters a great number of calcareous molecules; but they exist in much greater quantity in the anterior part of the body, sometimes called the shield, so as to form, especially in the grey limaces, the rudiment of a shell, which it is true, is very thin.

The viscous matter just mentioned serves to attach the limaces to the bodies over which they move. By means of this sort of spittle, become friable and shining, a limax may be traced, frequently many days after it has passed. Tobacco, common salt, and in general all irritants, produce so great an ejection of this matter, that the animal swells, stiffens, and dies, when a pinch of them is put upon its head.

These animals have four tentacula. The extremity of the anterior is swelled, translucent, and as it were gelatinous. That of the posterior tentacula presents a small disk, altogether black, which forms the organ of vision. The eye, very small, is nearly spherical. It has evidently a fibrous envelope, very thin, and allowing the black colour of the choroïd to be seen through. Behind, the sclerotica is applied against the nervous ganglion; in front it is continued with the transparent cornea, which appears also to be the continuation of the skin. The choroïd, very much coloured, is pierced by a pupil extremely small.

The organ of locomotion, as in all animals of the molluscous type, is in a great measure cutaneous, that is, the muscular fibres which compose it, are very adherent to the skin, confounded with the dermis, and disposed in all directions. Under the belly, however, where the locomotive disk exists, they are much thicker, and directed according to the length of the animal.

The tentacula are hollow in all their length, and formed by

an elongation of the dermoïd envelope, consequently some muscular fibres line the internal face of this hollow cylinder. These fibres are in a great measure annular, and of course their contraction suffices to elongate the organ.

The mouth is armed above with a small arched tooth, adapted for cutting the leaves of the plants on which they feed.

The limaces possess the sense of touch in more perfection probably than the helices, and especially in the anterior parts, and on the edges of the mantle. In taste, smell, and sight, there must be a strong resemblance. That they both taste and smell is certain, since they seek out and prefer certain substances to others. They do not seem, in reality, to perceive bodies, although they are provided with an organ of vision, and they are certainly deaf.

Their locomotion is performed pretty nearly like that of snails, by the successive contraction of the muscular fibres of the foot, and especially of those of its middle band. But it is more lively and more rapid, particularly when they are endeavouring to escape from a place where they have been detained.

Their food essentially consists of vegetable substances. These are, particularly young plants, paper, rotten wood, fruits, mushrooms, &c. They will also feed upon some animal substances, as cheese, meat, and other matter of this kind in a state of putrefaction. They are evidently most voracious animals, and eat more in the evening than any other part of the day. Their mode of eating is a sort of mastication, the lingual plate being opposed to the upper jaw, and then pushing the substance towards the œsophagus. As in all the molluscous animals, digestion appears to be very slow in the limaces, and accordingly they can support a very long fast. They are less capable of this, however, than the helices, unless they are in very favourable circumstances for it, in consequence of the nudity of their skin, which causes the dryness of the air, as well as the solar action, to be very pernicious to them.

They are animals, in fact, which never issue forth from the holes of old walls, from under stones, or half-rotten leaves, cracks in the barks of trees, mushrooms, and even from the interior of the earth, where they retire habitually, except at that period of the day when there is generally more humidity in the air, that is in the evening, and early in the morning. They are particularly to be seen in greater abundance after warm and gentle showers in spring and summer.

Like the helices, or snails, the limaces are afraid of cold; but although they can shelter themselves but very incompletely under their buckler, they appear to dread it less than the latter. Accordingly they are found to enter later into the state of hybernation. Nevertheless, for the purpose of passing the winter, they sink into the hollows of the ground. They appear, on this occasion, more particularly to prefer the humus or mould which is formed in the trunks of rotten trees. Individuals have been found at more than the depth of a foot in this substance. In this state of torpor, the limaces contract themselves as much as possible in length, so that they become almost hemispherical.

Their general activity increases according to the temperature. It is generally about the end of spring, and in summer, that they seek each other for the purpose of reproduction. We know but little respecting the particulars of their intercourse. They are hermaphrodites, and give and receive impregnation at the same time. The organs of generation are situated near the neck.

A very short time after intercourse, and generally in the month of May or June, the limaces lay eggs, more or less globular, and the size of which varies according to the species. They are deposited isolatedly in little heaps, more or less numerous, in humid places, and sheltered from the sun's rays, under stones, in dung, in the holes of walls, &c. At first perfectly transparent, they become by degrees, from the thickening of their envelope, opake, and of a yellowish colour. Finally, they disclose at the end of a period which seems to vary a little according to the external temperature. The young limaces are then extremely soft, and almost mucous. But they crawl with their tentacula, or horns, extended, as soon as they have come forth naturally, or even have been extracted artificially from the egg. We are not yet sufficiently informed as to the time when they become adult, nor as to the duration of their existence.

They are fond of rainy seasons. When they are forced to remain exposed to the rays of the summer sun, as they have not the resource of the helices of enclosing themselves in a shell, they begin by transuding from their bodies a greater quantity of the viscous matter, and end by perishing. In a very few hours after their death they are decomposed, and melt into a viscous matter, which deserves to be analysed with a greater degree of attention than, perhaps, has as yet been bestowed upon it.

The limaces are scarcely in any manner useful to the human race. Formerly many imaginary virtues were ascribed to the little shell of the grey slug *(L. griseus)*, and to the mucosity which issues from all parts of their skin. But those sorts of fantasies, like many others, have had their day, and will go down no longer. They are still, however, in some cases considered to be cooling, humectant, and pectoral. They are consequently sometimes ordered in consumptions, coughs, and spitting of blood. These properties are common to them with the helices, which are employed in preference, as being more easy to collect and preserve.

The limaces, as unfortunately is too well known, are very mischievous in our gardens, kitchen gardens especially, and in our orchards and fields. These animals principally seek out for their food the young shoots of esculent plants. They attack without distinction, like the snails, fruits, the young buds of trees, and vegetables of all kinds, when they are young and tender. They constitute a real scourge; for when circumstances are favourable to them, *i. e.* when the soil is rich and humid, planted with herbs of which they are fond, and exempt from the visits of animals which devour them, they multiply to excess. They have been known completely to devastate in a single night a very large seed plat, the plants of which had just begun to shoot forth. This evil is always to be apprehended in gardens infested by these animals. To prevent its happening, it is necessary to cover the earth, or the edges of the plants, with ashes, slacked lime, or simply with fine sand. These substances act mechanically upon the animal, and hinder it from walking, by attaching themselves to its body. But care must be taken to keep them constantly in a pulverized state.

Various methods have been devised to destroy these noxious animals, but we shall not describe them here; suffice it to say, that to prevent them from coming to any circumscribed spot, it is sufficient, as we have said before, to surround it with sand or dust, or some very agglutinating substances which they cannot pass.

The limaces appear to be found in all the northern zone of both continents, as well as in the temperate zone. Accordingly we find them in Norway, Sweden, Lapland, in all Russia, Denmark, and Britain, in all parts of Germany, in Greece, in Italy, in France, and in all the southern parts of the Mediterranean on the African coasts; whether they are to be found in the rest of Africa is doubtful. It appears certain that true limaces exist in North America; at least such is the assertion of M. Raffinesque. It does not seem so likely that the limaciform terrestrial mollusca, that are found on the coasts of the Mexican Gulf, in the American Archipelago, and in all the rest of South America, are true limaces. Perhaps they may be species of *veronicella*. It would also seem that the true limax does not exist on the Indian coasts, nor in Polynesia, nor even in Australasia. This would be a very curious subject of research.

It appears that the grey limaces, or slugs, seek human habitations in preference, more than the others do, from which Swammerdam has thought proper to separate the genus into domestic and rural slugs. The *L. agrestis*, though a very small species, is, nevertheless, the one that is most hurtful to agriculture, in consequence of its great multiplication. It has been observed that two individuals, after coupling, have laid seven hundred and seventy-six eggs, and that these eggs may be dried eight times in succession on a stone, without losing their property of disclosing the young.

A very curious species is the *phosphorescent slug, L. noctiluca*, which is only known after a description and figure, incomplete enough, given by M. d'Orbigny to the Baron de Férussac, and which the latter has published in his work on the mollusca. It appears particularly remarkable, because towards the posterior extremity of the shield, there is a small disk, or pore, covered with a matter which is luminous in darkness. This limax, which is fifteen lines long, and seven thick, was found under stones, in the Island of Teneriffe.

The limaces and terrestrial helices, for many years, have greatly occupied the attention of physiologists, and even of all persons who take any interest in scientific researches. Curiosity was excited to ascertain if, as Spallanzani had advanced, the head of these animals could be reproduced, after having been cut off. Observers were then busily engaged in various parts of Europe, in immolating to their (shall we not call it) cruel thirst for knowledge, myriads of these poor animals, and in spite of this sacrifice, many among them deny the reality of the fact, which others assert to have taken place. It is, in fact, impossible for any one who has not convinced himself of it by personal experiment, to believe in the reproduction of a part so important as the head. Nevertheless, even now, at the present day, many persons believe it, though it seems to be proved that it never takes place but when the horns, or that part of the head which is in front of the brain, alone has been removed. The animal infallibly dies, when the first ganglion, which essentially constitutes the head, has been taken away.

The motion of the limaces is in general very slow. It has even passed into a proverb. They have a great number of enemies among the birds, quadrupeds, and reptiles. Accordingly an immense quantity of them is destroyed every year, but the following spring amply makes up for the deficiency.

In treating of the HELICES, or *snails*, an interesting and well known genus of mollusca, we shall enlarge a little on the text.

The organization of these animals has many relations with that of the limaces, or slugs. To form an idea of it we may conceive one of the latter, *i. e.* an ovaliform, elongated body, convex above, plane underneath, in which the mass of the viscera of digestion, and a part of those of generation, should have formed a sort of hernia, for the extent of the middle third of the back, or rather in the space formed by the shield, and have drawn along with them the skin, considerably attenuated. This mass, in the front of which is the apparatus of respiration, is turned spirally, and is contained in a shell of the same form.

The body is pretty nearly cylindrical in all its anterior parts, and, behind, terminates in a sort of tongue, entirely muscular, which is only the elongation of the foot. This last name is given to a flatted and very thick portion of the under part of the body, by means of which the animal crawls. All the lower surface of the body is smooth, but the upper is rugose. The head, especially on the upper part, is not really distinguished from the body, except by the organs with which it is provided. There are two pairs of the tentacula, or horns, the posterior of which are the largest, and have a small black point at the extremity which has been considered as an eye. They differ much from those of other mollusca, by being retractile. At the anterior extremity of the head, is a folded aperture which forms the mouth. In this anterior part, at the root of the right horn, is a small cleft, where the apparatus of generation terminates. The visceral mass is entirely concealed by the shell. Of the latter, a description would be superfluous. It is attached to the rest of the body by the retractor muscles of the head and foot. Some of these shells are variegated, and very beautiful, but in general they are plain enough.

The skin of the snail, where it is uncovered by the shell, is of an extreme degree of sensibility, and accordingly it receives a great number of nerves. It must have a very considerable number of mucous pores, if we may judge by the great quantity of viscous or mucous matter which it ejects. This quantity, however, is less than in the slugs. That portion called the *collar*, which is a muscular ring, bordering the visceral mass, and which answers to the *mantle* of the other mollusca, has the greatest number of these pores.

Thus the general sense of touch, must be, and is, in fact, extremely delicate in those animals. We must add, that the skin which envelopes the tentacula appears to possess still greater sensibility than that of the rest of the body. It is more fine, less viscous, perhaps, but in particular more nervous.

With respect to the function of the tentacula, it is questionable if it be confined to this general sensibility, and whether they have not a peculiar office. The latter appears probable. Some writers are of opinion that the anterior pair may serve as olfactory organs. Others have imagined that the entire skin of these mollusca, was, as it were, pituitary, *i. e.* that it could transmit the sensation of odours, but from analogy, this appears to be extremely improbable. Be this, however, as it may, it is certain that snails can smell extremely well, for they are easily attracted by many plants, the odour of which pleases them.

The black points at the extremities of the last pair of tentacula already mentioned, all authors have agreed to consider as eyes, and very probably, with great justice. Swammerdam has even anatomized them, and declares that he has found there all the parts which compose a true eye. It must, however, be very imperfect, since, when we oppose a body to the first or second pair of the tentacula of these animals, they do not seem to perceive it more with one than the other.

There is no trace of any special organ of hearing in the snails, and in fact these animals do not seem to perceive noise, unless it be so considerable and so near them as to produce a sensible agitation in the circumambient air.

The apparatus of locomotion in the helices is general or partial. It is general in so much as the muscular or contractile fibre is not distinct from the skin of which it forms the internal stratum, directed in all ways. This is only more thick, and takes a more determined direction, when it belongs to the foot, where the muscular fibres are divided into little bundles, and disposed lengthwise. It is by means of this foot the animal moves, and very promptly, contracting and elongating successively each of these little bundles. There is a muscle which penetrates from one of the muscular bundles of the pillar, (another of these serves to retract the foot within the shell) into the interior of the tube of each tentaculum, and proceeds to fix itself at the extremity. This muscle, by its contraction, draws in the horn, turning it inside out, like the finger of a glove. These organs, on the contrary, are unfolded by the action of the annular fibres of the skin which forms them. The mouth and its appendages are drawn back by another muscular bundle from the pillar of the shell; and they are put forth by some small muscles, much shorter, which, from the circumference of the lip, terminate on the anterior edges of the mouth.

Of the interior anatomy, we shall merely notice the organ of respiration. This is situated in a large cavity, placed above the general mass of the viscera, occupying the last whorl of the shell. It communicates with the external air by an orifice formed in the right side of the thick edge of the mantle, or in the *collar*. All the inferior part of this cavity is smooth, and formed by a membrane evidently muscular; but the superior is almost entirely vascular.

Thus the snail, like all the mollusca of the same family, respires directly the atmospheric air in a cavity evidently pulmonary. The mechanism of this function is simple enough. The animal causes the air to enter, by drawing back the respiratory cavity into the last whorl, which is the broadest, at the same time putting forth from the shell, all the parts that can come out, and strongly dilating the pulmonary orifice; it expels the air, on the contrary, by withdrawing its body into a more narrow part of the shell, and by so much the more completely, as it draws in the head, foot, &c. But these movements of respiration are never regular. The fluid elaborated in the respiratory organ, or the blood, which is of a slightly bluish white, arrives to the heart, by means of the pulmonary vein.

These animals, like the slugs, are hermaphrodite, giving and receiving fecundation at the same time. The organs of generation are extremely complicated. Each individual has an ovary, two oviducts and a bladder; a testicle, an epidydimis, a deferential canal, and an excitatory organ.

The nervous system of the helices is very considerable. There is a central part above the intestinal canal, composed of two very thick ganglia. The nerves which it furnishes from all its external edge are very numerous and very considerable. There is a peculiar ganglion for the generative apparatus, which receives a thick thread of communication from the cerebral ganglion, and furnishes threads to the different parts of the apparatus.

The helices are found, as it would appear, in all parts of the earth. They are known in Europe, in Africa, in the two Americas, in Asia, and in Australasia. It is generally in humid situations that they are principally to be found; but they are also to be met with in warm and dry places, which is never the case with the limaces or slugs. They usually withdraw into the excavations of old walls, of rocks, under the bark of old trees, and even under ground. They sink more deeply during the winter season, at least in our part of the world; for in climates where vegetation is continual, it is probable that the snails do not hybernate, or if they do retire, it is during the intense heat, and especially at the season when no rain is falling. Before they enter into the torpid state, the helices of our climates withdraw their body entirely into the shell, and produce at its entrance a sort of momentaneous opercle, or lid, to which the name of *epiphragma* has been given. It is evidently composed of calcareous molecules, not very abundant, united by an animal gluten, and exuded in strata from the parts of the body which enter last into the shell, namely, from the external swelling of the collar. There are, however, a certain number of species, which, even in our climates, do not produce this epiphragma. Probably, at this period, they bury themselves deeper in the earth. It is at the end of autumn that the snails thus retire. During the whole of the fine season, they do not re-enter into the excavations which conceal them, except during the heat of the day, and particularly in dry weather; for as soon as rain falls, and especially in fine and gentle showers, they may be seen to issue forth from all parts, as they commonly do during the night. Their mode of locomotion, which may easily be observed by placing one of these animals in a transparent body, is a peculiar kind of reptation, in which the animal seems to glide on the plane which supports it, and to follow all its irregularities. But on closer examination, we find that this reptation is executed by means of the successive action of all the ranks of little muscular fibres of which the inferior face of the body is composed, something the same as in certain articulated animals, in which the number of articulations is very considerable. Although this mode of locomotion is very slow, yet the snails do not fail to advance more than one would imagine at first sight. As their resting point is always taken in front, it is always in this direction that they draw themselves along, and never backwards. The mucous matter which issues from all the parts of these animals, but especially from the foot, and which enables them to adhere more strongly even to the smoothest bodies, remains on the surface of these bodies, and through desiccation, which is very prompt, leaves a sort of silvery trace, which always betrays the route which these animals have taken. It is this matter which forms the epiphragma, as it is that which exudes from all parts of the skin, covering the visceral mass, and especially from the edges of the mantle, or collar, that produces the shell. No helices have been seen able to swim, or even crawl in an inverted posture on the surface of the water, as the lymneæ, and some neighbouring genera can do. It is generally in search of their food, or of some individual of their own species, during the season of reproduction, that the helices issue from their retreats. They are advertised of the presence of external bodies only by means of the fineness of their touch. In fact, at the slightest contact with any part of their body, but especially of their tentacula, they withdraw themselves more or less completely within their shell, and do not come forth again but by degrees, and with the greatest possible precaution. The choice which the snails make of certain herbs, does not permit us to doubt of their being provided with the sense of taste. It would appear that they do not perceive bodies at a distance, but by the assistance of the sense of smell, the seat of which is probably in the first pair of tentacula, and that in a manner complete enough, since it is known that these animals are attracted a good way by the odour of the plants which they prefer. It is not probable that the organ of vision found at the extremity of the large tentacula is of any great use to them. In the first place, it is by night that they are most active; then it is evident that the structure of the organ is very incomplete, and experiment proves, besides, that on approaching a body to these tentacula, the snail does not perceive it any sooner than when it is approached in the same manner to the first pair. Moreover, their extreme timidity, the precautions which they take in walking, to extend as much as possible the two pairs of tentacula in front of their body, to explore all obstacles, evidently indicate an animal pretty nearly blind.

The snails are nourished essentially with vegetable substances, with tender succulent herbs, and fruits of the same nature. But it seems that they also eat animal substances, such as cheese. They gnaw the leaves and fruits by means of their jaw, which is opposed to the tongue, and that with very considerable strength, and sometimes quickly enough to cause great devastation in our gardens. But it is especially in fine weather, and when they come forth from their hybernating state, that they eat more, and consequently occasion more mischief. In proportion as the autumn approaches they eat less and less, until at last they bury themselves in some excavation, and fall into a state of torpor.

Towards the end of spring these animals seek each other for the purposes of reproduction. The duration of the intercourse is almost two hours.

The rut of these animals continues for a very long time, and it is said that they can reproduce again at the end of six weeks. It appears that even fecundation does not take place until the third time of intercourse.

The helix has one organ peculiar to itself, which must not be passed over here. It would be difficult to say to what apparatus it properly belongs. It is formed by a purse or pouch, more or less elongated, obtuse, rounded at its posterior extremity, and the parietes of which are very thick and muscular. Its interior is a cavity of no great extent, with four furrows, and at the bottom is a nipple. It terminates in the cloaca by a narrow orifice above the origin of the female apparatus. The interior of this pouch, especially the nipple, excretes a cretaceous matter, which being disposed in strata in the cavity of the pouch, assumes its form and produces a sort of pointed and quadrangular dart, with a very narrow canal in its interior. The animal has the power of shooting forth this dart, which is an invariable preliminary to sexual intercourse.

If a helix be opened a little time after this intercourse, the male organ will be found diminished in volume, the multifid vesicles empty, and the pouch no longer containing the dart above mentioned, which, according to the observation of Swammerdam, is sometimes contained in the canal of the bladder. The ramifications of the oviduct in the interior of the ovary contain a fluid in which are swimming some little round membranes marked with a black point, which are eggs.

The first part of the oviduct, properly so called, presents some unequal dilatations, which seem filled with a calcareous matter. Its second part is manifestly augmented in all its dimensions, and contains a great quantity of white matter resembling the spawn of fish. This matter afterwards becomes gelatinous, and swells considerably in water. The testicle appears to be still filled with a tolerable quantity of fluid.

Some time after this, the embryos, detached from the ovary, come into this second or swelled portion of the oviduct. They are enveloped there in a considerable quantity of the matter of which we have been speaking, and which forms for the little animal the fluid, by the absorption of which it is nourished. Later, or in another part of this oviduct, is deposited an external membrane, sometimes rather calcareous, and then the egg is complete. After the end of about fifteen days these eggs are expelled. Duvernoy makes a curious observation respecting the manner in which these eggs are formed. "If we open a snail," says he, "a little before the eggs are laid, we find no eggs, but some little embryos swimming in a very clear fluid, and exhibiting very lively motions. They become eggs on the road which they are obliged to pursue to get out."

The eggs of the helices are usually rounded, tolerably big, and of a white colour. They are at first a little glutinous, and especially so in the species which deposit them *seriatim*, one after another, and in the form of a chaplet. Most frequently they are deposited one by one, or in an irregular mass, in holes which the animal excavates in a soft earth, but much more usually in natural excavations, fractures more or less deep in the ground, in the holes of trees, rocks, or old walls; in general in places where drought cannot reach them, and where moisture is permanent. The number of these eggs does not appear to be extremely considerable.

At the end of a time, which varies more or less according to the species, and, perhaps, also according to circumstances, the eggs disclose, and a young snail issues forth, which is already invested with a shell, extremely thin, it is true, and almost membranaceous. Accordingly at this time they are very much afraid of the drying action of the air, and especially of that of the sun, and do not issue from the holes in which they have been born, except during the night. Their growth at first is tolerably quick, but afterwards it becomes much slower, so that judging by the striæ of the shell, which mark the growth, these animals ought to live a long time. But with regard to this point we possess no certain information. The growth of their body, in fact, necessitates a proportional growth in the shell. When it is necessary that this shell should be increased, the helices assemble in troops. The animal remains in a state of repose, sinks into some cavity, and there issues from all the parts of the mantle, and especially from its thick edge, a stratum of calcareo-glutinous matter, which is applied within the preceding, out-edging it a little. It is this part of the junction of this new stratum which forms the stria that marks the growth; it is so much broader in proportion as the animal is better nourished and more vigorous. When the shell is arrived at the completest growth which it can attain, it merely becomes thicker, and forms, in most of the species, a sort of pad, more or less thick; and there is deposited on that part of the spire which modifies the aperture, a calcareous matter, generally not very thick, which can join the two edges. This is named a callosity. Though all this constitutes the complete or terminated shell, yet the animal was adult before, that is to say, was capable of reproducing its species. But it is very important to observe, that the shell of the same individual differs much according to the period of life of the animal which is examined. In general the spire is so much the less raised as the animal is younger, and consequently the last whorl is larger in proportion; the *umbilicus* is more uncovered, the aperture is wider, the edge more trenchant, and the shell is thinner. Accordingly, when it is terminated, the last whorl comes out from the line of the spiral, and becomes falling, which renders the aperture more narrow. These differences are important to be known, otherwise we might be liable to consider individuals of different ages as different species. We find also, in these animals, and especially in their shells, some anomalies. Thus we find some that are entirely *left*, that is, all the parts are reverted from right to left, and then the shell has its free edge at the left, and the columellary one at the right. We also find some in which that part of the body, which is in spiral, instead of being very compact, is twisted in a very loose manner, or merely a little recurved. The shell then assumes the form of a sort of cork-screw, and is termed *scalaria*. The cause of these irregularities is unknown.

The uses of the helices, or snails, are not very numerous. It appears, however, that the larger species, and especially the garden snails (*H. pomatia*, *L.*) serve for the aliment of man in many countries. The Romans, according to Pliny, consumed great quantities of them; and they must have been in great estimation for the table, since that author has thought it necessary to give, in his Natural History, the name of him who first thought of rearing these animals in sorts of parks, or depots, and fattening them with particular substances. The best came from the island of Astypalæa, one of the Cyclades, the smallest from Reate, in the Sabine territory, and the largest from Illyria. The Romans also greatly esteemed the snails of Sicily, of the Balearic Isles, and of the Island of Caprea. They shut them up in sorts of warrens, and fattened them there with cooked meat, flour, &c. It was Fulvius Harpinus who first conceived the idea of this previously to the civil war between Cæsar and Pompey. He carefully separated each species, and succeeded in obtaining individuals whose shell contained *octoginta quadrantes*, about ten quarts. All this history is taken from Pliny; but there would appear to be some confusion in it, especially with respect to the size produced by education; for Varro, after whom he writes, says the same only concerning the Mauritanian species, which naturally attained to these dimensions. It does not appear that this mode of educating snails was practised for any great length of time, for Macrobius says nothing about it. Some authors tell us that these animals are still eaten in different countries, and among others, in Silesia, Brabant, Liege, Switzerland, Italy, and many departments of France. It is said that in the environs of Rochelle, they form depots of them, putting them one above the other in layers, between each of which moss and other plants are spread. It is allowed that the individuals which live in elevated situations are the best, and that they acquire, in some measure, the flavour of the plants on which they feed. In general, it is very probable that their flesh must be rather hard, in consequence of the proportional size of the foot. It is certain that many half-civilized nations eat them dried in the smoke.

In Paris, and many other continental cities, a great quantity of snails is to be found in the market. But it is not as an article of food that they are employed. Mucilaginous broths are made of them for persons afflicted with maladies of the chest and lungs, and in such cases they have been found efficacious. It appears less certain that they may be employed with advantage in incipient hernia, and that they exercise an astringent effect upon the inguinal ring. Of this, however, we are assured by a French medical writer, M. Georges Tarenne. He employs for this purpose the blood of the animal, obtained by pricking it with a sharp instrument, and placing it in the form of a sort of cataplasm on the cushion of the bandage used in such cases. In a few months, he says, two or three hundred of these animals thus applied will work a complete cure. It appears much more likely that the cure, if effected at all, is effected by mechanical means. It is not necessary to detain our readers by enumerating all the properties which have been attributed to these animals, which for the most part have been conceived *à priori*, from the viscosity of the blood, and the calcareous matter which their humours contain. Those who are curious on this subject may find enough to gratify them in Gesner, and the older writers on the medical science.

We must, however, dwell a little more on the employment which physiologists have made of these animals, and which we have already glanced at in our notice of the slugs, to prove that reproduction is not confined to less essential parts, and to animals but little raised in the scale of being, but that it may equally take place as to the composition of numerous and very important organs, and even in the head itself. Spallanzani positively affirms the fact, grounded upon numerous experiments on the helices; but, as takes place in all such cases, many other authors have denied it, and founded their opinion equally on the basis of experiment.

The experiments of Trembley on the green hydræ, or fresh-water polypi, have established beyond a doubt, that in this degree of organization, an animal can reproduce not only the different parts of its body, but that when the latter is cut in pieces, each piece can become a perfect animal. He even succeeded in causing six or seven heads to sprout from a single body, by dividing the latter longitudinally into so many strips.

Bonnet, desirous of repeating the experiments of Trembley, and not being able to procure any of these green hydræ, tried whether some fresh-water worms, a species of naïs, could not also reproduce the parts which should be cut off, and as he could make no experiment on the appendages, he only tried upon the body, and found that it might be cut into twenty-six parts, and that each part reproduced a complete animal, so that by a very simple calculation he showed that from a single individual of two inches long, which might be cut into eight parts, and those successively into the same number, in proportion as they became perfect, there would be produced at the end of the fourth year 32,768 individuals.

This faculty which the worms possess of reproduction, when mutilated, was equally demonstrated in animals apparently more complicated, that is, in the actiniæ, by the Abbé Dicquemare. He proved, in fact, that their body might be divided into a sufficiently great number of parts, provided that in the strip which was cut off, a portion of the mouth should be preserved.

Up to this time, although these facts appeared sufficiently extraordinary, nevertheless, as they had not been observed, except in animals but little elevated in the living scale, and all the parts of which, to a certain point, were similar, a small number of persons remained incredulous, or rather, as the experiments were made upon animals not very common, nor easily to be procured, perhaps less attention was paid to them. But when, in 1764, Father Boscovich, in a letter to M. de la Condamine, announced that snails whose head was cut off, reproduced another altogether similar, as was proved by the experiments of the Abbe Spallanzani, a great number of persons massacred an immense quantity of snails with the idea of verifying these experiments. Voltaire himself, as may be seen in his questions in the Encyclopedia, article *Colimaçon*, commenced experimentalist. But he showed as little skill as a physiologist, as, on another occasion, he did in geology. In 1769, the celebrated Adanson, after having made trial on more than 1500 snails, denied that the individuals from which he had cut, not the entire head, but merely the tentacula and the jaw, but without leaving their roots, reproduced these organs, and he concludes that Spallanzani, in his amputations, only removed the cap or bonnet.

M. Cotti, an eminent meteorologist, declared, that after numerous experiments made from 1768 to 1774, he was also obliged to conclude, that the snail whose head was completely amputated, did not renew it, but perished. Valmont de Bomare tried similar experiments about the same time, and with similar success.

Nevertheless some persons were more fortunate, and among others, the celebrated Madame Bassi de Bologne, M. Lavoisier, Schœffer, &c. But the memoir which seemed to put the result of Spallanzani's experiments out of doubt, was one inserted by Bonnet, in the " Journal de Physique." He insisted on the precautions which were necessary to be taken that the experiment might succeed. He accompanied his memoir with figures, to show the parts retrenched, and the manner in which they are reproduced by a sort of vegetation. We there find that he never cut any part but tentacula as far as their base, as well as the entire cap of the head and the jaw, and that the reproduction, which has some variations both with respect to time, and the order in which the organs are reproduced, does, nevertheless, really take piace. Unhappily, he has not given us the anatomy of the organs which he retrenched, nor that of those which were reproduced, so that his experiments can scarcely yet be considered as conclusive.

In 1788, M. O. Müller published some experiments confirmatory of those of Bonnet. He made use of very sharp scissors, and placed them obliquely, so as also to take away only the upper half of the head, that is, the four tentacula, the upper lip, the jaw, and sometimes a small part of the foot.

Finally, some experiments on the same subject, and much

more conclusive, were made by M. G. Tarenne, of whom we have spoken before, in 1803. If we admit the testimony of this gentleman, we cannot refuse to believe that snails can reproduce the entire head, since he assures us that the piece which he suddenly cut off, with very sharp-edged scissors, placing them perpendicularly, a little behind the large tentacula, and under the foot, contained not only the tentacula, the jaw, and the upper lip, but also the buccal mass altogether, the brain, and the anterior part of the foot. He continues even to assure us that snails thus mutilated, at the end of a year and more have reproduced a complete head; but a fact so extraordinary, with reference to so great a length of time, requires much more evidence before it can be generally received. Nevertheless, Spallanzani says very positively that the head is regenerated, whether the section be made above or below the brain. Thus we find, that whatever repugnance we may feel in admitting this fact of the regeneration of the entire head in the helices, it is exceedingly difficult to deny it altogether. It took place in about two years after decollation, and the new head did not differ from the old, but in the circumstance of the skin which covered it being whiter and smoother. Sometimes also, there is a sort of furrow at its junction with the trunk. According to Spallanzani, it appears that the manner in which this reproduction takes place is rather variable, and that sometimes it even remains incomplete. But M. Tarenne says, that having cut the heads off of two hundred helices, and having put them into a humid thicket at the extremity of a garden, so that they might more easily find the nutriment which was suitable to them, he perceived on all the individuals that he could find again at the end of the fine season, a new head somewhat resembling a coffee-berry. It had four small horns, a mouth, and lips. At the end of the following summer the heads were perfectly reproduced, except that the skin was smooth, or cicatrized, in the same way as it is in partial amputations. How they could avail themselves of the nutriment which Tarenne placed at their disposal, till the new head appeared, does not appear. Although Spallanzani has given fewer details respecting the operatory process which he pursued, than M. Tarenne, we find that he had already really obtained the same results, as the latter is candid enough to avow. From all this, it would certainly appear established, that the entire head of the helices can be regenerated when cut off, and yet, notwithstanding, we cannot quite get rid of a certain degree of repugnance to admit this assertion as an indubitable fact. It is with difficulty that we conceive how the nervous threads, the muscles, the vessels which have been cut through the middle of their length, can re-accord with the portions which sprout from the head, now become a sort of germ; or, admitting that the regeneration should proceed from the nervous and muscular threads themselves, how should the nervous threads shoot forth and give birth to the brain. To render our conviction complete, it would be necessary that the reproduced head should be carefully dissected, and compared with the one which was cut off.

If the advantages which we derive from the helices are but inconsiderable, it is otherwise with regard to the detriment, or at least the inconvenience which they occasion us. All persons occupied in gardening regard them as a sort of plague. In fact, when these animals are abundant, they can lay waste in a single night an entire plantation of oleraceous herbs, a little time after they have sprouted from the earth, and are yet extremely tender. They also attack the finest, and especially the most succulent fruits, as they approach to maturity, and so produce their destruction, either of themselves, or by facilitating the destructive operation of other frugivorous animals, such as wasps and hornets, or that of rain, by which rottenness is produced. Many modes have therefore been thought of to destroy them as well as the slugs, or to hinder them from getting at the fruit. The best way of destroying them is, certainly, to hunt them morning and evening, or after a small rain, and crush them. But their propagation may also be hindered, by taking care always to avoid tufted borders, and especially those made with box, close hedges, and all those ancient ornaments of gardens, formed of yew and hawthorn. In general every arrangement must be avoided which can present humidity and shelter to the snail, unless it be used as a sort of snare in which they may be found collected in greater or less numbers, and destroyed at once. Their hurtful effects upon the fruits of isolated trees may be prevented by investing a part of the trunk with any very viscous matter, and among others with a sort of pitch or tar, the residue of distillation from coal or charcoal. The same effect may be produced by ashes, or any pulverulent substance, strewed at the foot of the tree, but the pitch is better, because it stops the progress of many other pernicious animals.

We now come to the *aquatic pulmonaceous mollusca*, which will not detain us long. The ONCHIDIA are without shells. The species called *O. typhæ*, by Buchanan, has an oblong body, and is about an inch and three quarters long in a state of repose, but when it walks it becomes linear, obtuse at the two ends, and its length extends to two inches. The foot is formed by a great number of transverse wrinkles, by means of which the animal walks, and adheres pretty nearly after the manner of an earth-worm. The head changes form considerably when the animal walks, becoming extended, flat and oval. From the upper part of the head arise two tentacula entirely similar to those of the snails, and having the appearance of eyes at the extremity.

This animal, says Buchanan, is not hermaphrodite, like

many other *worms,* but the sexes are separated in different individuals. This, however, is indicated by no external dissimilarity. The anus and sexual organs are placed in a common cloaca at the hinder extremity of the tail, and during intercourse the distinction of sex is very evident. This animal lives on the leaves of the *typha elephantina,* common on the banks of the Ganges.

The manners and habits of the PLANORBES are nearly the same with those of the next genus, (*Limnææ*) with which they are constantly found. Thus they are fresh-water animals, altogether aquatic, perhaps more so than the limnææ. In fact they are very seldom indeed to be met with out of the water. They crawl, like them, on the surface of bodies as well as on that of the water, with the foot upwards, and the shell downwards. They equally feed upon vegetable substances, which they eat in the same manner. Their mode of sexual intercourse is exactly similar: the individuals form long lines, beginning with a male, and ending with a female; but they are hermaphrodite. The eggs which they lay are in a gelatinous mass, applied against submerged bodies. In our temperate climates, and in cold climates, the planorbes, at the approach of winter, sink into the mud, and fall into a lethargic state, so that it is impossible to find an individual in a place where, some months before, there were myriads.

The species of this genus appear rather to belong to the temperate and boreal zones, than to the other parts of the world. They are very common in Europe and North America. Few are known in the Austral regions; many are found in the fossil state.

A viscous matter issues from the collar of these animals, when it is irritated.

The LIMNÆÆ appear to possess a greater degree of sensibility than most other mollusca, which is doubtless owing to their skin being more gelatinous and less tuberculous. They creep sufficiently fast by the assistance of the very broad muscular disk with which they are provided, not only over solid bodies, immerged or not, but also at the surface of the water. In this case they are inverted, the shell being below and the foot above. It appears that the contraction of the foot takes its hold of a very light stratum of the water, which they leave above them. Their strength, however, cannot be very great, and in fact the slightest wind is often sufficient to accumulate the limnææ, thus floating towards the opposite side from which it blows. On the least danger, they withdraw all their parts into the shell, become of a greater specific weight, and fall to the bottom. To return to the surface they are obliged to crawl along the bottom, as far as the bank, or to follow the stem of some aquatic plant. It is only in the water, and in the fresh water especially, that the limnææ are to be found, and as this fluid cannot serve them for respiration, they are obliged to come from time to time to the surface to respire the atmospheric air. Sometimes they are found altogether out of the water, but on some aquatic plant, and never even at the slightest distance from the element. They feed on vegetable substances alone, and especially on the leaves of aquatic plants, which they cut, like the slugs, with the tooth with which their mouth is armed. During winter, at least in our climates, they fall into a sort of torpor, and sink more or less deeply into the mud which is at the bottom of the ponds, marshes, rivers, or streams, that they inhabit. It is at the end of spring, when their activity becomes greater, that they engage in the grand work of reproduction. Although hermaphrodites, their mode of intercourse differs from that of the limaces and helices; for with the limnææ the concourse of three individuals is necessary, and as others come up, they thus form long cordons. At the end of a certain time after this, of the duration of which we are ignorant, the fecundated individuals deposit upon dead or living bodies in the water, small masses, glairy, translucent, and ovaliform, composed of a greater or less quantity of eggs. These eggs, at first, by no means distinct, become so by degrees. In each of them the little animal provided with its shell, is very distinguishable, and in a little time it separates from the others and proceeds in search of its food.

We are ignorant of the duration of the life of these animals, and of the time which they take to become adult. In certain localities they are accumulated in great abundance.

The limnææ are of no direct utility to the human species. They serve as food to aquatic birds, and particularly to fishes, which consume great quantities of them.

The species of this genus appear, along with the physæ, planorbes, &c. to be found in the fresh waters of all parts of the earth. They are known in the Northern Zone, in Europe, in Asia, and in America. The temperate zone assuredly contains them in the three quarters of the world. They are found in the tropical regions of America, Africa, and Asia. Finally, if we are not acquainted with them in the antarctic, or southern zone, it is probably owing to the want of direct observations.

We have nothing to add on the rest of this order.

The second order of this class is the NUDIBRANCHIA. In this DORIS constitutes a very numerous genus. Bohadsch first established this genus under the name of *Argo*, derived from this curious reason, that he thought the upper tentacula of many species, which are formed as it were of an aggregation of small rounded tubercles, were an agglomeration of eyes. The details of the organization are curious, but our limits oblige us to omit them, and rest contented with the brief description in the text.

The Dorides are hermaphrodites, the two sexes being united in each individual. The dorides are all marine, and live at variable depths, in rocky situations, where algæ, and marine plants in general abound. Their motion is slow; they creep along with their tentacula and gills very much spread out, by means of the broad disk, which occupies their entire abdomen, either on submarine bodies, or at the surface of the water in an inverted position. On the slightest contact they draw in their tentacula, and even in a great measure their gills, which have many relations of structure with the former organs, and gather up their bodies, much after the manner of the slugs. It was generally believed that they fed on living animal substances, and among others upon oysters, and other fixed conchyliferous mollusca, whose shells they pierced by means of their sort of tongue. But M. Dupont de Nemours assures us that their nutriment consists in sea-weed. Their mode of sexual intercourse is altogether unknown. Their spawn is in the form of gelly, adherent to submarine bodies.

As to the subsequent genera, down to scyllææ, their habits are either utterly unknown, or may be presumed to be analogous to those of doris.

The Scyllææ, like the preceding, are hermaphrodites; they are little pelagian animals, which seldom visit our coasts, but are not rare upon the masses of fucus in the Atlantic Ocean. They live, as it would seem, in the depth of the waters, where they creep upon the stems of fucus, as the canaliculated form of their foot would indicate. It is probable, however, that they can swim with equal facility, by means of the foliaceous appendages with which their body is provided, and especially the two branchial pairs. Their other manners and habits are altogether unknown, but, in all probability, they differ little from those of doris and other neighbouring genera.

Very little is known respecting the manners and habits of Glaucus, and the remaining genera of this order. We merely learn from Dupont and other observers, that they are found only in the high seas at a considerable distance from the shore, and that they often remain at the surface of the water, where they swim in an inverted position, like the planorbes, the limnææ, and many other mollusca, with the assistance of their small foot. André Dupont says that the middle line of what he names the back, but which in fact is the belly, appeared like a leaf of silver, and was in a continual undulatory movement. This little animal, little more than an inch in length, in consequence of its fine blue colour, silvery under the foot and at the extremity of the digitations, and especially from its form, appears extremely elegant when it is swimming in calm weather at the surface of the sea.

The little order of the INFEROBRANCHIA we must entirely pass over here, as nothing whatever is known concerning their habits or manners.

Of the first two genera of the order TECTIBRANCHIA, we know nothing, except what regards structure. The genus APLYSIA, by a typographical error, in the second edition of Linnæus, was called *Laplysia*, which misnomer, ridiculously enough, has been preserved. Pliny and Dioscorides speak of a species of this genus under the name of *sea-hare*, and depict it as a venomous animal, which not only should not be touched, but not even looked at. Following them, Rondelet has spoken in the same manner.

The Aplysiæ are supposed to possess the property of causing the hairs to fall from those parts of the body to which they are applied, and of causing strangury to those who are so foolish as to swallow a little of the sanies which flows from their bodies. But M. Cuvier has proved this opinion to be erroneous, at least as far as the first of these properties is concerned. But they exhale an odour so nauseous and fetid, that one would be more disposed to avoid than approach them.

The best known of the species resembles, when in a state of repose, a mass of unformed flesh. When it is in motion its figure approaches to that of the slugs. Its colour is a reddish brown; it has four horns in the head; and the eyes, which are very small, are situated between the two hinder ones.

This aplysia has a reservoir of ink like the sepia, and employs it for the same purpose, that is, it ejects it for the purpose of escaping the pursuit of its enemies. It inhabits, in preference, the muddy bottom of the water, and lives on small crabs, small mollusca, &c.

Draparnaud thinks we should not regard the two anterior prolongations of the head as tentacula or horns; if such then be the case, this genus in reality will have but two.

The remarkable molluscous animal, called OMBRELLA by Lamarck, and *Gastroplax* by de Blainville, exhibits many relations in its interior structure with aplysia. We know nothing respecting the manners of this animal, which lives in the Chinese seas. It is sometimes called the *Chinese parasol.* From the extremely anomalous position of the shell, it is not easy to conceive how it can crawl. Accordingly, M. de Blainville, considering that the back covered with a skin extremely thin, has need of being sheltered from the action of external bodies, has supposed that this molluscum was, as it were, compressed between two protecting bodies, one inferior, or the shell, and the other superior, which might be a sort of valve, extremely thin, and adherent, like the *anomia*, or even some rock. An hypothesis which may be yet further supported by a consideration of the cavity, at the bottom of which is the mouth, and towards which the pedicled tentacula might, by their movement, determine the arrival of nutritive substances.

In the order of HETEROPODA, we must pass the first two genera, and come to the FIROLÆ, of which we can say but little. Their manners and habits are but little known. They are found, as it would seem, pretty commonly in all the seas of warm regions, and even in the Mediterranean, where they swim with peculiar elegance, by means of their fin and tail. It often happens that they are mutilated, a circumstance which has occasioned some confusion in the distinction of species.

In the order PECTINIBRANCHIA, the animals of the first genus TURBO, are marine; they live on the sea-coasts, in the midst of rocks beaten by the waves, and consequently at no great depth of water. At low water, when the rocks are uncovered, these animals often remain fixed in the same place, but they are also sometimes seen to move, endeavouring without doubt to regain the sea. On the least touch they suffer themselves to fall, and thus easily escape from their enemies. Their locomotion is not rapid, which is certainly owing to the shortness of their disk. Their nutriment should be vegetable, if we may judge by the apparatus of their mouth; but of this we are not perfectly assured.

We are ignorant of all the particulars of their reproduction; we merely know that the sexes are distinct, and not combined in the same individual. The small species with corneous operculum are ovo-viviparous. Thus these animals do not deposit corneous eggs like the siphonobranchia. It yet remains a question whether there are any true turbines with calcareous operculum.

The animals of this genus are of some utility to the human species. In fact the inhabitants of the sea-coasts feed upon the smaller species, and from the larger a very fine *nacre*, or mother-of-pearl, is derived. Formerly, ornaments were made for cupboards of the shell of the *T. marmoratus,* completely stripped of its non-nacreous stratum.

The animal of the SCALARIA is but very incompletely known. From the figures given by some authors, we conclude that it does not differ much from the animals which inhabit other operculated shells of the same order. From a figure communicated by Dr. Leach to M. de Blainville, it would appear that the proboscis, or more probably the labrum, is, as it were, laciniated, or divided into little tongues swelled at the extremity, a disposition, which if true, is exceedingly anomalous. We also learn from Bianchi, that this molluscous animal suffers to escape from its body a great quantity of fluid which stains the fingers, and paper, of a fine purple colour, which suggested to him the idea that this might be one of the conchyliferous mollusca, which furnished to the ancients the purple with which they tinted the most precious stuffs. Linnæus, and some other writers, have adopted this opinion.

The scalariæ, like the turbines, appear to prefer the sea-coasts, where there are numerous anfractuosities and rocks. It is probable that some of them are to be found in all seas.

In the animals of the genus CYCLOSTOMA, the sexes are separate. They live on land, in somewhat humid places, under rotten leaves, or in the trunks of rotten trees, where they are sometimes found in great abundance, the common species especially, *cyclostoma elegans.* Its tentacula appear to be contractile, though M. de Férussac has asserted that they are retractile, which is contrary to all analogy, for tentacula of this kind are not known, except in the limacinæ. This little mollusca is remarkable for the mode in which it moves, making sorts of steps or strides, as M. de Férussac has observed, though this seems to be done by means of its foot, and its proboscidiform mass, and not by raising the two lateral parts of the foot alternately.

The females of the PALUDINÆ, which are always bigger than the males, present, in the apparatus of generation, a disposition which has been erroneously thought to be peculiar to these animals in the great development of the second part of the oviduct, to which the name of matrix has been given, and where the eggs are accumulated, until they are sufficiently developed to disclose the young, so that the little paludinæ come forth from the body of the mother in the living state.

This has occasioned the species most common in great rivers to be called *vivipara.* This singularity was observed long ago among many species of Turbo.

There is nothing very peculiar in the manners and habits of the paludinæ. They generally live at the bottom of rivers, on the aquatic plants which are found there. They appear to feed on substances of all kinds, but more especially vegetable. Their mode of sexual intercourse offers nothing very worthy of remark. We have said that the young issue forth in a living state from the interior of the mother; but they do not all come forth at once. The females of the paludinæ appear to lay during the entire of the fine season. The little ones, on issuing forth, place themselves on the shell of the mother, and appear to remain there for some time. They crawl pretty fast on a resisting ground, and sometimes come to the surface of the water, where they can also float after the manner of the limnææ. This, however, appears to take place very rarely.

The species of paludinæ appear to exist only in our northern hemisphere, and not in the torrid zone, where they are replaced by the ampullariæ. They seem to be very common in the rivers of North America.

The JANTHINÆ constitute a very distinct genus, established by M. de Lamarck on a molluscous animal, which Linnæus had classed with HELIX. The name of Janthina indicates the colour of the handsome shell, which is a fine violet. This genus has now been adopted by the generality of modern zoologists.

The body of the animal is globular, as the form of the shell would indicate, and the visceral part is very small, comparatively with the head, chest, and foot. The foot itself is not remarkably large, but it is very thick and muscular, and is distinguished by a vesicular mass, which forms the principal characteristic of this animal. It is said to be a sort of subcartilaginous froth, composed of small cells or utricles, which can swell or contract at the will of the animal. M. de Blainville, however, tells us that this mass, in an individual a long time preserved in spirits of wine, bore a strong resemblance to the cellular tissue, a little gelatinous, and did not present the least indication of being cartilaginous. Observers are doubtful as to its point of attachment. Forskal says that it is attached to the mantle of the animal, but M. Bosc that it issues from the anterior part of the foot, and is extended in length beyond that organ. M. Cuvier, as we see in the text, describes it as being situated under the foot.

The janthinæ inhabit the deep parts of the sea in all warm latitudes. When the water is calm they may be seen floating on its surface, with the shell underneath, and the foot and its vesicle just mentioned, upwards, and are sometimes in pretty numerous bodies. It is probable that they have no need of any muscular effort to maintain themselves thus on the surface of the water, and that the vesicles of their feet produce the effect of an hydrostatic bladder; but then they must of necessity be the sport of the lightest wind or the least current, which is not the case with the other mollusca, which can swim in this manner on the water, unless we admit that the foot should possess some action, or at all events that its lateral appendages should act as fins. Be this, however, as it may, on the slightest appearance of danger, or cessation of calm, the little animal withdraws its head into its shell, contracts its vesicles, either by forcing them to re-enter the testa with its foot, if they be attached to its posterior part, or by acting directly upon them, if their attachment be under the foot itself, and in the hollow which it forms; but in whatsoever manner this may be done, the air of these vesicles can neither be absorbed nor rejected. By the diminution of its volume the weight of the animal is increased, and it sinks more or less towards the bottom of the water. Whether it can creep on the ground, or fasten itself thereto, is doubtful. The first position does not seem at all probable; but the second is more likely to be founded in truth, from the peculiar form of the foot, in the style of a sucker.

The janthinæ, in all probability, feed upon animal substances, though we possess no certain information on this subject. Neither are we much better acquainted with their mode of propagation. According to Forskal it would appear that the female preserves her eggs in a sort of matrix, or inflated portion of the oviduct; at least he has several times seen young individuals issue from the body not larger than grains of sand, and which in the microscope, have appeared to him provided with a shell similar to that of the mother, with the exception of colour. According to the same observer, it would appear that the young animal presents some differences more considerable, by having towards the aperture of the shell two transverse laminæ, rounded, and ciliated in their circumference, which it might use as fins for locomotion.

Notwithstanding the probability of what we have just advanced respecting the ovo-viviparous character of the janthinæ, Sir Everard Home has published a contrary observation, that around a shell which he examined, there was a glairy and oviferous band, which he supposes to come from the body of the animal. His opinion too, founded on this observation, is that the animal never touches the ground, and that nature has thus given it the faculty of rolling its eggs around the shell.

According to M. Bosc, who has had occasion to observe many of these animals, the janthinæ are eminently phosphoric. They serve as food to fish and birds. The violet liquid which they produce might be employed with equal success as that of the purpura, with which doubtless it has much analogy.

In a supplement necessarily confined to such limits as ours, we are forced to pass in silence many subdivisions of the mollusca, even when some degree of popular interest may attach to them; but it would be unpardonable to omit all mention of the animal supposed to have produced the celebrated purple of the ancients.

M. de Lamarck was the first who established under the name of PURPURA, a distinct genus composed of several species of shells, previously classed by Linnæus and his followers, in the genera *buccinum* and *murex*. The name, however, was applied before by various authors, not only to the species of the last mentioned genera, but to many others, owing doubtless to this circumstance, that all these animals furnish in greater or less abundance, the materials which the ancients employed in dyeing. In fact, we must not even suppose that it is in the species of this genus that the purple is principally found. It exists equally in murex and buccinum, and it is even probable that the species of shell-fish, from which the ancients extracted the purple, did not belong to the present genus; this, however, is a convenient place for treating on the subject.

The purpuræ are marine animals, living in the fractures of rocks, in places covered with fucus, and also sometimes burying themselves in the sand. They creep, by the aid of their foot, like the other gasteropods. Their nutriment appears to be constantly animal, and obtained by piercing the shell, principally of the bivalve mollusca.

The mode of sexual intercourse is not known. The eggs are spheroïd, a little elongated, corneous, of a yellowish colour, and not deposited until towards autumn. They adhere to the rocks, and other submerged bodies, by means of a sort of paste. Their other extremity is closed by a sort of opercle, small, oval, thick, and transverse. In their interior, a thicker matter is found in the middle of one more fluid.

The species of this genus are found in all seas; but the

greater number and the largest come from the warmer latitudes, and more especially from the Austral seas. But without considering this genus more minutely, we shall seize the present opportunity of treating concerning the purple of the ancients.

The words πορφύρα, or *purpura,* were indifferently employed by the Greeks and Romans, to designate the colour itself, and the animal which furnished it. Aristotle is the first writer who has spoken of the purpura. He tells us that "with the exception of the head, all its other parts are contained in the shell; it is provided with a very firm proboscis, by means of which it pierces the testa of the animal on which it feeds. In the turbinated part of the shell lie the stomach, liver, and intestines; between the neck and the liver is the organ which furnishes the colouring matter; it has the form of a vein. The substance which fills the rest of the interval resembles alum." Here it is probable that Aristotle alludes to the cretaceous matter frequently found in the rectum of many of the mollusca. He continues:

"The purpuræ move but little; they remain concealed during the great heats of the dog-days. Fresh-water is positively injurious to them; but they can live as long as fifty days out of water altogether. They can perceive their prey from a very great distance. They assemble in spring in the same place, and there make what is named their wax, which is a production resembling a cake of wax, except that it is not smooth, or rather a multitude of white pease-cods joined together; no aperture is ever perceived in it. When the purpuræ, like the other testacea, commence to form this production, they produce a gluey mucosity, which serves to connect these sorts of pods. It is in this united mass that the young purpuræ are born, and that they are found attached, sometimes as yet imperfectly formed, to the shells of the old ones, when the latter are fished up. If the purpuræ be taken before they have cast this spawn, they do so in the baskets in which they are put, and the narrow space in which they are enclosed merely gives to the mass of the *wax* the form of a cluster of grapes."

Although Aristotle has pretended that in these animals, there is no generation, properly so called, but that they spring from the earth, it is evident that the above description agrees to the eggs of the purpuræ, which consequently resemble greatly those of the *buccinum undatum* in *buccinum* proper.

"We distinguish," continues the Stagyrite, "many species of purpuræ; accordingly there are some large, as those of the promontory of Sigetum, and small, as those of the Euripus and the coasts of Caria. In general those which are fished for in gulfs are large and of unequal surface; some of them weigh as much as a mina." The colour which they furnish, named by Aristotle flower, (ἀνθος) is most generally black, though sometimes red, and small in quantity. On shores, and around promontories, they are small, and their liquor is red. In places exposed to the north, it is generally black, and red in southern aspects. It is never of less value than when the purpuræ have cast their spawn, accordingly they are fished for in spring, at the very moment when they are getting rid of it. They are otherwise taken by means of baits composed of tainted flesh, or of small fishes, and without a net; but as they often fall back into the water, after having been drawn out of it, to avoid this inconvenience the fishermen place drag-nets underneath and around the bait, so that if they should happen to fall, which they do easily, when satiated, (and it is difficult to pluck them away) they are not lost. They then leave them in the drag-nets, with which they continue to take others until there is a sufficient quantity for use. To extract the fluid from them, they remove, at least in the larger species, the animal from its shell, and then take the part situated between the neck and the liver or vein; but as for the smaller individuals, they break them up with their shell, because it would be too difficult to separate them from it. But," adds Aristotle, "care is taken to do all this when they are living, without which, if they died naturally, they would shed their fluid in expiring."

Pliny considerably abridges what Aristotle says concerning the purpura, and even modifies it in a manner nearly unintelligible, which proves that he did not comprehend the text of Aristotle. He adds, "that they generally live seven years, though they grow much more quickly than other shell-fish, and attain their full size in about a year. They can live fifty days without eating. They remain concealed for thirty days during the dog-star heats; and it is especially on the coasts of Tyre in Asia, of Meninx, and Gitulor in Africa, and of Laconia in Europe, that the finest purple is to be found." But he gives many interesting details respecting the species of shell-fish from which different colouring matters are derived.

"There are two genera of shells which produce the purple colours, and the *conchylian* colours, colours which differ only in shade. The smallest is a *buccinum*, so called because it somewhat resembles that from which the sound of the horn is produced. Its aperture is round, and its edge emarginate. It is only found attached to stones, and in the neighbourhood of rocks. The other is named *purpura*; it is in the form of a club, and composed of seven whorls, which indicate its years, as in buccinum. But it is bristled with sharp points like needles, which do not exist in the latter. It is, moreover, provided with a projecting beak, (*rostrum*) and on the sides with small tuberculous spines, into which the animal can introduce its tongue."

"The purpuræ are also distinguished by the denomination of *pelagian*, among which many varieties are established according to the places which they inhabit and the substances on which they feed. Those which live in mud, or among the algæ, and feed thereon, are in very little estimation. Those of the coasts, which are gathered by the sea-side, are better, though the colour which they furnish is lighter and more clear. Another variety, which is called *gravelly*, from the gravel-beds of the sea, where it is found, is extremely proper for the *conchylian* colours; but the best for the purple colours is the *dialutensis*, that is, the one which finds its subsistence on different kinds of soil or strata.

"The purpuræ are taken with small nets, which are cast into the deep sea. Bivalve shell-fish are used by way of bait, which can open and close, or mussels, which, half-dead, are reanimated as soon as they are restored to the sea, and half open their shell. The purpuræ, eager to devour these, attack and thrust their tongue into them. But soon stimulated by this sting, the mussels close their shell, and retain the purpuræ, so that, victims of their avidity, the latter are carried off still suspended to their prey. The most advantageous season for this fishing, is after the rising of the lesser dog-star, or before the spring, because when the purpuræ have spawned, their juices are too liquid. But the workmen are ignorant of this, although it is very essential.

"In employing the purpuræ for dying, the artists begin by removing the vein already mentioned, and adding to one hundred pounds of this substance, twenty ounces of salt. The whole is allowed to macerate for three days exactly. It is boiled in a leaden cauldron, until greatly reduced. A moderate heat is then kept up by means of a long stove; after which, the flesh which necessarily remained attached to the veins, being skinned off, and the tincture being completely liquified on the tenth day, and afterwards strained, the wool is plunged into it. They continue to keep it warm until the desired point has been ascertained. A lively red tint is less valued than a blackish red. The wool is left to steep for five hours, for after being carded it is replunged into the bath, until it has imbibed as much of the liquid as possible. The buccinum is never employed by itself, because it produces a dye which will not hold, or rather, perhaps, because it does not preserve the lively red; but by mixing it with the purpura, it gives to the too dark tint of the latter the solidity and brilliancy of the scarlet, which is sought for. By this mixture, these colours mutually heighten or darken each other. The best proportion is for fifty pounds of wool, two hundred pounds of buccinum, and one hundred and eleven pounds of the purpura. Thus that superb colour is obtained, which is named *amethyst*. To obtain the Tyrian colour, the wool is saturated in a bath of the liquor of the purpura, still green and not black, and then it is transferred into one of the buccinum. Thus is produced the finest purple, the colour of coagulated blood, blackish when viewed in front, and of a brilliant hue when seen sideways. From this Homer applies the epithet of purple to the blood.

"The conchylian colour is obtained by similar processes, excepting that no buccinum is used, and half the bath is composed of equal portions of water and wine. Thus a palish colour is obtained, (aptly termed *conchylian*, from its resemblance to the usual colour of sea-shells), which is in high estimation, and more extensive, as the wool is less saturated.

"Another tint is yet obtained, which has been named *tyriamethyst*, by saturating a stuff which was at first amethystine, in a bath of the Tyrian purple, as its name indicates, so that they tint at first conchylian, to facilitate the Tyrian tincture, which then becomes, as is said, more agreeable and softer. Just as to obtain the deep poppy-colour, they retint in Tyrian purple what had been first tinted in Kermes."

The price of these colours varied according to the supply of animals; but, nevertheless, in consequence of the small quantity of juice extracted, and the length of the operation of dyeing, the purple was so dear, that in the time of Augustus, one pound of wool dyed with the Tyrian purple, could not be bought for thirty pounds. It is supposed that the opulence of the city of Tyre was much increased by the commerce of this precious dye. The purple decorated the magistrates of Rome; but as it grew scarcer, its use was reserved, under pain of death, to the emperors alone. The priests, when it was first known, assumed it as a colour agreeable to the divinity, and employed it in the public solemnities of religion.

For a long period nothing was known respecting the two juices which formed the Tyrian dye, excepting what we derive from the ancients, and, indeed, we may say more especially from Aristotle and Pliny, for no additional information can be said to have been communicated by other ancient writers; so that, although Aristotle and Pliny had given some intimations of their being primitively white, and Pliny had mentioned one of the intermediate colours, as we have already seen, a green, yet the other colours which they undergo on exposure to the sun, were not distinctly noticed, until the animals themselves were discovered at the end of the last century, and the beginning of the present; until then, no adequate conceptions could be formed of the changes they underwent before they became purple.

The buccinum was found in 1686, by Cole, in great plenty on some of the Irish shores, on the shore of Somersetshire, and the opposite shores of South Wales. Its juice was profitably employed to mark linen of a fine durable crimson. A small species of the buccinum was likewise found by Jussieu on the French coast. Cole found the juice of the buccinum, when taken out of the vein or reservoir, to be white and clammy; and if this viscid juice be then squeezed on linen or silk, it immediately, on being exposed to the sun, acquires a pale yellowish-green hue, then changes to a blue, and lastly to a deep purple red. These changes, though very rapid, are quicker or slower in proportion to the heat of the sun.

On washing the cloth with scalding water and soap, and again exposing it to the sun, the colour changes to a beautiful crimson, and then no further alteration takes place from sun or air, or any of the agents usually employed to try colours. The linen marked with the white juice, while drying, always yielded, for the first time, a strong fetid smell, resembling a mixture of garlic and assafœtida. A similar scent was attached to the purple of the ancients.

Duhamel informs us, with regard to the purpura, that its juice neither receives nor communicates colour without exposure to the sun, and that this colour is evolved, not by the heat, but by the light of the sun's rays; for when the silk or linen which is stained, is covered with thin opake bodies, which transmit heat without light, no colour is produced, which is not the case when transparent bodies are employed. Also the light of a fire, though concentrated by convex glasses, and concave mirrors, has no effect upon it.

Duhamel gives the colours which it assumed, on exposure to the sun, in the following order; 1. a pale green, or yellow; 2. an emerald green; 3. a dark blueish-green; 4. a blue, with an incipient redness; 5. a purple. These colours appeared in less than five minutes; but it only became green when it was not exposed to the light. This succession of colours is best observed when the sun is low; in the light of noon in summer, they come on so quickly as not to be easily distinguished.

Observers, however, are far from being perfectly agreed respecting the circumstances which cause the colouring matter to pass from the pale-green, or greenish-white to the purple-red. Respecting the permanence of this tincture, however, there is no dispute. Cole, Reaumur, Templeman, and particularly Duhamel, have proved, that when the stuff has been perfectly saturated, and all its parts completely exposed to the solar action, that the strongest lixivia, and most active tests, have no influence on the colour, except that when there remain many layers at the surface, the last having impeded the solar action upon the others, and its combination with the tissue not having taken place, the colour then grows very clear, so that Duhamel concludes, from his experiments on the subject, that the ancients must have had a peculiar process for extending the colouring matter, which is always pretty thick and viscous in the animals, and thus making it penetrate into all parts of the tissue. Perhaps it was for this purpose that water, urine, and salt, were employed by the ancient dyers. Templeman tells us, however, from his own experience, that the addition of salt has no effect. This is a subject for the analysis of chemists, who, nevertheless, appear to have bestowed but little attention upon it. It deserves investigation, not as regards the art of dyeing, for the purple of the moderns is as beautiful and as fixed as that of the ancients, more easy to be obtained, and consequently less expensive—but it is important to the science of animal chemistry. Duhamel thinks that the action of the sun in purpurification is somewhat analogous to what passes in the coloration of fruits which remain whitish, yellow, or green, in shady places, and are coloured only where the action of the sun is received. Here, however, the change is slow and gradual, but in the purple it takes place instantaneously.

From all that has been investigated respecting this subject, M. de Blainville draws the following conclusions.

1. That it is probable, that the molluscum from which the ancients principally derived their purple, is an animal tolerably large, known in the Mediterranean, perfectly described by Columna, and of which Linnæus, and the modern conchologists make their *murex trunculus*, or perhaps the *murex brandaris*.

2. That they also employed a species of buccinum, which was smaller, to obtain an analogous, though a little different colour, and this species is probably the *B. lapillus* of Linnæus.

3. That it is certain that a great number of species of this family furnish an analogous liquor; but it is probable that they do not all do so.

4. That it is even probable that all the individuals of the same species do not produce it. It is a query whether this depend on sex, or age, or the epoch of reproduction?

5. That we do not precisely know in what part of the animal this matter is found. Is it in the depurating organ, or in the generative apparatus? We might be induced to believe in the latter, as the eggs of the *B. lapillus* contain the same liquor in abundance.

6. That the process employed by the ancients in dyeing purple is as yet unknown.

7. That the chemical phenomena of the purpurification are as yet but very incompletely known.

We shall now pass at once to the last order of the gasteropods, the *Cyclobranchia*, on which we can afford but very little space.

The PATELLÆ are animals much more simply organized than the preceding mollusca. They live upon the shores of the sea, and constantly on those parts which are alternately covered and left dry by the waters. None are yet known belonging to the fresh water, nor have any been observed even in the mouths of large rivers. They are almost constantly fixed upon rocks and submerged bodies, and sometimes in excavations tolerably deep, which they have hollowed in the substance of the rock. They do not, however, always remain in the same place, as some persons have supposed. It is now a long time since Reaumur published his observations on their mode of locomotion. It takes place in the same way as in the other gasteropods; but, in spite of the size of their foot, their movements are extremely slow, which is occasioned by the thinness of the layer, composed of longitudinal fibres, which is the seat of locomotion. But, on the other hand, the patellæ can adhere to the plane of position in a manner truly astonishing. In fact, if previously to removing a patella from its rock, it has been first touched, and thus, as it were, advertised of the intent, it is impossible to get it away, and the shell might sooner be broken, unless a plate of iron be passed between the foot of the animal and the rock. It has been proved by experiment that a patella will thus support a weight of many pounds without falling. This faculty is owing to the great quantity of the vertical fibres of the foot, which, by raising the middle part form a hollow in the centre, and consequently a sort of sucker.

Patellæ may frequently be observed sunk two or three lines in depth in chalky rock, which is singular, considering that they do possess, though in a very small degree, the faculty of changing place. M. d'Orbigny, however, tells us that each individual constantly returns to the same place. From the dental apparatus of the patella, and their great abundance in places covered with marine plants, it is probable that they feed on vegetables. Concerning their reproduction nothing is known. They are eaten by some of the poor inhabitants of the sea coasts, but their flesh, though savory, is coriaceous and indigestible.

With respect to the OSCABRIONS (CHITON) though their organization is very remarkable, we know little or nothing of their manners.

SUPPLEMENT

ON THE

FOURTH CLASS OF THE MOLLUSCA.

THE ACEPHALA.

WE shall begin our review of this class with that well known animal the OYSTER, (OSTREA, *L.*)

Oysters have been pretty generally regarded as almost in the last rank of animality, but most erroneously so, since there are a great number of animals below them, most certainly their inferiors in point of organization, and consequently in its results. What has caused them to be so considered is, that they live, for the most part, fixed to submarine bodies, or to individuals of their own species, and that it is thought that they are not capable of changing place. This, however, is an error; certain species can move, if not by means of a foot, of which they have no trace, at least by abruptly opening and closing their shell, as many other bivalves do, so as to turn themselves, when by any chance they happen to be upside down. If their sensibility is nothing, or at least extremely obtuse in the greater part of their body, it is not so with the papillary edge of their mantle or cloak. At the slightest contact of an external body on the tentacular threads, at the slightest rough motion of the water, it contracts, and the animal closes its shell. It must be confessed, however, that the majority, being fixed more or less completely, according to age, and always by their inferior valve, they are obliged to live in the places where they are born. Some species form, by a successive accumulation of individuals, strata, or banks, often very much extended and very thick, while others remain more or less free and solitary.

They are found, as it would seem, in all seas, but never at very great depths, nor at a great distance from the shore. The gulfs formed by the mouths of great rivers, or those in which the waters are more tranquil, constitute their usual habitation. But it does not appear that oysters ever live entirely in fresh water, or delight in it, as Pliny is pleased to inform us. Certain species, it is true, live in those parts of rivers where the sea comes up, so that they remain dry during low-water. This especially takes place with the *Ostrea parasitica,* which attaches itself to certain shrubs and trees in the torrid zone, vulgarly called *Mangliers* in the French Antilles, *Conocarpus* of Linnæus. Then they close their shell exactly; but in their ordinary state, *i. e.* in the water, they leave it half open, the marginal line of their tentacular papillæ edging almost the entire of the cleft. On the slightest contact of a foreign body, with these tentacula only, they close the shell more or less completely, and can also enclose there some of the smaller crustacea, especially of the genus *Pinnothera,* but which do not serve them as food. In fact, the nutriment of oysters, is, in all probability, composed of animals much smaller, of infusoria, of animated molecules, and even of animal matters so abundantly spread in the waters of the sea; for, in spite of the size of their buccal aperture, the softness of its edges, and its position, will not allow us to believe that oysters can feed upon substances at all resisting; and the confidence with which we ourselves eat these animals, must induce the belief that their stomach contains no hard substances. Accordingly, it is generally admitted, that the water of the sea in which they live, continually attracted and rejected by the mantle, brings at once both the matter of respiration and that of nutrition.

The oysters having, like the other bivalve mollusca, only the female sex evident, they must be genuine hermaphrodites; and, in fact, it appears quite certain that a single individual can reproduce and continue the species. The eggs are expelled in the form of spawn, or white fluid, very similar to a drop of grease, in which may be perceived with the microscope, innumerable little oysters. This is called *spats* by our fishermen. This matter, in which they swim, doubtless serves to agglutinate them to the submarine bodies, or to individuals of their own species; then the new ones, in being developed, smother, as it were, the old ones, not permitting the water to reach them, or hindering them from opening their shell. It is thus are formed those immense banks of oysters which are found upon our coasts, and which, in spite of the consiant destruction wrought amongst them for centuries, never appear to be exhausted. The species which are not fixed, or are not so in a flat position, not being circumstanced so favourably for the adherence of the spawn, appear, in general, to be less multiplied.

We are entirely ignorant of the manner of growth in the oysters, and of the duration of their life. It would seem, however, that as the growth is rather slow, if an individual be placed in favourable circumstances, that is, without risk of being smothered by its progeny, it would live a very long time. But we have no certain data to go on here. However, if we give credit to what is asserted by the inhabitants of Marennes, on the coasts of the Atlantic, oysters do not live more than ten years. In three days after the deposition of the spawn, the shell of the little oyster is already three lines in breadth; in three months, or six, it is nearly the size of a half crown piece; and at the end of a year, as large as a dollar. The fishermen on the coast just mentioned, distinguish the age of the oysters by the striæ of the shell. When they approach the term of their growth, the shell is very large in proportion to the animal, which grows thin, and diminishes more and more. As the oysters can completely shut their shell, and thus enclose a large quantity of water in their interior, they can live a sufficiently long time out of this fluid, especially if the drying action of the air on their testa be prevented, and they be placed in their natural position. This faculty, which allows of their being transported to considerable distances, facilitates the extensive commerce to which they give rise.

The oysters which never cause us any injury, except it may be in occasionally contracting or diminishing the depth of a bay, are of very considerable utility, inasmuch as from time immemorial they have constituted the food of man, fresh, dried, or cooked, but particularly in their fresh state. The Greeks, and more especially the Romans, when they levied contributions upon land and sea, throughout the then known world, to cover the table of a Lucullus, or an Apicius, held them in very high estimation, and attached no small importance to the localities from which they were imported. Those of the Dardanelles, of Venice, of the bay of Cumæ, and of England, were those which they preferred; but they especially attached a very great value to those which, brought from these different places, and perhaps from places still more remote, were transported in large boats (*lacubus ligneis*) and deposited in the Lucrine lake, where they grew remarkably fat. The first Roman who entertained the notion of establishing this sort of *park*, or oyster-bed, was Sergius Orata, at Baiæ, in the time of the Marsian war. It appears that the Romans preferred those oysters which have the edges of the mouth a deep brown, almost black, and that they gave them a particular name, that of *calliblephara*, a word which is nevertheless supposed to be corrupted. Those are they which vulgarly, but most erroneously, are called males, because the oysters are all hermaphrodites. The Romans ate oysters raw, and also cooked, with various seasonings, into which entered pepper, the yolks of eggs, vinegar, oil, wine, &c. But it is not probable that they made so great a consumption of them as do the modern Europeans. Accordingly oysters, at the present day, constitute a very considerable article of commerce.

The best oysters in Europe are our own. The most esteemed in France are found upon the coasts of Bretagne, and the largest on those of Normandy, whence they are brought at great expense to Paris in the autumn and winter seasons. The places where oysters are chiefly caught in England, are the Pont Burnham, Malden, and Colne waters, near Chester. This brood, and other oysters, are carried to the creeks of the sea, especially on the coasts of Kent and Essex, near the mouth of the Thames, and then thrown into the channels which are called oyster-beds, or layers, where they grow and fatten for the supply of the London market. In two or three years the smallest will become oysters of full size.

The mode in which oysters are *parked*, as they call it in France, affords some curious details that may not prove uninteresting to our readers.

The oysters, which form a considerable object of commerce in the north of Europe, and especially at Paris, come from the bay of Cancale, in the British Channel, between the village of that name, Mount St. Michel, and St. Malo. The bottom of this bay appears even, solid, and without current, all favourable circumstances for the reproduction of these animals. It must be, therefore, very considerable, and the bank which the oysters have produced must be very much extended, to suffice for the continual fishing which has taken place there for so long a time, without any sign of diminution. Nevertheless, from 1774, to 1777, we carried off so great a number, with the intent of forming a bank upon our own coasts, that some diminution was perceived in the bay; but it was very speedily supplied. Although the French may certainly be said alone to possess the right of fishing in this bay, it is yet open to all nations, but not at all periods of the year. The fishing usually commences at the end of September, and finishes in April; during the other months it is severely interdicted, because then is the time of spawning, and the oyster is supposed to be of an injurious quality. This idea, which is probably erroneous, is nevertheless a useful prejudice; for, otherwise, the continual fishing would soon destroy the bank, not only by subtracting the adult individuals, but also by destroying the sources of reproduction. The mode of fishing is very simple. It is performed by means of what is called the drag, which is a sort of iron rake or harrow, behind which is attached a leathern pouch, and which is drawn along by a boat in full sail. By thus raking the surface of the bank, sometimes as many as eleven or twelve hundred are taken at a single haul. These oysters, shipped in the ports of Granville and Cancale, are then transported to the different places where *parks* or depôts have been established for them. These parks not only answer the purpose of preserving the oysters and facilitating the sale, but also of ameliorating them, as the ancient Romans well knew long ago. In fact, the oyster when it first comes from the sea, has a strong taste of the muddy bottom, is more or less hard, and of an indifferent flavour altogether. It acquires the desired qualities in those parks, or beds, which we have mentioned. They are simply reservoirs, more or less extensive, excavated in the soil, or even in the rock, as at Etretat, and into which, at will, the sea-water may be permitted to enter at full tide, and from which it may be suffered to run out. In general, these excavations, which are in the form of parallelograms, are only a few feet in depth, and their walls are sloping. They communicate with the sea by means of a canal, more or less long, and furnished with a small flood-gate. When the water is to be changed, the flood-gate is opened at the end of the low tide, and the reservoir is filled at high water. The bottom and sides of these passes are covered with pebbles. or very coarse sand, for mud must be carefully avoided, which is always very hurtful to oysters. Care must also be taken that the motion of the water be not considerable enough to cause any grains of sand to enter the shells. When the park is thus disposed, the oysters are placed there in their natural position, that is, horizontally, the gibbous valve underneath, on a part of the elevation of the slope, so deep, that it will be difficult for robbers to get at them, but not too deep, that we may avoid as much as possible the deposition of the mud. The more suitably the oysters are fixed, the more cautiously they are moved, and the more the deposition of mud is prevented, which may be done by washing the walls of the park, and frequently throwing water on the oysters, when left uncovered, the sooner will these animals become good and marketable. Those also which happen to be dead should be carefully thrown away, and they are easily recognized, as they remain half open when the water has retired. There are some doubts as to the preference which should be given to these parks, or depôts, according as the water which they contain is received every tide, or only twice a month. In the first case the oyster may be, perhaps, a little harder, and more coriaceous that in the second. But the water should always be very clear and limpid.

Fresh water is not good for oysters, at least when the quantity which is introduced into the parks, either by great rains or inundations, becomes too considerable. Experience has settled this question, beyond all doubt, in the case of the parks of the Courseulois, which are liable to be filled with the fresh water by the inundations of the Seule. This proves the necessity of more frequently renewing the water of the parks during great rains. As cold is equally injurious to oysters, it would be desirable to have them placed at a sufficient distance from the surface of the water; but another inconvenience would result from this, that they would be less easily inspected.

From all these considerations it appears, that to establish a park of oysters properly, it should be in places sheltered from the wind, to prevent the agitation of the water, and consequently the entrance of grains of sand into the shells. The bottom should not be muddy, or it should be well covered with pebbles and coarse sand, so that the animal may lose the muddy flavour, and not resume it. The mass of water should be pretty considerable, especially if not changed at each tide, to escape too great a proportion of fresh water from rains. The oysters should be placed sufficiently deep to escape cold, but not deep enough to prevent them from being easily viewed by the inspector, as in that case he could not throw out the dead individuals. Finally, the more it is in our power to renew, or not to renew the water at pleasure, the more influence can be exerted in modifying the oysters. If it be desired that the oysters should be very white, clear, and even bigger, the water should be changed at every tide, as is done at Etretat, and different points on the Atlantic. If, on the contrary, it is wished to have them more tender, smaller, and particularly to make them green, they must be left in the water a longer or shorter time, according to the season, and some other circumstances, probably atmospheric, of which not much is known.

It is quite certain that the green oysters are absolutely of the same species with the white, and come from the same places, and that the latter may be greened at pleasure. For this purpose a park is chosen, in general rather small, and sea-water is introduced into it, which is preserved for a longer or shorter time without being changed. When the pebbles with which the wells have been lined, commence to grow green, the oysters are put in. But they are obliged to be placed with much more precaution than the other oysters, so that they shall not be one upon the other. From this, it happens that in any given space, where oysters are placed for this purpose, a third of the quantity used in ordinary cases, cannot be placed. Sometimes three days are sufficient to give the oysters a light green tint; but a month is necessary to render it of a deeper colour. The oysters will not green either in winter, or during very hot weather. A moderate heat is necessary for this purpose, as in March, April, September, and October. Rainy and stormy weather is said to be unfavourable, as is the agitation of the water, especially by the north winds. Generally speaking, there are some years in which the oysters green easily, while in others they will scarcely do so at all. What can be the cause of this coloration, which other mollusca may equally experience? Certainly it is not that these animals feed upon fucus, nor even upon green matter. It might be a state of malady, and to a certain point this is probable, as the green oysters are in general smaller. But M. Gaillon thinks that it is owing to the penetration throughout their entire tissue of an animalcule of a fine green colour, which he terms the *vibrio* of the oyster. This penetration of the animalculæ may be owing to a state of atony in the animal, arising from unfavourable circumstances. The great precaution which is required in greening oysters, causes them to be very dear.

What we have now said on the art of greening oysters, is drawn from the observations of M. Lair, and other French gentlemen, on the coasts of Normandy; but it seems that this art is carried to much greater perfection on the coasts of Aunis, whence come the excellent green oysters of Marennes. The men who employ themselves in this sort of business do not take the oysters indiscriminately. They choose individuals only a year old, and especially those which come from oysters already greened. They catch them with the hand upon the rocks, or they detach some large individuals fished up with the drag, and more deeply; they also choose the best conformed individuals. The parks in which they place them are named *Claires.* These are situated on the banks, at the mouth of the river Seudre, and particularly on the right bank. Each park is enclosed by walls of three feet high at most, and can communicate with the river, or rather with the sea, at high tide twice a month only, by means of a small canal, with a flood-gate. All around, and within the enclosure, is a canal of three feet in depth, for the deposition of the mud. The middle is as smooth and level as the alley of a garden, and care is taken that no plant is left there. It is in this place, about a year after its arrangement, that the oysters are placed, very flat, and carefully insulated from each other; then the water is introduced, which is kept only six inches high, except during great heat, or great cold, when it is raised as much as possible. The oysters remain thus for more than two years, sometimes without being brought to market, and much care is necessary on the part of the superintendents, to change their places, and even to transport them into other parks, which accelerates their greenness. To prevent the deposition of mud, there must be a suitable mixture of salt and fresh water; crabs must not be allowed to introduce themselves into the park. With all these precautions, the green oysters which are obtained are of a very superior quality. As to the cause of this greenness, the people attribute it to a variety of circumstances, to the nature of the soil, to that of the waters, and particularly to the mixture of fresh and salt water; to the influence of the solar action, of the north-east wind; of a moderate temperature, but by no means to any single immediate cause. They are not at all disposed to coincide with the theory of M. Gaillon, above mentioned, against which they have many objections; but it would be too long to enumerate them.

The transportation of the oysters, thus become marketable, still demands some precautions. They must always be placed in their natural position, that is to say, horizontally, the hollow valve underneath, so that they may lose less of the water which bathes their gills. It is equally advantageous that they should be enveloped in fucus, or marine plants, to avoid the drying action of the air. The quicker this transportation takes place the better, especially when the weather is a little warm. But as this mode by land is very expensive, during the winter oysters are brought to Paris in rather large boats, which come from St. Vast, by the Somme. Then they are very cheap at Paris. Some years back, a notion was entertained of no longer sending oysters dry, but in a boat full of sea-water, after the manner of the Romans. But the attempt did not succeed, for so small a quantity of water, containing so many animals, not being capable of being renewed, soon putrified, as sea-water easily does, and caused the death of so great a number of oysters, that the police were forced to order the whole to be thrown into the river.

Oysters of good quality, are in general easy of digestion, but not very nutriticus, especially when they are eaten raw. They rather act as a provocative to appetite, which is occasioned by the agreeably subsaline water which they contain. Persons have been known to eat fifteen or twenty dozen at a meal, without experiencing any inconvenience. It would not have been the same if they had been cooked; they then become harder, more coriaceous, and consequently more indigestible. They are also eaten pickled with vinegar and fine herbs. In this state they are sent into places far remote from the sea, heaped on each other in small barrels without their shells.

If oysters almost always present a nutriment light and agreeable, it sometimes happens, though but rarely, that they produce grievous accidents. It is generally supposed that such accidents are more serious when they are eaten in certain months of the year, when they deposit their spawn, as in June, July, and August, but this is now considered erroneous.

The species of oyster most commonly eaten, is that which, from this very circumstance, has been justly named *Ostrea edulis*, and the one which is found in the greatest abundance. But there are also many others which serve for the nutriment of man. It appears, generally, that the oysters of hot climates are less agreeable than ours, whether this be attributable or no to the species or to the climate.

The oysters have a great number of enemies. It is reported that the crabs, to eat them with security, clap a small stone between their valves when they are partly open, to hinder them from closing. *Credat Judæus.* Among their enemies, there are many that introduce themselves furtively, and suffer themselves to be enclosed in the cavity of the valves. Others pierce them slowly, and all finish by killing the animal and living on it. It has been observed, that the oyster, to defend itself against the first, possesses the faculty of shooting forth very strongly the water which it holds in reserve in its body, and it can retard, and even hinder the action of the second by augmenting, at will, the thickness of the shell, in the place exposed to danger.

Some persons have, proceeding gradually, succeeded in making certain species of the oyster live in fresh-water.

It is not necessary to dwell upon the therapeutic qualities which have been attributed to oysters, because none of them will stand the test of an attentive examination. We shall only mention, that in some places, where limestone is scarce, their shells are employed with great advantage to make, by calcination, a most excellent lime.

The PECTINES are so called from the form of the canaliculations of their shell resembling a little the arrangement of the teeth of a comb. They evidently differ from the oysters in all their anatomical relations. Their habits are pretty similar to those of mussels, with this difference, that they are in general more free. They never sink in the sand, and, on the contrary, are always at the surface of the bottom of the sea, resting partly on the side, like oysters, and, as it would appear, at rather a small distance from the shore. The species which have a *byssus*, must, doubtless, never quit the place where they were born, and deposited in the egg-state ; but the others are said to be susceptible of a very singular species of locomotion, since they can raise themselves in the water, and even at its surface, by agitating the two valves of their shell, pretty nearly as birds do with their wings, and fishes with their fins. The small length of the ligament, its position, its great elasticity, and consequently the trifling separation of the valves, allow us to believe, that these animals can really move by contracting them suddenly on the fluid which they contain, and pushing themselves on in an opposite direction. But it appears by no means probable that a sort of flight can result from this operation. We know, indeed, but little respecting the habits of the pectines. On some sea-coasts the larger species are eaten ; but few, except the poor, have recourse to this kind of food, which is hard and indigestible. For a long time the hollow valve of the larger species of pectines was used in some places by the poor as a dish. It is capable of bearing the fire, and is sometimes adopted as a culinary vessel. The species of this genus are found in all seas, and pretty equally divided.

Passing over the intervening subdivisions as affording little or nothing of popular interest, we come to the genus AVICULA, a species of which the *mytilus margaritiferus* of Linnæus is celebrated as principally producing the fine " Orient pearl." A few short details on this subject cannot fail to be interesting to our readers.

The pearl is a body of a variable volume, and of different forms, composed of layers of the nacreous substance, extremely numerous and compact, which constitute a more or less considerable part of certain species of univalve, or bivalve shells, and which appears to be constantly accidental, and is supposed to be caused by a malady of the animal or its shell. This, however, is far from being an ascertained fact, as, were it so, the disease must extend to the far greater majority of individuals, so much so indeed, that every one is found to be accompanied by a certain proportion of minute particles, which are evidently the pearl in the first stages of formation; hence it may be fairly supposed that they are in some essential degree useful rather than prejudicial to the inhabitant of the shell.

When we treated of the structure of the coquillaceous envelope of the mollusca, we observed how it is produced and thickened by little and little, showing that the whole surface of the skin, which clothes the body, properly so called, is exhaled excessively from thin strata of calcareous molecules, dissolved in an animal mucus, and applied one within another, always outedging a little, from which results the augmentation of the shell, not only in thickness, but in breadth and length. We have likewise seen that the necessary modification in this formation, to produce the nacre, or mother-of-pearl, seems to consist in the calcareous molecules being placed so as to leave very small spaces between them, in which the light is decomposed before it is reflected back to us. Thus the pearls are an animal production, in which chemists recognize a great quantity of carbonate of lime united to an animal matter or mucus. They are capable of being dissolved in all acids stronger than carbonic acid. They are also necessarily formed of strata, which envelop one the other, at least when they are perfect. It has been suggested, moreover, that they always contain a little foreign body, around which the layers are deposited. Athenæus would have it that they are produced in the body of the animal, and compares this production to that of the hydatids in the flesh of a leprous pig. We certainly cannot admit the opinion of Pliny and Dioscorides, that they are a production of dew; nor without hesitation, that of Valentine, who thinks that they are the eggs of the females, an hypothesis, however, much more probable than that of Pliny.

In carefully studying a great number of shells, the interior of which is nacreous, we can easily see that at the places where the movements are most irregular, for instance, in the places where the fibres of the muscles of attachment are inserted in a univalve or bivalve, the nacreous substance is much less smooth, than at the places where the mantle only executes its ordinary movements of retraction and extension. Sometimes they even form pretty well marked swellings, and even sorts of irregular tubercles, as is very obvious in the haliotides. When a shell has received an external shock, considerable enough to cause it to lose substance, or even merely to occasion a depression of no great depth, we find in its interior that the nacreous matter, in depositing at first, has followed the inflexion produced by the depression, and afterwards was necessarily accumulated in this place in a greater quantity than if there had been no irritation, so as to form an irregular tubercle, more or less considerable. This is evidently the origin of at least one genus of pearls; for, once that the parallelism is lost in the deposition of the forming layers, the irritation produced by this sort of foreign or anomalous body, continues and determines this part of the mantle to deposit more nacreous matter, from which results a tubercle more and more thick, and even more and more regular. The original inequalities being necessarily effaced more and more in proportion as the new layers are deposited, the consequence is the production of a pearly mass ; but, that this should form a pearl, properly so called, that is, that the mass should assume a form more or less regular, either globular, oval, or pyriform, is a matter entirely owing to chance. Moreover, in this species of pearl, it is impossible to conceive that there should not always remain, at some part of the pearl, a pedicle more or less narrow, in proportion to the bigness of the swelling, and which is the place where the disturbing cause commences to act. A pearl of this kind cannot, therefore, by any means, be compared to an animal or vegetable excrescence ; for in the latter, the growth, the augmentation, takes place within, whereas the contrary is the case with the pearls.

Another genus of pearls is that, the observation of which we owe to Redi. On opening many pearls, he has constantly found in their interior a foreign body, like a little grain of sand, and in this case the formation is very easily conceived. Nevertheless, considering that it is only the mantle itself which produces the coquillaceous matter, nacreous or not, we must suppose that this grain of sand, which has penetrated by accident into the interior of the shell, being a foreign body, and in relation with a part of the mantle, has produced a point of irritation, and subsequently a continual deposition of layers of nacreous matter, somewhat like calculi in the bladder. Then we perceive why, in this genus of pearls, there should be no pedicle of insertion. They must be equally smooth in their whole circumference, and it is more conceivable why they should be spherical. As to their form, bulk, and even the beauty of their *nacre*, (*water*, as it is sometimes termed) these are matters pretty nearly inexplicable, although, in a certain degree, the first may be referred to the figure of the determining body, the second and third to the vigour and duration of life in the animal. M. de Blainville says on this subject,

"I have attempted several times, in the small pearls which merchants name seed-pearls, to find this little kernel in their interior, in which I succeeded, though the trial took several days, and I was obliged to employ an acid much stronger than the acetic, namely the sulphuric acid. I can even say that it cost me a good deal of trouble, so that we may entertain some doubts respecting the celebrated anecdote of Cleopatra, who, with the intent of expending a sum much greater than Antony had done in his most sumptuous repasts, where he had lavished the riches, and exhausted the luxuries of the East, took a pearl from her ear, of considerable bulk and value, put it into vinegar, where it was supposed to be dissolved, and swallowed it. To effect such a purpose, vinegar must be very strong, and many weeks, perhaps months, would be necessary to complete the dissolution. Some of these small pearls, in which, nevertheless, I could trace no pedicles of attachment, were equally formed of layers, although often not exactly concentric, but without any interior kernel; so that we must necessarily admit a third genus of pearls, which is formed independently of the shell, independently of a foreign body, and which would be but a sort of extravasation of the shelly matter, without doubt, in the mantle, perhaps even in its edge, that is to say, in the part which forms the greatest quantity of calcareous matter. The origin of this, however, may be in some accidental or external irritation produced on the edge of the mantle, or, in fine, in some internal malady of the animal, and in that case, it would be nearly impossible to determine the formation of these pearls. It is not so with the first two genera. We know, in fact, that certain shells may be brought to produce pearls by artificial means. Linnæus has effected this in the shells of the genus *Unio*, in the rivers of Sweden. By piercing the shell while the animal is living, the formation is induced, at the corresponding part of the interior, of a pearly mass, which will present the desired size and form. The Swedish government at first made a secret of this invention, and established artificial pearleries; but at the end of a few years it was obliged to abandon them, the advantages of the undertaking being very far from counterpoising the expence; because, in the number of pearls which were formed, it was very rare to find any of the slightest value."

The inhabitants of India appear to pursue a system pretty nearly analogous to this. We find sometimes in collections, though not often, the large shell which produces the most numerous and the finest pearls, crossed by a thread of brass-wire, in a part of its length, and we see in the interior, some crumplings, or blisters of the nacre, which indicate that pearls might be found there. A fluviatile shell, from China, has been seen penetrated by a brass-wire, rivetted externally like a nail-head, while the part which penetrated inside had a very well formed pearl, soldered, as it were, to its extremity.

Pearls are also said to be obtained by depositing in a living shell, a very small piece of spherical nacre, in a place where the mantle is raised up, and the nacreous paries scratched. This mode is reported to have been tried with success in Finland. It has been even said, that certain Asiatics introduced into shells little works of handicraft, which, in the course of time, are invested with the substance which forms the pearls.

We have already observed, that it is easy to imagine that all the nacreous shells may produce pearls. Thus we find them sometimes in the patellæ, the haliotides, and the pinnæ; but in general, they are more especially produced by the thick bivalve shells, such as the *uniones*, and more particularly by the pearl-mussel, vulgarly called the pearl-oyster, *M. margaritaceus* of Linnæus, and the old conchologists, *avicula margaritacea* of Bruguières, and finally, the *pintadina margaritacea* of Lamarck. In all the countries in which it exists, as in the Gulf of Mexico, (if indeed it be the same species) in the Red Sea, in the Persian Gulf, on the coasts of Japan, and especially in the Indian ocean on the coast of Ceylon, it has given rise to regulated fisheries, and to a very considerable commerce, as it would appear, from time immemorial.

This fine species of shell exists in considerable banks, attached by its *byssus*, or beard, to the submarine rocks, somewhat like the mussels, and as it would appear, constantly at considerable depths. There are many such banks in the Gulf Manaar, island of Ceylon, at Arippo, Condatchy, and Pomparippo. The most considerable is said to occupy a space of twenty miles opposite Condatchy. To prevent the useless destruction of a great number of individuals, the bank is divided, as it were, into regular cuts, pretty nearly like the banks of coral on the coast of Sicily, that is to say, it is divided into seven parts, which are worked successively every year, because it is supposed that these animals, in that space of time, attain the full size of which they are capable, and that if they were left longer, the pearls would become troublesome to the animal, and that it would finally expel them from the shell. Be this as it may, in the commencement of February, when the fishing begins, to end in April, all the boats which are to be employed in it, assemble in the bay of Condatchy, whither they come from different parts of the continent, and from the isle of Manaar. At ten o'clock at night, on a signal given by cannon, the boats set out together, so as to be upon the bank where the fishery takes place, at break of day, which is the hour of commencement. Each bark is manned by twenty men, besides the master, ten of whom are rowers, and ten divers. The latter, who are from infancy habituated to this trade, and the most skilful of whom come from Colang, on the coast of Malabar, and from the island of Manaar, divide themselves into two bands, of five each, which dive and rest alternately. Each man is provided with a net, in the form of a sack, to put the pearl animals in, with a cord, to which a stone is attached to facilitate his descent, and finally with another cord, one extremity of which remains in the bark, and of which he makes use to give notice that he wishes to come up. At the moment when he is about to dive, he takes between the toes of his right foot the cord with the stone, between the others, the net, and seizes the cord of call with his right hand, at the same time that he stops his nostrils with the left. Arriving quickly at the bottom of the water, sometimes to the depth of from four to ten fathoms, he hooks the net to his neck, and works with the right hand in plucking away the shells, with which he fills the net. At the end of two, and sometimes of four, five, or six minutes, which last, however, is very rare, and depends upon the skill of the diver, he gives the signal for ascent, with his cord of call, and is drawn up by the men who remain in the bark. Each diver can repeat this operation fifty times in the same day, bringing up about fifty shells each time, but sometimes the blood will stream from his nose and ears. The fishery continues thus until noon, when a second discharge of cannon recals the barks to the point of departure. There the proprietors of the fishery, or the government, if it has reserved to itself the right, cause the shells to be deposited in pits of one or two feet in depth, or on mats, in square places, surrounded by palisades. At the end of some time, when the animals are dead, which is judged of by the opening of the shell, they search attentively in the latter, and in the animal itself, that is, in the lobes the mantle, sometimes even by boiling, for the free pearls, which may be found there. They choose, moreover, the finest shells, most proper for the production of nacre, and leave the rest. In spite of the pestilential exhalations which result from such a considerable mass of dead mollusca, the poor of the country come afterwards to glean what the rich by chance may have left behind.

The free pearls thus obtained are then sifted and selected carefully, and even drilled, and strung by negro workmen, who are extremely adroit in this sort of operation. As to the adherent pearls, they must be first detached and then rounded, and polished at the place of their adherence, which is likewise done in the same country, by means of a powder, furnished by the pearls themselves.

The commerce of pearls appears to be of the highest antiquity. History, in fact, apprizes us, that from time immemorial, the princes and princesses of the East have sought after this kind of ornament with a sort of passion, and employed it in all parts of their dress, and even in instruments, furniture, &c. The Persians, according to the report of Athenæus, paid for pearls with their weight in gold. The pearl mussels, therefore, must be like our common mussels, which, in spite of the prodigious quantity which have been eaten for so many ages, do not appear to suffer any sensible diminution. It is, nevertheless, said that the Dutch having fished too often on the coasts of Ceylon, the pearl fishery is not now so lucrative in the hands of our government as it was formerly.

We should hope that the facility for this business, given by the invention of the diving bell, may not contribute to hasten the destruction of those pearl-banks, but that it will be rather employed to regulate the fishery still better than has hitherto been done.

In Europe also a certain number of pearls are derived from the *Unio margaritifera*, a large species, which is found in the great rivers of the north, and in many lakes. Those of the Tay, in Scotland, are in tolerable estimation.

From what we have said concerning the origin and nature of pearls, it is evident that it must be very rare to find any which possess all the requisite qualities, that is, great regularity in form, whether that be round, oval, or pear-shaped; a fine water, or a tint white, lively, with brilliant reflections, similar to those of the opal, which is called a fine *Orient*, and finally, a considerable bulk. Accordingly, those which combine all these qualities are of an excessive price, and it requires some art in the jeweller to unite, in the formation of a necklace, or any other ornament, pearls which are well matched in size, and still more in tint or colour.

The irregular pearls are termed, in commercial language, *uneven*, and those which are extremely small, and seldom very regular, are called *seed pearls*.

The pearls of Europe, or the *Union* pearls, are sometimes very fine and rather large. But the majority have little or no brilliancy. They are found of all forms.

Pearls are of different colours; most generally they are white and nacreous, but sometimes they are yellowish, or greenish, or even livid or lead-coloured. Tavernier informs us that he had six in his possession that were jet-black. It is supposed that these colours are owing either to the pearl having remained too long in the putrified animal, or perhaps to the latter having lived in mud which was more or less fetid.

The jewellers sometimes make very large ones, but which are composed of two knobs of pearl applied one upon the other; these are only hemispherical tubercles, rounded, and which have been taken from the interior of a pearly shell.

Pearls have the great disadvantage of changing, and being deteriorated by time. They more especially alter quickly when worn immediately upon the skin. They tarnish and lose their brilliancy. It has been proposed, for the purpose of restoring them, to cause them to be swallowed by pigeons. Supposing that the art of jewellery possessed no other means than this, attention should be paid to prevent them from remaining too long in the crop of these birds, otherwise they would soon diminish in bulk. Redi, who made this experiment, informs us, that having caused a pigeon to swallow twelve grains of pearl, they diminished one third in weight. This author also relates, that on the opening of the tomb, where the daughters of Stilicho had been interred, with all their ornaments, eleven hundred and eighteen years before, every thing was found in high preservation, except the pearls, which were so brittle as to be very easily crushed with the finger.

The genus PINNA approximates very much to that of the mussels, the only difference of any import being in the thickness and fineness of the byssus. In manners and habits they are also very similar. They live, as it would appear, constantly fixed by their byssus, in a vertical position, the summit of the shell being undermost, and the base, or posterior extremity, uppermost. But it is particularly in a sandy, or even muddy bottom, that they fix themselves in considerable troops, and by attaching the filaments of the byssus to surrounding bodies, and even to grains of sand, so as easily to resist the movements of the sea. The ancients relate many things respecting these animals, which have not been confirmed by modern observations; among others, they tell us that these mollusca have many enemies, the presence of which is indicated to them by a small crustaceous animal, thence named *pinnothera*, and which shelters itself in their shell. The most common species exists in certain places in the Mediterranean, at the depth of from five to six fathoms. The inhabitants of Sicily and Calabria seek after them, not merely for eating, like the mussels, but also to gather their byssus, of which is manufactured in certain places, stuffs remarkable for their suppleness and warmth. The fishermen, to procure these pinnæ, make use of a large kind of iron rake, called a *cramp*, the teeth of which are a foot long, and the handle proportioned to the depth of the water, where the shell-fish are found. By dragging the rake strongly along, either by force of arm, or with the assistance of the motion of the bark, which carries the fishers, the pinna is torn away, the filaments of the byssus breaking in some point of their length, nothing then remains but to cut at their origin such as are long enough, and to spin them when they have been dried, to form tissues of different kinds, such as gloves, stockings, caps, and even larger clothing. The filaments of this byssus being extremely fine, of a perfect equality of diameter through their whole extent, of very great strength, and of a very brilliant and unalterable reddish brown colour, they produce a stuff extremely supple, smooth, warm, and solid, the colour of which never changes. The ancients were acquainted with this sort of stuff, and it is still made in certain parts of Calabria and Sicily; but its great dearness, resulting from the number of pinnæ marinæ which are necessary to make even a pair of gloves for example, has caused it to be scarcely any thing more than an object of curiosity, and the number of families that devote themselves to this sort of industry is diminishing every day. Were it an object to continue this manufacture, the best mode, perhaps, (and it has been proposed) would be to form depôts for the pinnæ, as has been done both for oysters and mussels. By placing them in favourable circumstances, they might be made to multiply prodigiously, and they might be gathered when they had attained the requisite size.

The pinnæ are found in all the seas of warm climates, and in the Mediterranean; one alone has been found in the British Channel.

We now come to those well known animals the MUSSELS, MYTILUS, *L.*

The mussels do not appear to possess a greater general or special sensibility than the other testaceous acephala; perhaps, indeed, their sense of touch may be less fine, in consequence of the absence of tentacular filaments at the edges of the mantle.

Their locomotion is nothing, or very trifling, according to some observers, who even assert that the mussel never totally changes place, and that the linguiform appendage of their abdominal mass serves only to spin the different threads of the byssus, or to place or fix these threads on submerged bodies. According to Reaumur, on the contrary, the mussel can change place when it has been accidentally detached by the cutting of the fibres of the byssus. He tells us, that in the saline marshes on the coasts of the ocean, where the fishermen throw the mussels at hazard, they are found at the end of some time, united in packets. By putting them into glass vessels, he observed that their mode of locomotion consisted in drawing their linguiform appendage out of the shell, curving it, hooking it to some bodies, and thus drawing themselves towards the point of attachment. It is certain, that in ordinary circumstances, the mussel does not change place, fixed as it is by means of a greater or less number of its fibres, to all the surrounding bodies of whatsoever nature they may be. When we examine how this fixedness takes place, we find that it is by agglutination, each fibre being often a little widened at its extremity. The linguiform appendage of the foot may also conduce to a similar purpose, and this organ, (as has been ascertained by experiment) when cut off, will sprout again.

The mussels undoubtedly feed upon very small animals, or their spawn, as is proved by the property which they acquire of being poisonous, when they have eaten that of the asteriæ.

It is certain that the mussels are hermaphrodites, like the others of the same tribe; that is, all the individuals are similar, or a single one constitutes the species.

The female product of generation does not come from the mother in a perfect state. It is rejected in the form of a gelatinous substance, in which are contained the germs of the young mussels. The latter, no larger as yet than a grain of millet, already have their byssus, which, most likely, is born with them, and which, probably, serves to attach them, with the assistance of the linguiform appendage of the mother.

The species of this genus live in assemblages, more or less numerous, usually placed in a compact manner, one against the other, fixed more or less solidly by their byssus, in an oblique situation, the top of the shell being undermost and behind ; the base, or broadest part of the shell, is uppermost, the two valves being a little open. Through this semi-aperture, the fringe which edges the mantle comes out behind, and the filaments of the byssus pass through the emargination of the ventral edge.

Some species are thus placed at the superficies of bodies ; others, in preference, seek the excavations which may exist in such bodies. Finally, some species excavate a lodge for themselves, like the other lithophagous bivalves. It also appears that some live sunk in mud, like the pinnæ. It is probable that they have a thicker byssus, contrary to the lithodomous species, in which it is very small, and even does not exist except in youth. As their shell closes exactly, they can very well support the alternation of the flux and reflux upon our coasts, and thus live during six hours out of the water, but in general they are more constantly submerged.

Mussels are almost always found in salt, or at least in

brackish waters. According to the observations of Adanson, some species can remain six months of the year in salt water, and the other six months in fresh. Moreover, it appears certain that there are true mussels constantly existing in river waters. One species, in fact, belongs to the Danube, and another to the lakes of North America. M. Beudant succeeded in causing the common mussel to live in water altogether fresh, by taking suitable precautions.

Mussels are to be found in all the zones of the earth, whether frigid, temperate, or torrid. The identity of our mussel with the *mus* of Aristotle, is allowed by all naturalists and commentators.

The flesh of mussels, which is a sufficiently agreeable food, though less so than that of oysters and certain species of *Venus*, very often produces, not only serious, but even very alarming symptoms of illness. This would appear to take place more in certain places, and at certain times of the year than others. The symptoms produced by the hurtful qualities of cooked mussels, are said to be an uneasiness, and universal numbness which seize the body in about three or four hours after the repast. These are followed by a constriction in the throat, a sensation of heat and swelling in the whole head, and especially in the eyes, an inextinguishable thirst, nausea, and sometimes vomiting. If the patient has not the good fortune to vomit up altogether, or in part, the mussels, the constriction of the throat, the swelling of the face, of the lips, of the eyes, and of the tongue, increase to such a degree as to render speech difficult. The colour of these parts becomes so red, that they seem to be excoriated; it extends externally, first to the face, then to the neck, the breast, the belly, and in fine, over the whole body. This eruption is the most characteristic symptom of the malady. It is constantly accompanied with delirium, a singular degree of inquietude, an insupportable itching, and sometimes with a great difficulty of respiration, as well as an extreme stiffness, as in catalepsy. It cannot be compared to any other cutaneous eruption. Although the skin is redder than in cutaneous complaints generally, it is thickly sown with points of a still deeper red, which are infinitely smaller than a grain of millet, and which, viewed with the microscope, appear distinctly to be the apertures, or pores of the skin, leaving the subjacent tissue quite uncovered.

Sometimes, this malady is accompanied with nervous phenomena, such as convulsions, spasms, and insupportable pain. At other times the inflammation of the throat is so great that gangrene supervenes.

Notwithstanding the alarming character of these symptoms, they are not so formidable as might be supposed, and if suitable remedies are administered, the cure will be effected in about three or four hours, though the numbness will sometimes last for several days. There have been instances of persons who have suffered horribly for three or four days, and even of some patients who died of this malady.

The cause of this singular disorder has been attributed to the orange colour of the mussels, to their putrefaction, to their leanness, to the phases of the moon, to a particular disease of the animal, to the little animals which introduce themselves between its valves, and particularly to a small species of crab of the genus pinnothera. But it would seem that all these suppositions are erroneous; at least we are told by M. de Beunie that the mussel never produces this effect but when it feeds upon the spawn of the asteriæ. This spawn, observed in the microscope, appears at first to be nothing but a dead and formless mass of jelly; but after a few days of hot weather, it appears living, and filled with animalcules, which become developed and metamorphosed into little asteriæ. It is from the end of April, or the beginning of May, to the middle of July, or the commencement of

August, that the asteriæ spawn, which explains sufficiently a vulgar opinion, that mussels are poisonous only during the months into the name of which an *r* does not enter. This spawn is so poisonous, so caustic, that, according to M. de Beunie, it immediately causes swelling and inflammation, with insupportable itching in the hand of the person who touches it, and stiffens this part to such a degree, that it appears ready to fall into a gangrene. But this accident is followed by no evil consequences, especially if the part be rubbed with vinegar.

It is not merely to man and quadrupeds that this spawn is deleterious, it is also hurtful to some fishes, and among others, to sturgeon, salmon, &c. The little star-fish themselves are equally poisonous, according to the experiments of M. de Beunie. Several of these animals, raw, were wrapped up in meat, and given to dogs and cats; these animals either died in consequence, or were seriously ill, except when they were made to swallow a quantity of vinegar, or when the star-fish were cooked.

From all these considerations, M. de Beunie thinks that the mussels owe their injurious quality to the spawn of the asteriæ, which is very abundant during the months of May, June, July, and August, on the mussel-banks which are on the coasts of Flanders, and, in fact, it was at this period alone, that he witnessed the malady in question, especially at Antwerp, where it appears more frequent, since every person there, even down to infants of three years old, eats raw mussels. He thinks, indeed, that cooking will remove this dangerous property of these mollusca, which, however, unhappily, does not appear to be the fact; at least M. Durondeau, a physician of the same country, relates that he has seen this malady constantly produced at Brussels by feeding on these animals. He even cites examples of it in the months of April and September, and in fact, in all the rest of the year; so that the opinion of M. de Beunie does not appear to be divested of all shade of doubt. It appears that this accident is more common in cold and humid countries, than in dry and warm climates, and more prevalent in Flanders than elsewhere. We know of no example of it reported by travellers on the coasts of the Mediterranean. This may certainly be referrible to the mussel-banks of Flanders being more in relation with the asteriæ and their spawn, as being less deeply situated in the sea. It must also depend a little on the idiosyncrasy, or individual constitution, since, among many individuals who have eaten of the same dish of mussels, and pretty nearly in the same quantity, some experienced very serious effects, while others suffered no inconvenience whatever. The curative means, however, are very simple. They consist in making the patient vomit, with the assistance of ipecacuanha, and then, after having recourse to a general bleeding, to make him drink plentifully of some refreshing diluent, and three ounces of vinegar, a little diluted with water, every hour. Vinegar appears to be essentially the antidote to this poison. Accordingly, all persons who have observed the effect, agree that raw mussels are more dangerous than cooked, but that they seldom cause any accidents, when, in either of those states, they are seasoned with vinegar alone, or vinegar mixed with pepper.

The mussels being a source of nutriment to the human species, means have been sought to cause them to multiply, and impart to them some qualities which they do not possess habitually.

The fishing of mussels presents no great difficulty. Women and children are usually employed in it, on the north coasts of France. An indifferent knife is sufficient for the purpose, and they gather the mussels, breaking the filaments of the byssus, which attach them to submerged bodies, or to one another. In places where the banks of mussels are open to sea, they are seldom very fine. They are much larger, and even of a more delicate taste, when they come from banks which are not uncovered, except at the great monthly or annual tides. Notwithstanding the great destruction which is made of them, the multiplication of mussels is so considerable, that no sensible diminution is observable on the coasts above mentioned, and especially none in those which are not continually explored.

On the coasts of the ocean more art and industry are exercised than on those of the British Channel. There the mussels are *parked* pretty nearly in the same manner as we have already described the oysters to be. As it happens with the latter genus of mollusca, it appears that the mussels are rendered more tender, and the quality of their flesh is improved, by putting them in places where the saltness of the sea-water is tempered by rain or river-water. Pliny has made the same observation respecting that species of bivalve which he names *myas*, in saying that it was better in autumn, because a greater quantity of rain is then mingled with the sea-water. Accordingly, on the coasts of the Atlantic, the fishermen throw into the saline marshes the mussels which they have taken in the sea. In the port of Tarentum, in the kingdom of Naples, in the month of March, long poles are sunk into the mud, on which the spawn of the mussels is fixed. In the month of August, a period at which they are as large as almonds, the poles are transported to the mouth of those streams, which fall into the gulf. In October, they are replaced in the port, and it is only in spring that they are eaten, although they have not yet arrived at their full growth. In the neighbourhood of Rochelle, the mussels fished in the sea, are deposited in sorts of ditches or ponds, to which they give the name of *bouchots* *, and in which the salt-water is

* The literal meaning of the word *bouchot*, is a fishing-hurdle, or crawl.

stagnant, and where a greater or less quantity of fresh-water can be introduced. These *bouchots* are formed with two ranks of stakes, intertwisted with poles, and united so as to form an angle, the summit of which is opposite to the sea. They are placed on a muddy bottom of great depth at the mouth of the Seine, and to the west of Aunis. The mussels which are there attached, deposit their spawn, which is sheltered in the branches of a sort of coralline which abounds very much on the wood of the bouchots. At the end of some months, part of the mussels are detached from among those which are too much accumulated, and they are distributed in such places as were unfurnished with them. To facilitate their adherence, they are carefully engaged in the hurdle-work, and even, for greater precaution, they are enveloped with a net, without which they would be soon carried off by the waves. The mussels multiply in these bouchots in the proportion of ten to one in the same year. They are gathered from the end of July, for more than six months, either at low water, or by the assistance of a machine named *acon* *. The product of these bouchots would be considerable, were it not for the expense of repairing the wood, which has been gnawed by the teredo since 1720, at which time this species of mollusca was introduced by a vessel wrecked upon the coast.

Besides the human species, mussels have a great number of other enemies; many sea-birds detach them from their situation, by breaking the shell, and then feeding on them. Many cephalous mollusca, and among others the *turbo littoralis*, according to the observation of Reaumur, pierce the shell with their proboscis, and suck the soft parts.

The *fresh-water mussels*, (ANODONTES) are, some of them, very large, particularly the species called *mytilus cygneus*. The heart of this animal is remarkably large, and the foot

* *Acon*, or *accon*, signifies a sort of flat-bottomed boat.

composed of three layers of fibres, directed according to length, breadth, and depth, imparts to the animal the faculty of changing form in these three dimensions. The anodontes are viviparous. Thousands of young ones are frequently found, in winter only, between the plates of each gill. They may be seen distinctly with a good microscope, opening and closing their shell. All the anodontes are fluviatile. When they want to walk, they place their shell level, put out their foot, and crawl something like snails. It has been said that they swim by striking the water with their valves. They are most commonly found sunk in mud, the aperture of the valves, and the obtuse extremity which corresponds to the mouth, directed upwards. They may serve as food, but are in no kind of estimation.

There is nothing to induce us to mention the UNIO here, except the circumstance of some of the species producing pearls, on which we have already sufficiently enlarged. The species are very numerous.

The mollusca of the genus CHAMA live attached to rocks, like the oysters, by one of their valves, also to corals, and even to other shell-animals. This often injures their development, and renders it more or less irregular. They are frequently found in numerous groups, pressed one upon the other. They are thus cemented so strongly, that they cannot be detached without breaking the shell. The adherent valve assumes, like that of the anouriæ, and oysters, to a certain point the form of the bodies on which it reposes. It is usually rather irregular, and much less coloured than the other. From this mode of existence, we may easily conceive that the chamæ make but little use of their foot.

Almost all the CARDIA, to which we now pass, live in the sea; a very small number of species inhabit the fresh-water. Some of them remain remote from the coasts: but the majority prefer sandy shores, where they remain concealed in the sand at the depth of three or four inches. They come forth from this retreat, re-enter it, raise themselves briskly, advance or recede, by means of their long fleshy pedicle, which is moveable in all ways, and capable of contraction and elongation, in a variety of directions. If they are desirous of sinking in the sand or mud, they elongate it, and make it penetrate as much forward as they can, hook themselves by its extremity, which they curve, then shorten it, and force the shell to approach its point, cutting the sand with the edge. To cause it to spring back, they curve it into an arch, then straighten it quickly, and thus raise the whole body with agility. The same manœuvre enables them to raise themselves above the soil with a sort of jump. It is easy, therefore, to conceive how they can advance or recede by similar operations.

The cardia are found in all known seas. The species are numerous, and the individuals exceedingly multiplied and spread in very different latitudes. In many countries of Europe, such as Italy, England, Holland, and the coasts of France, a very great quantity of them are consumed. Their flesh may not be very delicate, but it is, nevertheless, wholesome and agreeable enough. Their abundance causes them to be very cheap.

The *Cardium rusticum* is frequent on the maritime coasts of the two Sicilies, where it lives sunk in sand, covered by about fifteen feet of water. The fishery, which is prohibited in summer, is made with an iron rake, with which the sand is worked to discover these animals. But, though the flesh is pretty good, none but the common people eat it. It is dressed with oil, crumbs of bread, pepper, and aromatic herbs. This species is equally found in the European Atlantic.

The *Cardium edule*, which is our common *cockle*, is fished up in immense quantities during winter, on the coasts of England, Ireland, and Holland, where they are used as food, during that season. According to Pali, those which are consumed in Naples, are found in great abundance, sunk in the mud of the lake Fusoza, which is partly fed by the waters of the sea. Their flesh is good, but not very savoury. The foot, if we are to believe Pali, is not of the same colour all months of the year. In October it is whitish, in December and January it grows yellow, and afterwards becomes a fine vermillion.

The CYCLADES are constantly found in fresh-water, and have the habits and manners of *Venus*, of which we shall treat presently. They walk by means of the abdominal appendage, which is sometimes formed of a widened base, from which arises an elongated, flagelliform appendage. They live in mud, but not in a vertical position, and put forth their tubes through the posterior part of the shell, a little open.

The mollusca of the genus VENUS are animals found in all parts of the world. They constantly live on the sea coasts, sunk in the sand, but at no great depth, so that they issue forth easily, and can walk very well with the assistance of their foot. It is even said that they can jump, and as it were vault, striking the interior of the water repeatedly with their valves. This same faculty has caused Pali to name the whole class of the bivalves, *subsilientia;* but all are very far from possessing it.

Very little is known concerning the habits and manners of the venus, which cannot differ much from those of other bivalves.

At different sea-ports these animals are eaten, and esteemed a delicacy, instead of, and indeed preferably to, oysters. The taste is stronger than that of oysters, and one must be habituated to it to relish it.

The mollusca called SOLEN, all live at but small distances from the shore, sunk vertically in the sand, the mouth underneath, and the anus upwards. The holes which they make there are never lined with a calcareous deposit, as in certain neighbouring genera, the reason of which may be very well conceived, as the mantle is entirely covered by the shell. The movements of the solens are usually confined to an ascent or descent in their hole, which is sometimes two feet in depth. This movement is doubtless produced by the action of the foot, which becomes attenuated at its extremity for the purpose of descending, or being widened, assumes a resting point from which to ascend, and cause the tube, and even a part of the shell, to pass the orifice of the hole, at the surface of the sand, and to rise more or less in the water which covers it. It does not seem probable that the animal ever comes forth altogether from this retreat of its own accord, though we can very well conceive it possible that it should do so. But it is certain, according to the observations of Reaumur and Adanson, that if by any cause it has been drawn out of it, it can enter there again. The first of these observers has described the mode in which the animal proceeds for this purpose. Curving and sinking the extremity of its foot, it commences by raising its shell more or less obliquely to the horizon; then a new impulse, by straightening the foot, begins the sinking of the shell, at the same time that it makes a still shorter angle with the horizon. The same action renders it vertical and a little sunk. Then it extends its foot as directly as possible, and giving it the form of an angle, by then drawing it back, the shell to which it is attached descends. By repeating these motions, it sinks very quickly. The ascent, on the contrary, is made by drawing back the foot strongly, and widening it much; the resisting point is on the swell of the foot, and the motion takes place in the shell upwards.

We know little more concerning the natural history of the solens. Aristotle, however, tells us, that these animals seem to hear, when a noise is made near them. The only ground for this opinlon is, that if a sudden and somewhat loud noise be made near them, they will sink down into their holes. But this may proceed simply from the immediate shock upon the water, and on the cirri which terminate their tubes.

We are ignorant as to how the solens reproduce, and how their germs, or eggs, are placed by the mother. Aristotle maintains that they reproduce in the sand, which is conceivable, if he means that the eggs are deposited, at a very small depth, in the sand itself. The distinction found in Pliny, and subsequently in Rondelet, and other writers of the same æra, of male and female solens, does not rest upon any thing positive.

The ancient authors, and Pliny among others, tell us, that the solens are essentially phosphoric. But this was, doubtless, because they comprehended under this name, animals of the genera Pholas and Lithodoma, for Reaumur does not say that the true solens possess this property.

The inhabitants of the coasts where the species of this genus are to be found, proceed in search of them, either for food, which, however, is only the case with the poorest sort of people, or to use them as baits for the catching of certain fish. It is when the sea is considerably withdrawn, especially after high tides, that they are able to procure them in great abundance, and with more facility. They recognize the place where any are to be found, by a transverse aperture, widened at each extremity in the form of a key-hole, above the hole which they inhabit. To draw them out, which is often rather difficult, the animal being sometimes sunk very deeply, they throw some pinches of salt into the hole. The salt produces such an irritating effect on the extremity of the tube of the animal, that it immediately ascends out of the hole to get rid of it; then it is seized, but still some address, and quickness, more especially, are necessary, without which the animal would re-enter as rapidly as it came forth, and fresh pinches of salt would no longer produce the same effect as the first. The animal, warned by the danger it had escaped, would prefer suffering the irritating action of the salt, to the certainty of being taken. Then the fisherman is obliged to have recourse to a long iron crook, which he sinks pretty deeply, and drawing it out obliquely, carries away the sand and the solen contained in it. In some places an iron rod is employed for the same purpose, terminated by a conical button. The solens are the inhabitants of almost all known seas.

The organization of the PHOLADES has scarcely any thing to distinguish it from that of others of the same family, but the particular disposition of the mantle, which is closed in almost its entire extent, except in front, and underneath, where it presents a very small and oval cleft, for the passage of the foot. They are all marine, and inhabitants of the shores. It appears, however, that they can live in fresh-water, for Adanson tells us that he has found them in the Niger, at a height to which the sea does not ascend. They live constantly buried, with the mouth and foot under, and the tubes upwards, in argillaceous soils, or in calcareous stone, so that all their locomotion consists in mounting or descending in their hole, that their tube may reach the water in which they are immersed, a little above the bottom. They probably, too, excavate their lodge. They are of the number of terebrant, or *lithophagous* mollusca, which last expression, though almost consecrated by the usage of naturalists, is, nevertheless, erroneous.

This is a convenient place to notice the opinions of writers, in explanation of the modes by which these animals perforate the substances in which they lodge. The complex denomination of *lithophagous*, which signifies *stone-eating*, is employed in the natural history of several molluscous animals, to designate the habit which they have of living more or less deeply in the interior of stones and rocks, and not because they actually feed upon them. Lithophagous species are found in almost all the families of the bivalves. The majority live in our European seas, and especially in the waters of the Mediterranean; yet, notwithstanding the facility of observation, we are still ignorant of the process pursued by these lithophagous animals, to penetrate thus into the interior of the stones. Some persons have thought that it was only when the stones were soft that this could be done, because, in fact, the pholades are found in a sort of white soft argile, which has been regarded as a sort of incipient stone. But they are also found in the true lime-stone, more yielding and softer, no doubt under the water, than when it is exposed to the air. This opinion was supported by Reaumur, and by Lafaille, of the Academy of Rochelle. M. Fleurian de Bellevue, who made his observations in the same places as the last mentioned naturalist, was convinced that the pholades, however small they may be, pierce the calcareous stone itself; and M. de Blainville has seen, on the coasts of Normandy, the same species of pholas in the clayey depositions of the mouth of the Seine, and in the tolerably hard calcareous mass of the chalk formation, which borders the sea for a great portion of its extent. Moreover, pholades and lithophagous mussels are sometimes found in marble on the coasts of the Mediterranean. The direction which the lithophagous mollusca take, in the substance where they conceal themselves, varies according to the genera. The pholades place themselves vertically; but not so the Saxicava, and some approximating genera; these animals pierce the stone in all directions, so as sometimes to meet one another. If we should admit that the pholades, whose shell is pretty thick, and furnished with asperities at its anterior extremity, can excavate their stony lodge, by mechanical means, in turning on themselves, the thing is conceivable enough, because they are free. But this can hardly be the case with others, which fill the cavity almost completely, so as to be unable to move there, an impossibility which is often augmented by a ridge of the stone, which fills the furrow, formed by the crooks of the two valves. When we add that these shells are often very smooth, and that one species is so thin as to be merely membranaceous, we are led to reject every idea of the possibility of motion, whether of rotation or vibration, by means of which these animals could file the stone to introduce themselves. We must, therefore, have recourse to the theory of a corrosive, or solvent fluid, which would act upon the stone, soften it, and convert it into a sort of liquid, which the movement of the foot of the animal would subsequently expel from the cavity. But what is the organ of the animal producing this sort of humour, and what is its nature? It is probable, according to the opinion of M. de Bellevue, that it is the foot, or abdominal appendage, which furnishes the greatest quantity. In the pholades, this part constantly passes the coquillaceous envelope. As to the nature of this liquid, the same observer is induced to think that it must be an acid strong enough to decompose the calcareous salt, of which the shell is formed, but not sufficiently so to attack the animal matter, which also enters into its composition. He has, in fact, observed that when the *rupellariæ*, which pierce the stone in all directions, come to meet, they make an irregular wound, one in the shell of the other, but without destroying the membranaceous part. He has also observed that the pholades in their cavity are bathed in a sort of black slime, tolerably abundant, which even penetrates some distance into the substance of the stone, when the latter is somewhat soft. He has made the same observation on the other lithophagous mollusca, and even on certain worms which also lodge in stones. This black slime appears to be the result of the corrosive humour of the animal mingled with the earthy matter of the stone. M. Fleurian having, besides, observed that the pholades and *modioli* possess the property of shedding a phosphoric light, appears inclined to believe that the fluid which serves the lithophagous mollusca to soften and dissolve the calcareous stone in which they lodge, contains a greater or less quantity of phosphoric acid. Whatever probability may attach to this view of the subject, it is far from being beyond the reach of doubt, the more so, as it would seem, from Spallanzani, that pholades are also lodged in rocks which are not calcareous, as, for instance, in lava. A chemical analysis of this black liquor of the pholades would be highly desirable, to ascertain whether it is acid or not. That it is so, is not very probable, as the patellæ, which excavate pretty deeply the calcareous stone of the coasts of the Channel, where they live, have no trace of acid in the humour which issues from their foot; so that it appears a tenable opinion, that the excavations, more or less deep, formed by the mollusca in stones, are owing to a simple, constant maceration, produced by the mucous fluid which issues from the foot. It is probable that it is the same way with worms, which possess the same faculty; for though, unfortunately, we are as yet but very insufficiently acquainted with them, we may nevertheless presume that their mouth is not armed with organs or instruments by means of which they could act mechanically upon the stone; were it so, they would be no longer worms properly so called, but species of the family of the *nereïdes*, and the problem would be less difficult to solve.

The pholades present another singularity still more inexplicable than their mode of lodging in the stone. This is their phosphorescence. It appears that there are few mollusca which are so luminous; and it is said that persons who eat them raw, and in an obscure or dark place, seem to be swallowing phosphorus. No one appears to have attempted any explanation of this phenomenon. The pholades feed, probably, like others of the family, on small animals, which the water brings to them by penetrating to the mantle. Their reproduction, in all probability, is similar to that of other bivalves. Their eggs, however, must be agglutinated pretty near the parents, if not placed by the parents themselves; for the place occupied by the *pholas candida,* in the horizontal banks of clay, appears to augment in every direction.

Mankind use many species of this genus as food, and particularly on the coasts of the Mediterranean, where the largest are found. It is even probable that the ancient Romans were fonder of them than the moderns, which explains why the columns of the temple of Jupiter Serapis, at Puzzuoli, are pierced by phalades at a level, much superior to the actual level of the sea; and in fact, it appears that it served as a *piscina,* or reservoir for sea-fish, as was first remarked by M. Desmarest, the elder. This goes fairly to overturn all the hypotheses of geologists on this subject, for it seems very probable that the pholades were placed there artificially.

Though the pholades are not considerably numerous in species, it appears, nevertheless, that some exist in all seas. It does not seem, though, that the Australasian expeditions have brought any from that part of the world.

The manners and habits of the TEREDINES (vulgò *ship-worm*) have been studied with considerable care, particularly by the inhabitants of the sea-coasts, who have been forced to construct dykes, to resist the invasions of these animals. It is generally known that they live constantly buried, pretty nearly vertically, with the head down, and the anus upwards, in such pieces of wood as are constantly immerged most frequently in salt-water, but sometimes also in brackish, and even in fresh-water, according to the observations of Adanson. The highest point at which they commence to bury themselves, is always some feet below the lowest waters, so as to be constantly immerged. By means of one of the tubes of their mantle, they let in, as far as the mouth, and traversing the branchial cavity, the fluid which is to serve at once for the purposes of nutrition and respiration, while, through the other come the excrements, and the product of generation. This double movement is produced by the action of certain organs, to which the name of *palettes*, or *palmulæ*, has been given. They vary in form in each species, but are always similar to each other, and symmetrically placed, one at each side of the extremity of the mantle. In the common teredo, each is composed of a calcareous piece in the midst of a membranaceous part. These palettes are opposed to each other at their internal face, and, by being applied together, can close the orifice of the hole which the animal inhabits, after it has withdrawn the tubes of its mantle. These parts therefore serve as kinds of opercula, but they cannot be better compared than to the antennæ of certain male insects. They are, in fact, composed of a considerable number of calcareous articulations. The hole inhabited by the teredo commences with an orifice extremely small, which is often even difficult to perceive. Its beginning is always a little horizontal or oblique, but beyond that point it curves, and becomes more or less vertical or straight. The nature of the wood has a great influence on the regularity, and the flexions of the canal, which is hollowed in its interior. The neighbourhood of other teredines exercises a still more manifest effect upon it. In fact, they seem to endeavour to avoid each other, which sometimes produces very considerable flexions. The depth of the holes is necessarily proportioned to the size of the teredo, and to the duration of its life, which appears to be rather short and rapid. But by what process does this animal, which, very probably, is ovo-viviparous, and rejects the product of generation, already provided with the shell, thus excavate the hardest woods, and that in so very prompt a manner? The form of the shell, so well adapted for boring and filing, its solidity, the nature of the ligneous body, and the perfectly smooth state of the hole which is excavated, will not permit us to doubt that it is done by mechanical action, aided, however, by the maceration of the wood. It appears that the thick adductor muscle, which passes from one valve to the other, is in reality the power which puts the shell in play, supports it at the bottom of the hole, and causes it to turn in the wood like a real gimlet. There is here no necessity to have recourse to the presence of a solvent acid, which we have already observed is so doubtful as to the other terebrating shells. The relation of cause and effect is too evident. In proportion as the little animal grows, it excavates its hole, which it also carpets in proportion as it goes on, with the cretaceous exhalation of the parts of its body not covered by the shell, and there buries itself until it has arrived at its full size.

From the singular habits of the teredines, it is evident that they must be very formidable animals to the human species, for they very considerably hasten the ruin of all constructions in wood, established in or near the sea, such as piles, dykes, moles, barricades, and even vessels. Piles are in general completely untouched by them, as far as they are discoverable at the lowest sea; but from a certain point which is constantly immerged, to a very considerable depth, the holes of the teredines are so numerous, that the thickest piles give way to the slightest efforts. Holland is the country where the results of this are most fatal. Vessels have been known to spring leaks in consequence of the holes made in them by teredines. Means have been devised to obviate such serious accidents, either in the selection of the wood, or in securing it from the attacks of these animals. No European wood is certainly secure against the terebrating action of these mollusca. It has been asserted that there are some American timbers which they will not attack, either in consequence of their hardness, or of some resinous gum which they contain; but what these woods are, and even whether the assertion itself be true, are points not ascertained. It is said that the previous carbonization of piles, to the depth of a few lines, will be sufficient to protect them completely; but for vessels, the best method undoubtedly is coppering the bottoms.

For the great injury done us by the teredines, there appears to be some slight compensation. They are said to constitute an article of food with some of the inhabitants of the coasts of the Atlantic, and their flesh is reported to be very delicate, and of a more agreeable flavour than that of the oysters. It is probable, also, that they may have their use in removing decayed vegetable matter, serving to somewhat the same end at sea, as the *termites*, and many other insects, do by land.

According to Seba, who made his observations in Holland, certain species of nereides are mortal enemies to the teredines, penetrating into their tube and devouring them. The species of teredo appear to exist in all parts of the globe.

We now proceed to the second order of Acephala, or *those without* shells.

The BIPHORES of the text, THALIA SALPA, &c., are curiously constructed animals. These gelatinous, transparent mollusca, sometimes free, and sometimes united in long lines, or cordons, like fire in consequence of their phosphorescent quality, must long ago have attracted the attention of navigators. But it must have been somewhat difficult to form an exact idea of them, in consequence of their remoteness from ordinary conformation. Had they come under the inspection of some ancient naturalist, which is very probable, they would have been placed among *the purgamenta maris*, that is, in the very numerous section, where old writers placed all animals that do not come under the ordinary forms.

The Salpæ float constantly immerged at variable depths in

the interior of the sea; but it appears that whether free or aggregated, they can move, probably, without any determined direction, by means of the water which they cause to enter into their mantle for respiration and nutrition. This water penetrates through an aperture provided with lips and valvules, and issues forth by an opposite one. This alternative action has been sometimes called *systole*, and *diastole*, and its result is, that the body is carried in an inverse direction to that of the rejected water.

We have no knowledge respecting the digestion of these singular animals. It is probable that it must be easy, in consequence of the fluid form in which their nutriment is conveyed. It is not likely that some foreign bodies which are often found in the cavity of their mantle are digested there; for this does not constitute their stomach. The function of respiration takes place by the introduction of water into the cavity of the mantle.

Circulation appears to be somewhat singular. The motions of the heart are made spirally; they take place by a twisting of its parietes, and always begin from one of its extremities. If it be that which touches the *nucleus,* a name given to the little mass of the digestive apparatus, the motion of the blood is made into the aorta, and into its principal ramifications; if it be the other, the march of the fluid proceeds in an opposite direction. These motions of the heart are easily perceived, first pushing the blood in one direction, stopping, contracting, and then impelling it in an opposite direction. We then see this fluid fall back, as it were, by its own proper weight, and assuming a direction opposite to what it had at first. But as the two systems of vessels which issue from the heart communicate together, it happens, after a certain time, that these sorts of oscillations send the blood into all parts of the body.

These motions of the blood are more perceptible, as it is

composed of small whitish lumps, visible through the parietes of the vessels, sometimes of a reddish brown, and transparent. These observations may be still more assisted, by holding the animal vertically, with the nucleus downwards; then, as the blood pushed into the aorta, is forced to ascend against its own weight, its march is much less rapid, and the movement of the globules can be very well followed.

These animals are essentially marine, and are seldom found, except in the high sea. It appears that they exist in all the seas of warm climates, and even in the Mediterranean, on the African coast. Farther northward they do not seem to be found, or if so, they have been driven thither by currents or tempests. They are sometimes found in great abundance in the deep water, at a great distance from the shore, either solitary, or united according to a particular method for each species, so as to form long cords or ribands, which float in a serpentine manner at no great distance from the surface. It is especially during the night that they are most perceptible, in consequence of the phosphoric property with which they are endowed to so high a degree. All navigators are agreed upon this subject, and say that the salpæ, when chained together, produce the effect of long ribands of fire, drawn along by the currents.

Although they really possess the locomotive faculty, it is extremely probable that they are the sport of the waves and winds, which drive them along in their own direction. This appears most likely with regard to the individuals which are linked together.

Their nutriment, without doubt, is entirely animal, and composed of animalculæ, and even of the amorphous matter which is found in such great abundance in the sea-water, which traversing the cavity of their mantle, serves at once for locomotion, nutrition, and respiration. The salpæ are true hermaphrodites.

There are very great singularities in the product of generation, or the offspring of the salpæ. It may be solitary, or united with a great number of individuals similar to itself, the union of which takes place in a constant manner by means of those organs to which the name of suckers, or *spiracula,* has been given.

The single or solitary fœtuses appear considerably to differ from the individual from which they come; so much so, that according to Messrs. Quoy and Gaimard, without being apprized of the fact, one might be disposed to set them down as distinct species. They are suspended in the cavity of the mantle by a sort of cord, which M. Chamisso names the umbilical cord. MM. Quoy and Gaimard also speak of a pedicle attached to a sort of placenta, filled with mucous matter.

M. Chamisso informs us that the species of this genus present themselves under a double form, a race entirely dissimilar to the mother, during the whole course of its life, nevertheless producing young ones all similar to her, so that a salpa which differs equally from its parent, and its own children, resembles its grand-parent, its grand-children, and its own brethren. In both states, the salpa is androgynous, or to speak more correctly, female, and equally viviparous; but in one the product of generation is a solitary animal, and multiparous; in the other, it is a *stirps* composed of individuals united together in a determined manner, and oviparous. Thus each species presents a solitary, and an aggregated race, equally capable of reproduction.

Many animals of this genus produce eggs enchained together, and from each egg comes an animal entirely similar to its parents. But the solitary race, instead of eggs, produce animals thus enchained, from each of which, as from an egg, issues a solitary salpa, similar to the first mother; so that we might say, that the solitary race is an animal, and the enchained race is nothing but a mass of aggregated and living eggs. M. Chamisso discovers some analogy between this singular disposition of the salpæ, and the metamorphoses of batracian reptiles and insects. What seems to confirm this opinion is, that there are extremely important differences between the two forms of the same species of salpa, not only in external conformation, but also in the disposition of the muscles, and the position of the viscera.

In the solitary, and consequently multiparous individuals, the body presents none of the appendages, or protuberances which are proper to produce a junction. The orifices are terminal, the first bilabiate, with unequal lips, and the second truncated. In the aggregated or uniparous races, there are in different parts of the body appendages, protuberances, or spines, by the assistance of which the individuals are linked together in a determined order. All the individuals of one and the same group are perfectly similar in bulk and length, though they may differ much in these respects, in different groups.

The mode of life in the ascidiæ, very much resembles that of oysters, to which they approximate also in internal organization. Most of the species have the habit of grouping together in numbers on the same body, where they remain fixed all their lives. Sometimes the young are fixed and grow on the body of the mother.

The ascidiæ furnish an abundant nutriment to fish, and even man feeds on several species. It is thought that they place themselves, in preference, at a certain elevation on the coasts, to escape becoming the prey of the former. Their only mode of defence consists in shooting out the water contained in their sac, through its two apertures, which is done whenever they are touched. Some of them can squirt this fluid to the height of three feet.

The little animals called BOTRYLLI, were for a long time very imperfectly known. They were considered as belonging to the division of Zoophytes. They are most generally grouped, either in a circle or ellipsis, sometimes pretty regularly; but some are also found which may be said to be scattered, two or three only being together. They have much of the exterior of radiated animals.

In the curious compound genus called PYROSOMA, the organization of each little animal composing the whole, has a very great analogy with that of the salpæ. In fact, we can scarcely be assured even yet, that the pyrosomata may not be young salpæ, not yet adult, and which disunite themselves later. According to the actual state of our information, the pyrosomata form cylindrical masses, more or less elongated, soft, gelatinous and bristling at their surface, with a great number of sorts of spines, or pointed tubercles, a little more hard, or more cartilaginous than the rest. These masses, which float horizontally in the interior of the sea, are, doubtless, abandoned to its movements, and cannot resist them, nor direct themselves. Accordingly, they are never to be met with but in the open sea, and often in troops, composed of a great number of these masses. Nothing can be more brilliant, sparkling, and lively, than the phosphoric light which these animals emit. They often form long trains of fire, from the manner in which the masses are disposed in cordons. But a more singular phenomenon attached to this phosphorescence, is, that the colours vary instantaneously, passing rapidly from the most lively red to the principal tints of the solar spectrum, to the crimson of the morning, to orange, to greenish, to azure-blue, and finally to an opaline yellow, when the mass is, to all appearance, in a state of absolute repose.

SUPPLEMENT

ON THE

FIFTH CLASS OF THE MOLLUSCA.

THE BRACHIOPODA.

In the first genus of these bivalve mollusca, the animal has altogether the form of the shell; that is, it resembles pretty nearly a large claw, pointed at one extremity, widened at the other, which is nearly straight, with a short, obtuse, and middle point. The shell is moderately hollow, and is only curved in the direction of its breadth. It is formed, like all other shells, by imbricated strata from point to base. The two valves are not completely similar. The upper differs from the lower in having a sort of internal pad tolerably long and projecting, which corresponds to an excavation of the latter. At its base are two muscular impressions. The lower is a little larger, and more pointed behind. It gives an attachment to the tube, or ligament, in a small fossette, hollowed at its external face.

The tube is very elastic, transparent, and striated transversely through its entire extent. It adheres to the lower valve by a thinner part. It is hollow in its entire length, and contains in its interior a soft pulpous body of the same form as itself. This does not appear to be contractile.

There is a considerable byssus of the same structure as that of the mussels. It appears to proceed from the adductor muscles, and not from the tube.

The body of the animal exactly fills the two valves of the shell, and is so placed that one valve corresponds to the back, and the other to the belly.

Seen from above, the body presents a posterior or visceral cavity, covered with a very thin, transparent membrane. In raising this from the back, a sort of regular figure is perceptible, anterior and surrounded with branchial laminæ. One altogether similar is on the other side.

The body, properly so called, is comprized between two cutaneous laminæ forming the mantle, whose whole circumference, thicker and more evidently muscular, presents no trace of papillæ or tentacula. This membrane is very thin, and altogether adherent on the mass of the viscera, which it allows to be seen in almost the entire posterior half of the body. In all this half are the fasciculi of muscular fibres, which pass from one valve to the other, and which are five in number. From one of these seems to spring the byssus just mentioned.

Beyond the first pair of muscles, the lobes of the mantle are entirely free, as far as their adherence to the trunk. Their form is altogether that of the extremity of the shell. On their internal face is a disposition evidently branchial.

On raising this part of the mantle from front to back, we find the mouth and tentacular apparatus. The mouth is very small, but quite visible, transverse, and at the extremity of a sort of point or flatted nipple. There are really four tentacula, to the first pair of which organs the name of arms has been given, whence the denomination *brachiopoda* for the class.

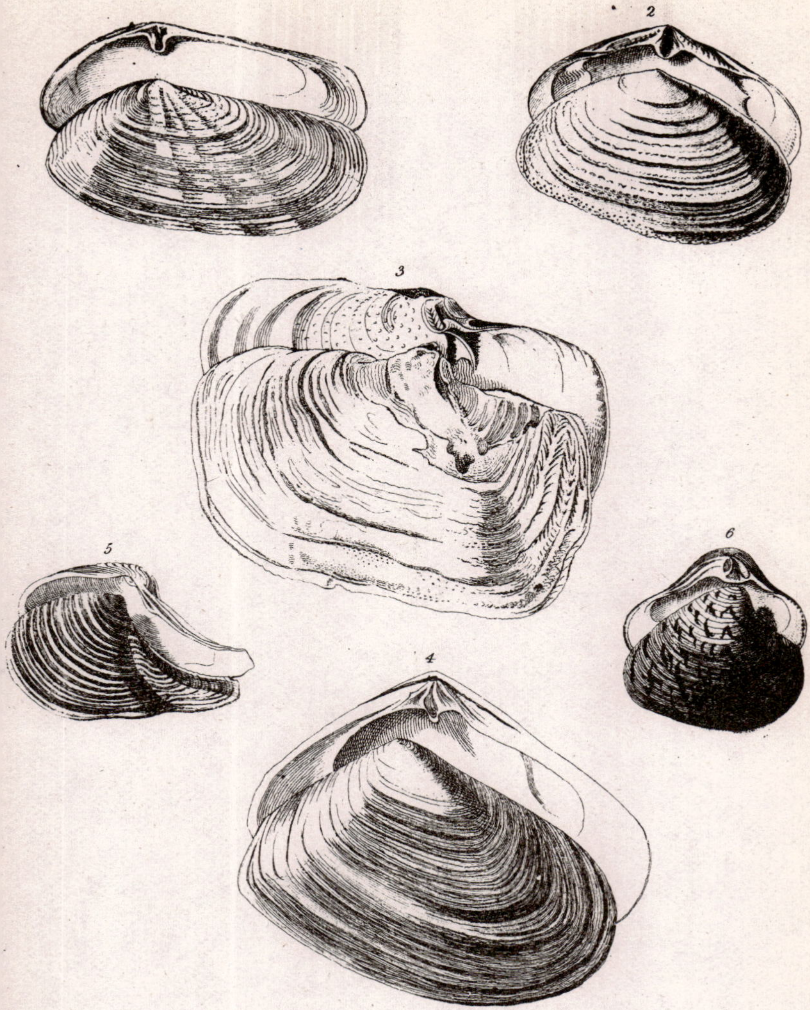

1 *Mesodesma solenoides.* 4 *Mesodesma subtriangulata.*

2 *Mes. denticulata.* 5 *Neæra Chinensis.*

3 *Annatina elliptica.* 6 *Mesodesma ornata.*

In the anterior cavity are two orifices placed symmetrically at the lower face of the upper lobe, in front of the mouth, and even of the tentacula. These two orifices appear similar, one at the right, and the other at the left of a medial canal. It seems not improbable that they are the termination of the organs of generation.

In the visceral cavity, on each side, externally to a granular mass, which occupies all the interval of the muscles, is seen from the point of the shell to the termination of the external pair of muscles, a gelatinous body, pretty large, and thick. This, in all probability, constitutes the ovaria. Another organ, much smaller, is placed at the right side, and formed by kinds of little leaves, joined by a common and longitudinal pedicle. This may or may not belong to the generative system.

The rest of the visceral cavity is filled by a mass subdivided into two, and as if composed of grains, which very probably is the liver, and at one of the sides of which is a part of the intestinal canal, which may be the rectum.

The heart seems to be placed at the middle of the anterior part of the anterior visceral mass immediately behind the middle pair of muscles.

This description of the anatomy of the lingula is taken from M. de Blainville. The Baron's view of its anatomy differs in some points; but it is proper to observe that the animal is but imperfectly known. There is but a single species, which comes from the Indian Ocean, near the Molucca Islands.

The TEREBRATULÆ constitute a genus of mollusca, established, and perfectly defined by Linnæus, under the name of *Anomia*, but into which many unsuitable species were introduced. The shell may be considered horizontal. The valves are equilateral, but the under one is always the smallest. The upper is generally more gibbous, and its constant character is to have its top more or less elongated into a sort of hook. This hook is always perforated by a trench or furrow, proportional to its length, and whose edges, in touching, terminate the canal, which produces at the extremity a hole of a form somewhat variable, whence this genus of shells has received its name of Terebratula.

The shell appears to be doubled in extent by a mantle, which must be very thin, as Pallas describes it under the name of periosteum. As to the body itself of the animal, it is contained altogether in a small part of the shell, pressed against the summits, and sustained by the double forks, when there are any. It appears that the belly is applied upon the complex valve, and the back upwards; for it is not placed in its shell, as is the body of the bivalves generally, one valve on each side. The body of the terebratulæ is depressed or flatted from bottom to top, and perfectly similar on the right and left. What constitutes its greatest part is what is now named its arm, and what were termed gills by Pallas. The mouth is medial, and very large. The intestinal canal, which is probably the stomach, of a compressed and conical form, reposes in the angular sac of the plane valve, where it receives the œsophagus. It is entirely surrounded be a certain black matter, which is the liver.

The gills, or arms, (for authors differ on this point) form a considerable fasciculus on each side, almost altogether behind the abdominal mass. Each is composed, not only of a single triangular ligula, very narrow, pinnate in its entire extent, which, attached by its base on the sides of the mouth, would turn and be free at its extremity, but also of a ligula longitudinally pinnate, wider in the middle, and attenuated at the extremities. This, attached by one extremity on each side, in front of the mouth, after having gone backwards and underneath, reascends upwards, convoluting, directs itself anew towards its origin, and, uniting itself in the middle line with its fellow of the opposite side, forms a vaulted mass. All its external edge is furnished with flatted, triangular, and flexible filaments, longer in the middle, and diminishing as they approach more to each extremity. Thus each branchial mass forms in reality two planes placed one above the other, and which, commencing at the mouth, finish very probably at the anus. These singular organs are very firm and solid, and probably capable, by their movements, of partly opening the shell, and coming out to a certain extent, not, however, in the way of arms, which the animal would agitate externally to attract the water for the purposes of nutrition and respiration.

The muscular apparatus of the terebratulæ is not less remarkable than their respiratory system. All the muscles surround the visceral mass. Some even appear to belong to it, and it is in their interstices that the intestines are situated, while the largest proceed from one valve to another. Pallas describes three pairs, but M. de Blainville could distinguish clearly only two. The first, or most considerable, proceeds from the bottom of the most convex valve, and is inserted near the middle of the anterior edge of the flatted valve. The other two, conical, proceed obliquely into the sinus of the former. They are very probably the fibres of one of the more internal muscles, which issue forth through the orifice of the pierced valve, or rather attach themselves to the membrane which closes its orifice; for it seems probable that the adherence of the terebratulæ takes place by means of this membrane, and not immediately through the muscular fibres.

We may observe, that, though the anatomy of these animals is not completely known, yet, from all that we have now said touching their organization, they seem to hold an intermediate place between the lingula and the ordinary bivalves, perhaps nearer to the latter. In fact, the gills are certainly not attached to the mantle, and the mouth is not provided with extensible labial appendages, as in the former; but also, the gills are not distinct from the labial appendages, and are, moreover, solid, resisting, and perhaps even altogether mobile, instead of being soft and flexible, as in the ordinary bivalves. All this might lead us to form a distinct order or perhaps even a particular class of the terebratulæ.

The definition of this genus may be thus summed up. Body oval, or rounded, compressed, horizontal, provided with a double pair of gills, free, resistent, extensible, turned in the form of a double comb, contained in a regular, dorso-ventral, equilateral, and inequivalve shell; the upper valve in general longer, more gibbous, and lengthened into a top pierced with a hole, or emarginated; the lower, shorter, more flat, provided internally with a system of support, or a very diversiform apophysis; the hinge ginglymoïd, composed of two tubercles, or teeth, more or less separated, for each valve; no ligament; muscular impressions not apparent.

The terebratulæ, which are so extremely abundant in the fossil state, have as yet been found but in a small number in the living, probably because they remain fixed at tolerable depths, to immoveable bodies, and more especially to rocks that are always submerged. Accordingly, we know scarcely any thing concerning their manners, and their organization, but very imperfectly. We know, however, that some species of them exist in all seas. They are, in fact, to be found at the most remote points of the two hemispheres; that is, in the Norwegian seas, and those of New Holland, as well as in the seas of the warmest latitudes. Their habit of living fixed to rocks, and as it would seem, to other bodies, will account for the number of fossils to be found in this group of organized bodies.

The last tribe of this class which we shall notice is that of Orbicula. This was established by M. de Lamarck, on a small shell of the northern seas, of which Müller, and subsequently Gmelin, made a patella; because, not having observed the adherent valve, they took it for a univalve shell. The genus may be thus characterized. Body very much compressed, rounded; the mantle open in its entire circumference; two tentacular ciliated appendages, as in lingula; shell orbicular, very much compressed, inequilateral, or a little irregular, very inequivalve, and without hinge properly so called; one very thin valve, more or less perforated by a fissure; the other patelloïd, with a top more or less inclining towards the posterior side; four muscular impressions on each valve, two of which are larger, and more approximating to the centre; the others more separate, and more anterior.

These animals live in the northern seas, contained in the excavations of rocks, whence they often assume rather an irregular form, or they are often rooted by some fibres of the central pair of adductor muscles, which traverse a fissure of the flat valve.

SUPPLEMENT

ON THE

SIXTH CLASS OF THE MOLLUSCA.

THE CIRROPODA.

THE ANATIFA is a genus distinguished, as well as the Balani, from all the other animals of this division, by articulated and ciliated tentacula, ranged by pairs, to the number of twenty or more, on each side of their body. They were the more deserving of forming a separate order, or class, as they now stand, inasmuch as they approximate in many respects to the articulated animals.

The name of anatifa is an abridgment of anatifera, and owes its origin to a strange fable. Believing some analogy to exist between the most extended species of the anatifa, and the numerous wild ducks, or *bernacles*, which abound on the maritime shores of the west and north of Europe, their inhabitants imagined that it gave birth to these birds, and it therefore received the name of *anatifera*, or bernacle.

The shell of the anatifera has the form of a flatted cone. It is usually composed of five or seven principal valves, and sometimes of a great number of small ones, for the most part triangular, or trapezoïd. These valves are not joined by means of any hinge. They hold together by means of the mantle of the animal, which lines their interior, and opens anteriorly by a longitudinal cleft, as in the bivalves. The base of the shell is united to a fleshy tube more or less long, fixed by its other extremity to the rocks, to the timber of vessels, to fuci, to gorgons, and even to the sandy bottom of maritime coasts. This tube is composed of three strata of fibres, of a circular form in the two external strata, and much harder, and more consistent than in the internal stratum. The fibres of which this last is formed extend in a parallel manner, from one extremity of the tube to the other. It is by its base, or its posterior and inferior part, that the body of the anatifa adheres to the tube. The mouth, placed towards the middle of the belly, is directed forward, when the animal extends itself out of its shell, and upwards when it is folded back. It presents a very remarkable apparatus of organs. Six parallel thin leaflets, of a form almost triangular, denticulated at their lower edge, which is free, surround it superiorly, and on a part of the sides. The two external leaflets are attached a little to another organ, situated at the inferior part of the mouth, hard, very projecting, of a triangular form, and which much resembles a lower jaw. A pointed proboscis, at the base of which a small aperture may be observed, protected on each side by two triangular lamellæ, terminates the upper extremity. Between it and the mouth, are, on each side, five or six trunks, which support an equal number of pairs of tentacula, of a corneous substance, whose length augments in proportion as they are removed from the mouth, composed of numerous articulations, ciliated and curved forwards. The animal puts them forth from its shell and agitates them at every instant. It cannot be doubted, that it endeavours by this manœuvre to attract towards its mouth the alimentary corpuscula. It has an intestinal canal, which runs along the back, and ascends towards the proboscis, and two tubes winding at its sides, which hold the place of testicles. The ovaries are towards the base of the body. We sometimes find between this base and the mantle, an immense quantity of eggs, forming a thick stratum of yellowish matter, which covers the back of the animal. The gills are little conical leaflets, of a soft substance, attached to the base of the tentacula in such a manner, that there are as many pairs of threads as there are pairs of tentacula. But these threads are directed in a contrary way, that is, towards the back, and inclined against the body under the mantle. The nervous system likewise presents some peculiarities. The brain is situated cross-wise on the mouth. Four nerves repair to the muscles which border this organ, and to the stomach. Two other cords embrace the œsophagus, approximate to form a ganglion, and then rise, pressed one against the other between the tentacula, each pair of which receives a nervous filament detached from the principal cord, without forming any enlargement.

We have purposely entered into some detail respecting the organization of the anatifæ, because it presents many peculiarities, when compared with that of the other mollusca. Those concerning the nervous system, and the gills, are taken from the Comparative Anatomy of M. Cuvier.

Fixed by its fleshy tube, and unable to move but by the elongation and contraction of this tube, and by its flexion in all directions, the anatifa presents but little that is remarkable in the history of its manners. The medium in which it is immersed, is filled with substances, which serve it as aliment. This was necessary for its preservation, since it is incapacitated to go in search of food. To obtain it, it is sufficient that the animal should determine towards its mouth, a current of the surrounding fluid, by means of its multiplied arms.

Most of the species live in groups of fifteen or twenty, more or less. They prefer such situations as are beaten by the waves, and when attached to the keel of a vessel they place themselves at some inches from the line of water, and especially near the helm.

The POLLICIPES, which now forms a subgenus of *anatifa*, derives its name, as is said, from the resemblance which the ancients imagined they could perceive between it and the nail of the thumb, or the great toe. A strong, thick, and short tube, of a conical form, shagreened, as it were, by a great number of little valves which cover it, and five large valves, constitute the character of the species on which this subgenus is founded. Its individuals are found united in groups of twenty, or more, of different size, in the Mediterranean, on the coasts of Spain, Normandy, Brittany, &c. They serve for food in many countries, after having been boiled in water. A notion has even been entertained that they are aphrodisiac. Their flesh, according to Rondelet, becomes red by boiling, like that of the lobster.

The *lepas aurita*, or *leporina*, is our author's subgenus OTION. This species has a very remarkable conformation. The tube is dilated into a coriaceous sac, which contains the animal. Five very small valves are, as it were, dispersed over the surface of this sac, which has two appendages formed like an ear at the superior and posterior part. It has been found in several of the European seas.

The BALANI have also been named *sea-acorns*, from some sort of resemblance to the fruit of the oak. They are mollusca which have the greatest analogy with the anatifæ; accordingly Linnæus united them together under the same generic name, notwithstanding the great differences existing in their envelope. We observe in the balani, as in the anatifæ, twelve pairs of articulated tentaculæ, a transparent tube between the bases of the two most elevated pairs; a mouth similarly placed both in one and the other, and also surrounded by analogous organs; and an internal organization absolutely similar in all that is essential. The eggs, as in anatifæ, are placed at certain periods in the folds of the mantle. The latter is formed of an extremely thin membrane, which lines the interior of the shell, and in which wind innumerable vessels. Two appendages of this membrane, placed on each side of the animal, in which is remarked a wide canal, filled with a lacteal humour, which ramifies into a body fringed with a purple colour, form, according to Poli, the gills of the balani. We have some difficulty in believing that they differ so much from those of anatifæ, which we have just described after the Baron Cuvier. Four teeth arranged in pairs, one above the other, and surrounded by an equal number of crustaceous palpi, bristling with setæ, arm the mouth externally. The intestinal canal makes the circumference of the body, reascends behind towards the base of the proboscis, and terminates in this place, by an oval aperture, at the bottom of which the heart may be seen to beat. It is also towards the proboscis that we discover the small end of the testicles, which are two claviform sacs placed on each side of the body. They are continued with a canal which winds into the proboscis, which Poli has observed filled with a whitish and opake fluid, similar to that contained in the testicles, and from which the same fluid trickled, when he compressed the latter. The same author thinks that the fecundation of the eggs takes place out of the body, by means of the proboscis, which inclines downwards, to shed on them the spermatic fluid.

The shell of the balani has, in the majority of species, an oval or rounded form, which gives it some resemblance to an acorn. From thence the generic denomination under which they are described. Six valves, which touch each other by their base, and separate towards their summit, essentially compose this shell. But a single species is known in which there are but three valves. The interval which they leave between these and their summit, is filled with testaceous laminæ, of which the furrows, canaliculations or striæ, are usually in a direction opposite to those of the first. They hold together by the teeth of their edge, which are fastened one in the other, or by a scaly suture. Their structure is sometimes tubular, and very remarkable. They rest, in the greater number of species, on a base of the same nature, more or less thick, flatted, or hollowed, which holds strongly to the bodies on which the animal has the habit of fixing. This base is of a coriaceous nature in a small number; sometimes it appears to be wanting altogether. Finally, an operculum, usually pyramidal, and formed of two or four valves, closes the superior aperture. It is raised, or lowered, by five different muscles, and holds to the valves by a tendinous membrane, ample enough to permit the play of the muscles. The form of the valves, that of the operculum, and of the base itself, vary in the different species. The majority of the balani unite, like the anatifæ, in groups more or less numerous, in which individuals of all sizes are to be met with. Frequently this agglomeration does not allow all of them to be equally developed, and hinders the shell from assuming its natural figure. The *lepas testudinaria*, or balanus of the tortoise, and that which lives parasitically on the whale, are the only species which are isolated. Some of them are to be found in all the known seas, and many are extended through very remote latitudes. Their fecundity is almost beyond what imagination can conceive. They lay their eggs in summer, and the little ones which issue from them are filled at the end of four months, according to the observation of Poli, with similar eggs ready to disclose. Their manœuvres to draw to them their nutriment are the same as those of the anatifæ. They raise their operculum at every moment, put forth their arms, move them with great swiftness, and draw them in with equal velocity. This is always done on the approach of the slightest danger. The ancients were of opinion that they hooked themselves more firmly to the rocks when it was attempted to pluck them away; and the difficulty thus experienced, has served Aristophanes as a comparison to express that of detaching an old woman from a young man, with whom she is in love. The balani serve as food in many places.

We shall conclude by briefly noticing a few of the principal species, just premising, that they constitute subgenera in the text, and will be found named in the notes.

The *lepas balanus* has a conical shell, with six valves, furrowed longitudinally, and marked transversely with fine striæ. The operculum terminates in a curved point. It is of a yellow colour, the valves of a pale rose, and the furrows whitish. These last are united by teeth, which are fastened one in the other; and a similar mode of articulation unites them to the base, which is very thick, and pierced with several ranks of quadrangular cells, which communicate together. Some canals hollowed longitudinally on the internal face of the valves abound with the most superficial of the cells which surround the base. This remarkable structure, which is far from being the same in all the species, has been described with much detail by Poli. The animal presents no particularity. It lives in numerous groups, fixed to rocks, to shells, and other marine bodies, in the Mediterranean and the Atlantic, on the coasts of England, Holland, and Greenland.

The *lepas balanoïdes* has a truncated shell, with six smooth valves, rose colour, marked in their length with purple lines. This species is more rare, says Poli, in the Mediterranean, than the preceding. Fabricius has found it much more frequently on the coasts of Greenland. It is also to be found on our coasts, and those of Holland, united in groups intermixed with those of the preceding species.

The *lepas tintinnabulum* has a shell with six valves, rose

coloured, and spotted with white; the aperture is quadrangular, and the operculum prismatic. The largest individuals are usually an inch and a quarter in height. They are found agglomerated in great numbers on the rocks, on shells, zoophytes, and other marine bodies in the Mediterranean, the European Atlantic, the coasts of the Island of Amboyna, and those of Jamaica. Rumphius relates that the Chinese make of these animals a delicate viand, prepared with salt and vinegar. They are red, and grow white by boiling. Their taste is similar to that of our lobster. This species attaches itself, in preference, to vessels, the course of which it sometimes slackens, by accumulating there in innumerable quantities.

The *lepas spongites*, has a base hollowed in the form of a chalice, the structure of which is spongy. It is pierced with many longitudinal series of pores, and embraced by six triangular valves, of a purple colour, and wrinkled transversely. The least effort is sufficient to disunite them. This balanus remains fixed by its base in the cellules of the softest sponges. It is found in the common sponge of the Mediterranean.

The *lepas diadema* (not the subgenus DIADEMA of the text), is vulgarly termed the whale-louse. The shell is almost conical, with six valves marked externally, with three or four raised ribs forming a sort of cone, in relief, on each valve. The membranous operculum opens anteriorly, and is provided at its posterior part with two small testaceous teeth. A horizontal partition, pierced in the centre, divides the shell into two cavities, one superior, which contains the animal; the other, inferior, divided into eighteen principal cellules, by testaceous lamellæ, which are fixed, as well as the circumference of the valves, on the skin of the whale. Fabricius has seen, in the largest individuals of this species, some small membranous sacs come out through the orifice of the cellules.

Each of these little sacs enclosed an embryo, which perfectly resembled the adult animal. The latter does not appear to differ from other species of balani. It fixes itself, in preference, in the furrows of the chest, and near the pectoral fins of the *balæna boops*, L. Its whiteness, similar to that of ivory, causes it to be easily recognized by the whale-fishers. It withdraws promptly into its shell when it perceives the harpoon, and produces on the skin of the whale a sensation, which, in the opinion of Fabricius, may serve to give it timely warning of the danger with which it is menaced. If this observation, made by an eye witness, be exact, we cannot help thinking it somewhat singular, that an animal provided with good eyes should be warned more tardily of the approach of a body, than one which possesses no means of information but from the touch alone.

FOURTH AND LAST GRAND DIVISION

OF

ANIMALS.

THE ZOOPHYTES, OR ANIMALIA RADIATA,

COMPREHEND a considerable number of beings whose organization, always manifestly more simple than that of the three preceding divisions, also exhibits more gradations, and seems to be constant only in this point, that the parts are disposed round an axis, and on two or several radii, or on two or several lines, proceeding from one pole to the other. The intestinal worms themselves, have at least two tendinous lines, or two nervous filaments, proceeding from a collar around their mouth. Several among them have four suckers round a prominence, in the form of a proboscis; in a word, in spite of some irregularities, and with very few exceptions, (such as the planaria, and most of the infusoria) we always discover some traces of the radiating form, very much marked in the great majority of these animals, and especially in the asteriæ, the echini, the acalephæ, and the innumerable polypi.

The nervous system is never very evident; when any traces of it have been at all discoverable, they were also disposed in radii; but most frequently there is not the slightest appearance of it.

Nor is there ever any true system of circulation. The

holothuriæ have two vascular apparatus; one attached to the intestines, and corresponding to the organs of respiration; the other serving only to fill the organs which occupy the place of feet. This last alone is distinctly seen in the echini and asteriæ. We can see through the gelatinous substance of the medusæ, some canals, more or less complicated, which proceed from the intestinal cavity. All this indicates no possibility of a general circulation; and in the great majority of zoophytes, it is easy to convince one's self that there are no vessels of any kind.

Some genera, such as the holothuriæ, the echini, and several intestinal worms, have a mouth, and an anus, with a distinct intestinal canal; others have an intestinal sac, but with only a single opening, representing both mouth and anus. In the greater number, there is nothing but a cavity, hollowed in the substance itself of the body, which sometimes opens by several suckers. Finally, there are many in which no mouth is perceptible, and which can only be nourished by the absorption of their pores.

The distinction of sex is observable in several intestinal worms. The greater number of the other zoophytes are hermaphrodite and oviparous; many have no genital organs, and are reproduced by buds, or by division.

The composite animals, of which we have already observed some traces among the last of the mollusca, are greatly multiplied in certain orders of the zoophytes, and their aggregations form trunks and expansions of every sort of figure. This circumstance, united to the simplicity of organization in the majority of species, and to that radiating disposition of the organs, which reminds us of the petals of plants, has obtained for them the name of *zoophytes* or *animal plants*, by which these apparent relations only are meant to be indicated; for the zoophytes, possessing sensibility, voluntary motion, and nourished for the most part, on matters which they swallow, or suck, and digest in an interior cavity, are assuredly animals in every sense of the word.

The greater or less degree of complication of the zoophytes has given occasion to their division into classes; but as we are not yet perfectly acquainted with all the parts of their organization, these classes cannot be characterized with as much precision as those of the preceding divisions.

The echini, and asteriæ, to which the spines that they are usually furnished with, have caused Bruguières to apply the name of ECHINODERMATA, have a distinct intestine, floating in a large cavity, and accompanied with several other organs for generation, for respiration, and for a partial circulation. It has been found necessary to unite to them the holothuriæ, which have an analogous internal organization, perhaps even still more complicated, although they have no moveable spines on the skin.

The INTESTINAL WORMS, which form the second class, have no very evident vessels in which a distinct circulation might be carried on, nor separate organs of respiration; their body is, in general, elongated or depressed, and their organs disposed longitudinally. The differences in their nutritive system will probably cause them one day to be divided into two classes, which we already indicate by establishing two orders. In fact, in some there is an alimentary canal suspended in a true abdominal cavity, which is wanting in the others.

The third class comprehends the ACALEPHÆ, or *sea-nettles*. They also have neither true circulating vessels, nor organs of respiration. Their form is generally circular, and radiating, and the mouth almost always serves as anus. They differ from the polypi only in a greater development in the tissue of their organs. The hydrostatic acalephæ, which we leave at the end of this class, may, probably, at some future time, constitute a separate one when they shall be better known; but it is only by conjecture that we can judge of the functions of their singular organs.

The POLYPI, which compose the fourth class, are all those little gelatinous animals, whose mouth, surrounded with tentacula, conducts into a stomach, sometimes simple, sometimes followed by intestines in the form of vessels. It is in this class that those innumerable composite animals are found, with a fixed and solid stem, which for a long time were regarded as marine plants.

It is customary to place subsequently to them the thethyæ, and sponges, although no polypi have yet been discovered in them.

Finally, the INFUSORIÆ, or fifth and last class of the zoophytes, are those little beings which have been discovered only by the microscope, and which swarm in stagnant waters. The majority of them present nothing but a gelatinous body, without viscera. Nevertheless, we have at their head some more complex species, possessing visible organs of motion, and a stomach. Of these, too, in all probability, at some future day, a separate class will be formed.

THIRD CLASS OF THE ZOOPHYTES.

THE ECHINODERMATA.

THE Echinodermata are as yet the most complicated animals of this division. Invested with a well organized skin, often supported by a sort of skeleton, and armed with points, or with articulated and mobile spines, they have an interior cavity, in which distinct viscera are floating. A sort of vascular system, which, in truth, does not extend to the whole body, keeps up a communication with divers parts of the intestine, and with the organs of respiration, which, most generally, are also very distinct. We even observe in several species, some filaments which might perform the nervous functions, but which are never distributed with the regularity, and in the fixed order, which exists in the other two divisions of invertebrata.

We divide the echinodermata into two orders: those which have feet, or at least vesicular organs, to which this name has been given, and those which are destitute of them.

FIRST ORDER OF ECHINODERMATA.

THE PEDICELLATA

ARE distinguished by organs of motions altogether peculiar. Their envelope is pierced by a great number of small holes, placed in very regular series, through which pass some cylindrical, membranaceous tentacula, terminated each by a small disk, which performs the office of a cupper. The part of these tentacula which remains in the interior of the body is vesicular. A fluid is spread through all their cavity, and is carried, at the will of the animal, into the external cylindrical part, which it extends, or it re-enters into the internal vesicular part, and then the external part sinks in. It is by elongating and contracting in this manner, their hundreds of little feet, or tentacula, and fixing them by the cuppers, which terminate them, that these animals execute their progressive movements. Some vessels proceeding from these little feet, repair to the trunks, which correspond to their ranges, and which end at the mouth. They form a system distinct from that of the intestinal vessels which are observed in some species.

Linnæus makes of these animals three genera, very natural, but sufficiently numerous, and comprehending species sufficiently varied to be considered as three families.

ASTERIAS, *L.* Vulgò, *Sea-stars*, or *Star-fish.*

Have received this name, because their body is divided into radii, most frequently five in number, at the centre of which, underneath, is the mouth, which at the same time serves as anus.

The frame-work of their body is composed of small osseous pieces, variously combined, the arrangement of which would merit investigation. They have a very great power of reproduction, and not only reproduce the radii, which have been taken away separately, but a single ray with the centre being preserved can reproduce the others, which is the cause that we so frequently find them irregular.

In the

ASTERIAS, *Lam.*,

Or Asterias, properly so called, each ray has, underneath, a longitudinal furrow, on the sides of which are pierced all the little holes which allow the feet to pass. The rest of the inferior surface is provided with small mobile spines; the entire surface is also pierced with pores, which allow some tubes much smaller than the feet to pass, which probably serve to absorb the water, and to introduce it into the general cavity for a kind of respiration. On the middle of the body, a little towards the side, is a small stony plate, to which corresponds

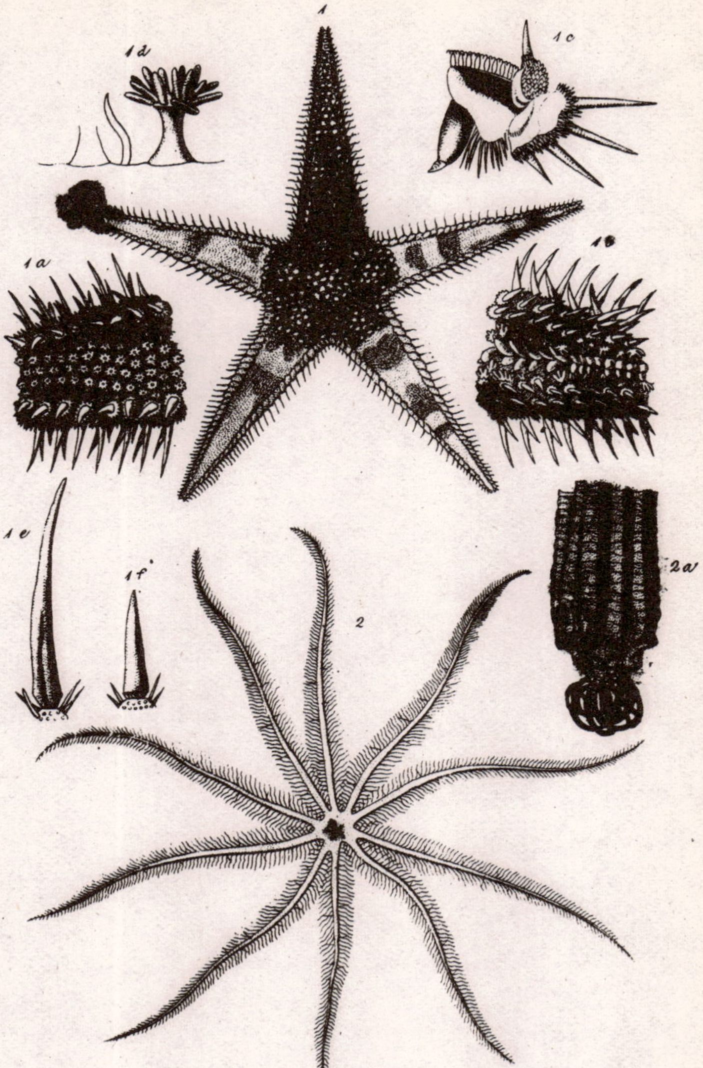

1 *Asterias*

2 *Comatula carinata.* Lam.

internally, a canal filled with calcareous matter, which is believed to contribute to the growth of the solid parts. Internally, is seen a large stomach immediately upon the mouth, from which proceed, for each ray, two cœca, ramified like trees, and each suspended to a sort of mesentery. There are also two ovaries in each ray, and it appears that the asteriæ fecundate themselves. A peculiar vascular system corresponds to their intestine, and there is another for the feet.

M. Tiedemann considers as their nervous system, a very fine filament which surrounds the mouth, and sends a branch to each arm, which proceeds externally between the feet, and gives forth two ramuscula internally.

Their osseous frame-work consists principally, for each branch, in a sort of column predominating along the inferior face, composed of vertebræ, articulated one with the other, and from which proceed the cartilaginous branches, which support the external envelope. Between the roots of these branches are the holes through which the feet pass. Other osseous pieces, to which mobile spines are frequently attached, furnish, in many species, the lateral edges of the branches.

Certain asteriæ have the form of a pentagon, with rectilinear sides, rather than that of a star. The radiation is marked externally only by the sulcus of the feet. (*Asterias discoïdea*, Lam.)

Others have, on each side of the pentagon, a slight re-entrant angle. (*Asterias membranacea*, Link.)

In others, the sides are concave, which begins to give them the figure of a star. (*Asterias tessalata*, Lam., &c.)

In these different species, the cœca and the ovaries are not as much elongated as in the majority of the others, which have their radii elongated, and separated by well marked re-entrant angles.

Such are

Ast. rubens, L. Encyc. cxiii. 1, 2. which is exceedingly

common on all our coasts, so much so, that in some places it is employed to manure the grounds.

Ast. glacialis, L. Link. xxxviii. 69. Encyc. cvii. and cviii. is frequently more than a foot in diameter; the spines which invest the upper part of its body are surrounded with a multitude of little fleshy tubes, which form sorts of cushions around their bases.

Ast. aurantiaca, L. Link. vi. vii. xxiii. Encyc. cx. Echin. pl. iv. 1., is our largest species. The edges of its branches are furnished with pieces, like mosaic work, on which some strong mobile spines are articulated. All the upper part is covered with some other small spines, terminating in truncated and bristling heads.

Some have a number of rays above five. Their cœca and ovaries are very short. (*Ast. paposa*, Link.)

We must separate from the other asteriæ, the species in which the rays have no longitudinal furrows underneath, for the purpose of lodging the feet; in general, these rays are not hollow, and the stomach is not prolonged into cœca, but its prominences remain in their intervals. Locomotion is effected principally by the curving and the movement of the radii, and not by the feet, which are too few in number.

M. de Lamarck names *Ophiures* those which have round a central disk five radii not branched. But we should still distinguish

Those in which these radii are furnished on each side, with mobile spines. The small fleshy feet also issue forth on each side from between the basis of these spines. (*Ast. nigra*, Müll., &c.)

And those in which the radii, having no lateral spines, but being furnished with imbricated scales, resemble the tails of serpents. The central disk, has, in each interval of the rays, on the side where the mouth is, four holes which penetrate into the interior, and serve, perhaps, for respiration, or, according to others for the issue of the eggs. There are no feet, except in five short furrows, which form a star around the mouth. (*Ast. ophiura*, Lin., &c.)

The GORGONOCEPHALA, *Leach*, named *Euryale*, by M. de Lamarck, are those in which the radii are divided into a double point. There are some in which this division commences from the base of the radii, and which present the appearance of a parcel of serpents. They have been vulgarly named *heads of Medusa*. The base of each ray has two penetrating holes. (*Asterias caput Medusæ*, &c.)

But there are also some in which the division commences only at the end of the ray, and is but little repeated. (*Euryale palmiferum*, Lam.)

We should still separate from the other asteriæ

The ALECTO of *Leach*, which M. de Lamarck calls *Comatula*. They have five large articulated radii, divided each into two or three, which support two ranges of articulated filaments. These five radii are attached to a stony disk which again supports, on the side opposite to the mouth, one, two, or three ranges of other articulated filaments without branches, shorter and thinner than the large radii, and which, it is said, enable them to hook themselves to bodies. The sac which contains the viscera is at the centre of the large radii, opened by a star-formed mouth, and another tubular orifice, which may be the anus. (*Asterias multiradiata*, &c.)

It is near the comatulæ that we should place

ENCRINUS, *Guett.*,

Which may be defined to be comatulæ, with a disk prolonged into a stem divided into a great number of articulations. Their branches themselves are articulated, and divided into pairs of branches, supporting a range of filaments, all articulated, and the stem itself supports some smaller ones, at various heights. In the centre of the radii is the mouth, and the anus on one side.

There is in the seas of Europe but one very small species, ***Pentacrinus Europæus,*** Thomson, Monogr., which is attached to divers lithophytes.

The seas of the warmer climates produce larger and more complex species, such as ***Encr. asterias,*** Blum., ***Isis aster.,*** Linn.

But the fossil encrini are very numerous, and vary sufficiently in detail to be divided into several subgenera, according to the composition of the central body, placed at the summit of the stem, and from which the large radii proceed.

This body may be formed of pieces articulated with the stem, and supporting radii by similar articulations. Then if the stem is round and enlarged at the top, these are the APIOCRINITES, *Mill.*

If it is round, but not enlarged, the ENCRINITES.

If it is pentagonal, the PENTACRINITES.

Or this body may be formed of angular laminæ, joined together by their edges, and forming several ranges.

Among those

The PLATYCRINITES have but two ranges, one of three laminæ, the others of five.

The POTERIOCRINITES have three ranges, each of five plates.

The CYATHOCRINITES have also three, each of five, but the last has some intercalary laminæ, which may increase it even up to ten.

The ACTINOCRINITES have several ranges, the first of three, the second of five, the others more numerous. The first two have radiated crests.

The RHODOCRINITES have also several ranges, the first

of three, the second of five, the third of ten, all three with crests; then follow some with more numerous laminæ.

Finally, the central body may be all of one piece, but which appears to be composed of five cemented together. These are the EUGENIACRINITES.

The fossil productions, known under the name of Entrochi, are pieces of the stem and of the branches of animals of this genus.

ECHINUS, *Lin.* Vulgò *Sea urchins.*

Have the body clothed with a testa, or calcareous crust, composed of angular pieces, which join exactly, and are pierced with several very regular ranges of innumerable small holes, through which pass the membranous feet. The surface of this crust is armed with spines articulated on little tubercles, and moveable at the will of the animal, to whose motions they administer, conjointly with the feet, which are situated between them. Other membranous tubes, much finer, and often divided at their extremity, probably serve to introduce and to expel the water which fills the interior of their shell. The mouth is furnished with five teeth, enchased in a very complicated calcareous frame-work, resembling a lantern with five panes, furnished with divers muscles, and suspended in a large aperture of the testa. These teeth, in the form of long bands, grow hard towards their root, in proportion as they are worn at the point. The intestine is very long, aud attached spirally to the interior parietes of the testa by a mesentery. A double vascular system runs along this canal, and extends partly over the mesentery, and there are also particular vessels for the feet. Five ovaries situated around the anus empty themselves each by a particular orifice. They form the eatable part of these animals.

The echini subsist more especially on small testacea, which they seize with their feet. Their motions are very slow.

Their testæ are preserved in great abundance in ancient strata, especially those of the chalk formation, when they are usually filled with silex.

The echini should be divided into regular and irregular. The regular

ECHINI, (properly so called) *Lam.* CIDARIS, *Klein.*

Have the testa generally spheroidal, the mouth at the middle of their inferior face, and the anus precisely opposite. The little holes are ranged there on six bands approximated by pairs, which proceed regularly from the mouth to the anus, like the meridians of a globe.

Certain species have large and thick prickles of very various forms, supported on thick tubercles of the testa, and the bases of which are surrounded with other smaller prickles. (*Echinus mamillatus*, L., &c.)

It is among these species that are ranged, as M. de Luc has discovered, those whose prickles, in the form of olives, are frequently found petrified in chalk or other ancient formations, and have received the name of *Judaic stones.*

The most common species, and especially those of our coasts, have only slender spines, articulated on small tubercles, much more numerous. Such is

ECHINUS ESCULENTUS, *L., Klein., Lesk.*, I. *A. B.*, *Encyc.* 132.

Of the form and size of an apple, all covered with short, striped tubercles, usually of a violet colour. Its ovaries are eaten in spring uncooked, they are reddish, and of a flavour sufficiently agreeable.

The neighbouring species are difficult enough to distinguish, being marked by the greater or less approximation of the bands of holes, by the equality or inequality of the tubercles, &c. (*Ech. miliaris*, Kl., &c. &c. &c.)

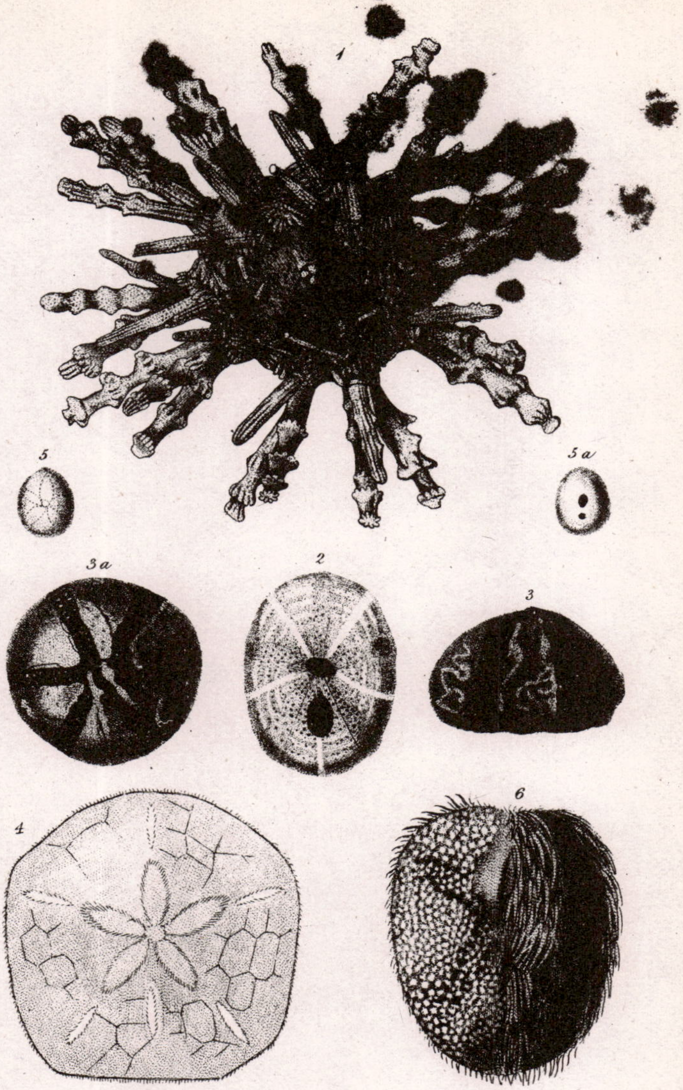

1 *Echinus verticillatus.*

2 *Echi. semilunaris.*

3 *Galerites sexfasciata.*

4 *Scutella hexapora.*

5 *Fibularia ovulum.*

6 *Spatangus pilosus.*

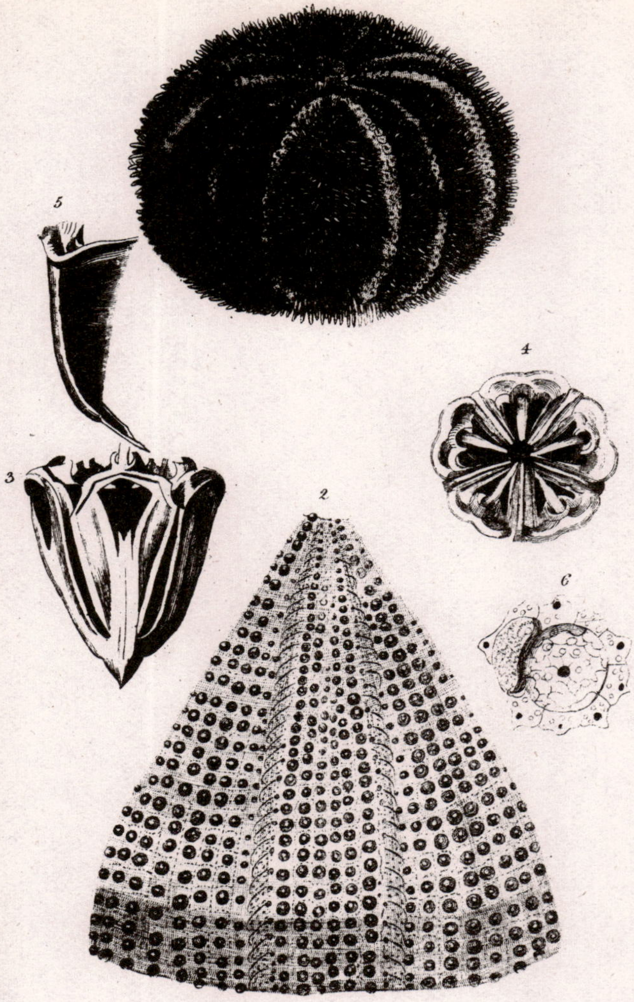

1 *Echinus esculentus.* 4 *Top of same.*

2 *Part of the Head.* 5 *A single Teeth.*

3 *Dentary apparatus.* 6 *Orifice of the ovary.*

Some round and depressed echini lose something of their regularity by a wide sulcus, with which they are furrowed on one side. (*Ech. sinuatus*, Kl.)

There are also some of these echini with mouth and anus opposite, which, instead of having the spheroidal form on a circular plane, are transversely oval, that is to say, one of their horizontal diameters is larger than the others. (*Ech. lucunter*, Kl.)

They also differ among themselves by the equality or inequality of the prickles, and by the relative proportions of the tubercles.

One species should be distinguished,—*Echinus atratus*, L. Encyc. 140. 1—4, in which the prickles, widened, truncated, and angular at their extremity, touch each other there, like pavement. Those of the margin are long and flatted.

We call irregular, all the echini in which the anus is not opposite to the mouth. It appears that they are furnished only with short and slender prickles, almost like hairs. Among them some still have the mouth at the middle of the base. They may be subdivided, according to the extent of the bands of holes for the feet. Sometimes these go, as in the preceding, from the mouth to a point directly opposite, or they unite after having embraced the entire testa. In the former, the

ECHINONE, *Phelsum* and *Leske*,

Have the round or oval form of certain regular echini, the mouth at the middle of the base, and the anus between the mouth and the margin, or near the margin, but underneath; oval species, *Echinus cyclostomus*, Müll., &c.; round species, *Ech. depressus*, Walch, &c.

NUCLEOLITES, *Lin.*,

Have, with these characters, the anus near the margin, but above.

The known species are all fossil. (*Spatangus depressus*, Leske.)

Others,

GALERITES, *Lam.* CONULUS, *Kl.*,

Have a flat base, on which their body is raised like a cone, or a semi-ellipsoid figure. The mouth is at the middle of the base, and the anus near its margin.

They are very common in rocky strata, but no living species are known.

The most extended is *Ech. vulgaris*, L. Encyc. 153. 6—7. Klem. Ed. Fr. VII. D. G.

Some have not their bands of holes distributed in the quinary number. *Ech. quadrifasciatus*, Walch, &c.

SCUTELLA, *Lam.*,

Have the anus between the mouth and the margin, the testa exceedingly depressed, flat underneath, and of a form approaching to orbicular.

Some have it entire, and without any other holes than the series of little pores which are seen in all the echini.

Others have the testa equally, without large holes, but divided by two emarginations. (*Echinus auritus*, Seb., &c.)

Others have it entire, and pierced from interval to interval, by some large holes which do not penetrate into its cavity. (*Echinus hexaporus*, Seb., &c.)

Others have it at once emarginated, and pierced with three large holes. (*Echinus tetraporus*, Seb.)

There are some, in fine, ROTULÆ, *Kl.*, in which a part of the posterior margin is festooned, like a dentated wheel, and these are divided according as they have large holes. (*Ech. decadactylus*, Encyc.;) or according as they want them, (*Echinus orbiculus*, Encyc.).

CASSIDULUS, *Lam.*,

Are oval, and have the anus above the margin, like the

nucleolites, but they are distinguished by their incomplete bands of pores, that is to say, not proceeding from one pole to the other, and figuring a star. (*Cassidulus Caribæorum*, Lam.)

Other irregular echini have not the mouth at the centre of their base, but it is towards one side, often transversely, and directed obliquely. The anus is towards the other side. They are subdivided according to the extent of their ranges of holes.

Thus the ANANCHITES, *Lam.*, GALEÆ, *Kl.*, have a little the form of the Galerites, and their complete bands, their greatest difference, consists in the position of the mouth. They are only known in the fossil state. Such is

ECHINUS OVATUS, *L.*, *Cuv. et Brong.*

(Envir. de Paris, 2d. edit. f. v. 7. A. B. C. D.)

A species extended in innumerable quantities in the chalk formations of our environs.

Some have four bands. (*Ech. quadriradiatus*, Kl.)

We may make a particular subgenus of certain species, in which the four lateral bands are disposed by pairs, and do not rejoin at the same point. (*Ech. bicordatus*, Kl.)

At other times these irregular echini, with central mouth, have bands of pores which do not lead to the mouth, but which form on their back a sort of rose. Such are

CLYPEASTER, *Lam.* ECHINANTHUS, *Klein.*,

Which have the anus near the margin, and whose body is depressed, with oval base concave underneath. They have sometimes the contour a little angular. (*Ech. rosaceus*, and its varieties, *Kl.*)

Sometimes their back is elevated in the middle. (*Echinus altus.*)

There are some also whose contour is not angular. (*Ech. oviformis*, Seb.)

And even where it is almost orbicular. (LAGANUM, *Klein. Echinus orbiculatus.*)

FIBULARIA, *Lam.* ECHINOCYAMUS, *Leske.*,

Have, with the rose of the clypeaster, the body almost globular, and the mouth and anus approximated in the middle of the under part. They are usually very small. (*Echinus nucleus*, Kl., &c.)

On the contrary, SPATANGUS, *Kl.*, have, with the lateral mouth of the ananchites, some incomplete bands of pores, forming a rose on the back. There are usually but four; that which is directed from the side of the mouth is obliterated.

Some, BRESSOIDES, *Kl.*, have the testa oval, without furrows. (*Ech. teres*, Seb.)

Others have a large sulcus more or less marked, in the direction of the obliterated band. (*Ech. spatangus*, Seb., &c.)

When, besides this, they preserve the oval form, they are BRISSUS, *Kl.*, but sometimes this sulcus grows deep, and the testa widening at the same time on this side, assumes the figure of a heart. (*Ech. purpureus*, Müll, &c.)

We have some in our seas of the last two forms. Branched tentacula, like those of the holothuriæ, have been observed round the mouth.

HOLOTHURIA, *Lin.*,

Have the body oblong, coriaceous, and open at the two ends. At the anterior extremity is the mouth, surrounded by very complicated branched tentacula, which can be completely retracted. At the opposite extremity opens a cloaca, where terminate the rectum and the organ of respiration, in the form of a hollow tree, very much ramified, which is filled with water, or emptied at the will of the animal. The mouth has no teeth, and is furnished only with a circle of osseous pieces. Some appendages, in the form of pouches, pour a saliva into it. The intestine is very long, diversely plicated, and attached to the sides of the body by a mesentery. A sort of partial circulation takes place in a very complicated double system of vessels, exclusively relative to the intestinal canal, and in a part of the meshes of which, is interlaced one of the two respiratory trees of which we have just spoken. There also appears to be a nervous cord, but very much attenuated, around the œsophagus. The ovary is composed of a multitude of blind vessels, partly branched, which all terminate at the mouth by a small common oviduct. They assume, at the time of gestation, a prodigious extension, and are then filled with a red matter, which appears to be the eggs. Some cords, of an extreme sensibility, attached near the anus, and which are developed at the same time, appear to be the male organs. These animals must, then, be hermaphrodites. When they are disturbed, they frequently contract themselves with so much force that they tear and vomit up their intestines[1].

The holothuriæ may be divided according to the distribution of their feet.

In some, they are all situated in the middle of the under part of the body, which forms a softer disk, on which the animal crawls, elevating the two extremities where the mouth and anus are placed, which they contract more than the middle. The anus finishes almost in a point. Their tentacula are very large when developed.

We have one in our seas whose envelope is almost scaly, *Hol. phantapus*, L. Müll., Zool., Dan. cxii. cxiii. Mem.

[1] For the anatomy of Holothuriæ, consult M. Tiedemann's excellent work already referred to.

de Stok., 1767. The feet of its central disk are on three series.

Others have the inferior face altogether flat and soft, furnished with an infinity of feet, and the superior face gibbous, supported frequently by osseous scales, and pierced on the front, with a star-like orifice, which is the mouth, and from which the tentacula issue; and on the back part with a round hole, which is the anus.

We have a small one, *Hol. squamata,* Müll., Zool., Dan. x. 1, 2, 3 ; but there are some of these of a tolerable size in the warmer seas.

Others have the body cartilaginous, flatted horizontally; trenchant at the edges; the mouth and feet at the inferior face, and the anus at the posterior extremity.

Such is in the Mediterranean,

Pudendum regale, Fab., Column, Aquat. xxvi. 1. *Hol. regalis,* Nob., a species more than a foot in length, three or four inches broad, and crenulated all round.

Others, again, have the body cylindrical, susceptible of enlargement in every direction by the absorption of water; all the under part furnished with feet, and the rest of the surface variously bristled.

Our seas, especially the Mediterranean, produce in great abundance one of blackish colour, which is more than a foot long in its greatest extension. Its back is bristled with conical and soft points; its mouth is furnished with twenty branched tentacula; it is the *Holothuria tremula,* Gm., Bohatch., Anim. mar. vi. and vii.

Some are found in which the feet are distributed in five series, which extend like the ribs of a melon, from the mouth to the anus, which has caused them to be called sea-cucumbers.

Such is in our seas.

Hol. frondosa, L. Gunner, Mem. de Stok. 1767. pl. iv. f. 1,

2. and cxxiv. The body is brown, and a foot or more in length.

Finally, there are some whose body is equally furnished with feet all round. (*Hol. papillosa*, &c.)

THE SECOND ORDER OF ECHINODERMATA,

OR

ECHINODERMATA APEDICELLATA,

COMPREHEND but a small number of animals, which present great relations with the holothuriæ, but which want the small vesicular feet of the preceding order. Their body is clothed with a coriaceous skin, and without armature. Their internal organization is not yet cleared up on all points.

MOLPADIA, *Cuv.*,

Have, like the holothuriæ, a coriaceous body, in the form of a thick cylinder, open at the two ends, and their internal organization is pretty similar to that of those animals; but, besides that they want feet; their mouth has no tentacula, and is furnished with an apparatus of osseous pieces, less complicated, however, than that of the echini.

I know but a single species belonging to the Atlantic Ocean, the extremity of whose anus is finished in a point. (*Molpadia holothurioïdes*, Cuv.)

MINYAS, *Cuv.*,

Have also the body without feet, and open at the two ends; but its form is that of a spheroid depressed at the poles, and furrowed like a melon. I can find no armature to the mouth.

There is a very fine species of a deep blue, in the Atlantic ocean. (*Mynias cyanea*, Cuv., Règ., An. iv. pl. xv. f. 8.)

PRIAPULUS, *Lam.*,

Have a cylindrical body, marked transversely with deep annular wrinkles, terminated in front by an elliptical mass, slightly wrinkled longitudinally; pierced with the mouth, and behind with the anus, from which issues a thick bundle of filaments, which may be the organs of generation. The interior of the mouth is furnished with a great number of very sharp corneous teeth, placed like a quincunx, and directed backwards; the intestine goes straight from the mouth to the anus; the muscular system resembles that of the holothuriæ.

We are acquainted with but a single species from the north seas, *Holothuria priapus*, L. Müll., Zool., Dan. xcvi. 1. two or three inches in length.

LITHODERMIS, *Cuv.*,

Have the body oval, compressed behind, and the surface is as it were incrusted with a stratum of small stony grains, which form a very hard crust. The mouth is surrounded with tentacula, and the intestines appear to have resemblances to those of the holothuriæ. I do not find any anus.

We know but a single species from the Indies, *Lithod. cuneus*, Cuv., blackish, two inches in length.

SIPONCULUS, Gm.,

Have a cylindrical elongated body, with a thick skin, wrinkled in both directions; the mouth at an extremity in the form of a proboscis, which may be retracted or put forth, by means of large interior muscles, and the anus more or less near the base of this proboscis. The intestine originates from the mouth, proceeds as far as the opposite extremity, and returns, convoluting spirally round its first part. We find

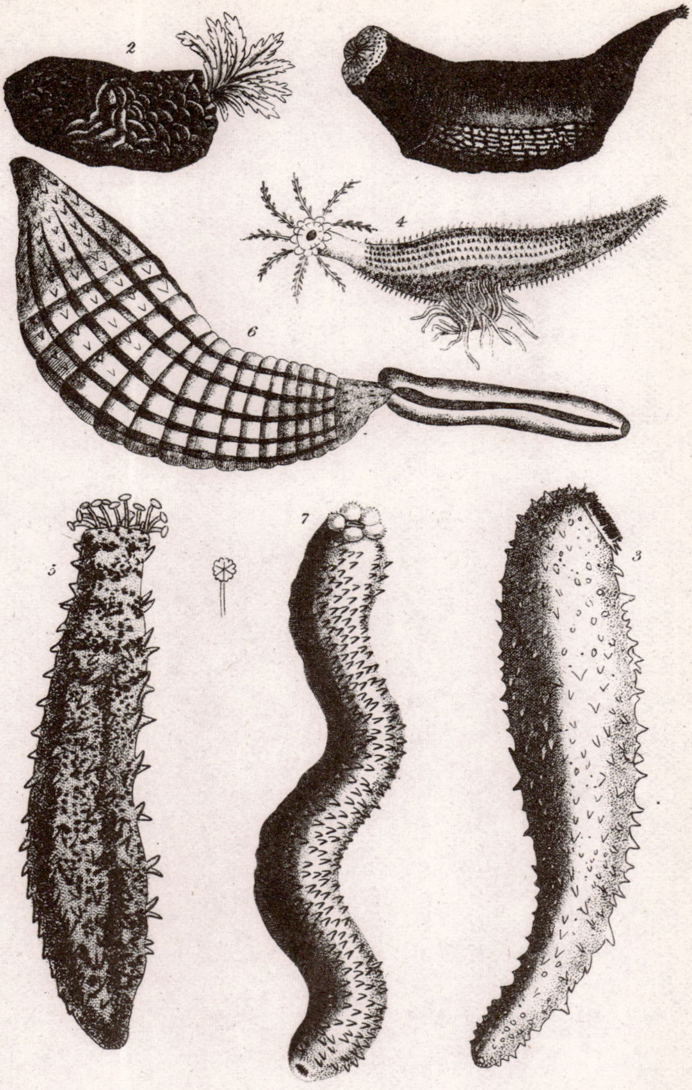

1 *Holothuria phantopus.* 4 *Hol. fuscus.*

2 *Hol. squamata.* 5 *Hol. cucumer.*

3 *Hol. elegans.* 6 *Hol. eacuari.*

7 *Hol. edulis.*

nothing in it but sand and fragments of shells. Numerous vessels appear to unite it to the external envelope, and there is, moreover, along one of their sides, a filament which may be nervous. Two long pouches situated in front, have their external orifices a little under the anus, and internally, near this last orifice, is sometimes seen a packet of branched vessels, which may appertain to the respiratory function.

These animals remain in sand under the sea water, like the arenicolæ, thalassema, &c., and they are sought after like those, for the purpose of being used as bait.

There are several species of them as yet but badly distinguished.

One of them, *Sip. edulis*, Nob., *Lumbricus edulis*, Gm., Pall., Spic., Zool. x 1. 7. serves as food for the Chinese who inhabit Java, and who go to seek for it in the sand, with little bamboo sticks prepared for the purpose.

Others, sufficiently small, *Sip. levis, Sip. verrucosus* Cuv., pierce the submarine stones, and lodge in their cavities.

BONELLIA, *Rolando*,

Have the body oval, a proboscis formed of a plicated lamina, susceptible of an extreme elongation, and forked at its extremity. The anus is at the opposite extremity of the body; the intestine is very long, replicated several times, and near the anus are two ramified organs, which may serve for respiration ; the eggs are contained in an oblong sac, which has its issue near the base of the proboscis.

These animals live deeply in the sand, and protrude their proboscis as far as the water, and even sometimes into the air itself, when the water is low.

We have one species in the Mediterranean, *Bonellia viridis*, Rol. Ac. de Tur. t. xxvi. pl. xiv.

THALASSEMA, *Cuv.*,

Have the body oval, or oblong, and the proboscis in the form of a plicated lamina, or small spoon, but not forked. But one abdominal filament is discoverable.

We distinguish

THALASSEMA, (properly so called,)

Which have only two hooks placed very much in front, and the posterior extremity has no setæ. (*Thalassema neptuni*, Gertner, &c.)

ECHIURUS,

Whose posterior extremity is furnished with some transverse ranges of setæ.

One is known, *lumbricus echiurus*, Gm., Pall., miscell. Zool. xi. 1—6, which inhabits our coasts, on the sandy bottoms. It serves as bait to fishermen.

STERNASPIS, *Otto*,

Which, beside the setæ of echiurus, have under the anterior part, a disk, a little corneous, surrounded with ciliæ. (*Thalassema scutatum*, Ranz., &c.)

SECOND CLASS OF ZOOPHYTES.

INTESTINA, ENTOZOA, *Rudolphi*,

Have, for the most part, this peculiarity, that they are unable to live, or to propagate, except in the interior of the body of other animals. There is almost no animal that does not nourish several kinds of them, and those which are observed

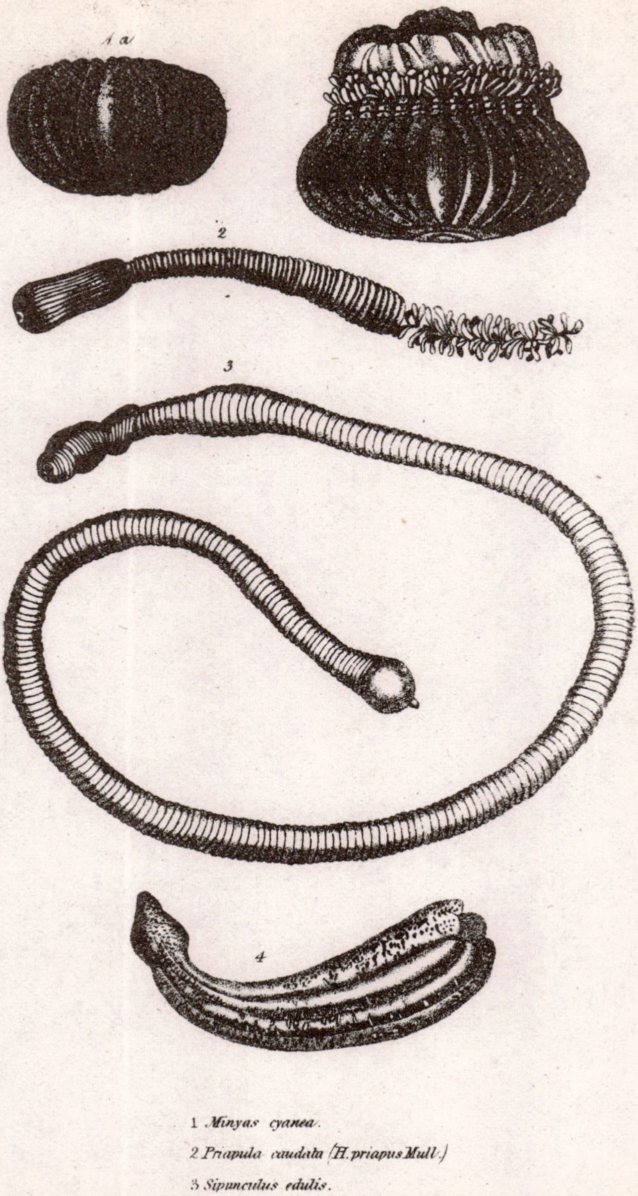

1 *Minyas cyanea.*

2 *Priapula caudata (H. priapus Mull.)*

3 *Sipunculus edulis.*

4 *Sip. tigrinus.*

in one species but rarely extend to many other species. They are not only found in the alimentary canal, and the canals which conduct thither, such as the hepatic vessels, but even in the cellular tissue, and in the parenchyma of the best enveloped viscera, such as the liver and the brain.

The difficulty of conceiving how they get into such places, joined to the observation that they never appear out of living bodies, has caused some naturalists to imagine that they engender spontaneously. It is now, however, quite certain, not only that the majority manifestly produce eggs, or living young ones, but that many of them have separate sexes, and couple like ordinary animals. We may then believe that they propagate by germs, sufficiently small to be transmitted through the narrowest passages, or that frequently the animals in which they live bring the germs into the world with them.

We discover in the intestinal worms, neither tracheæ nor gills, nor any other organ of respiration, and they must receive the influence of oxygen through the medium of the animals which they inhabit. They present no trace of a true circulation, and only vestiges of nerves, so obscure, that several naturalists have doubted their existence.

When these characters are found united in an animal, with a form similar to that of this class, we range it here, although it does not inhabit the interior of another species.

Every one knows to what an extent the intestinal worms injure the animals in which they are too much multiplied. Against those of the alimentary canal many remedies are employed, of which the most efficacious appears to be animal oil, &c., mixed with oil of turpentine.

We divide them into two orders, perhaps sufficiently different in organization to form two classes, if adequate observation could affix their limits.

Intestina Cavitaria. (**Entozoa Nematoidea,** *Rud.*,)
Which have an intestinal canal floating in a distinct abdominal cavity, and a mouth and anus.

Intestina Parenchymata,

Whose body encloses, in its parenchyma, some viscera, but imperfectly terminated, and most frequently resembling vascular ramifications, even sometimes being not at all perceptible.

THE FIRST ORDER OF INTESTINA.

Cavitaria. Nematoidea, *Rudolphi*,

Comprehend those whose external skin, more or less furnished with muscular fibres, and in general striated transversely, contains an abdominal cavity, in which floats a distinct intestinal canal, proceeding from the mouth to the anus, and where also are generally observed some distinct organs for the two sexes. The intestine is united to the neighbouring parts, and to the general envelope by numerous filaments, in which some have imagined that they discovered nutritive vessels, others tracheæ, but without any proof. It is impossible to observe in these animals any true circulation, but there appears to be in many, one or two nervous cords, proceeding from a ring which surrounds the mouth, and running the entire length of the body, on the internal face of the envelope.

The intestine is generally straight, and tolerably broad; the œsophagus is frequently more slender, and in some species we remark a stomach more ample and more robust. The

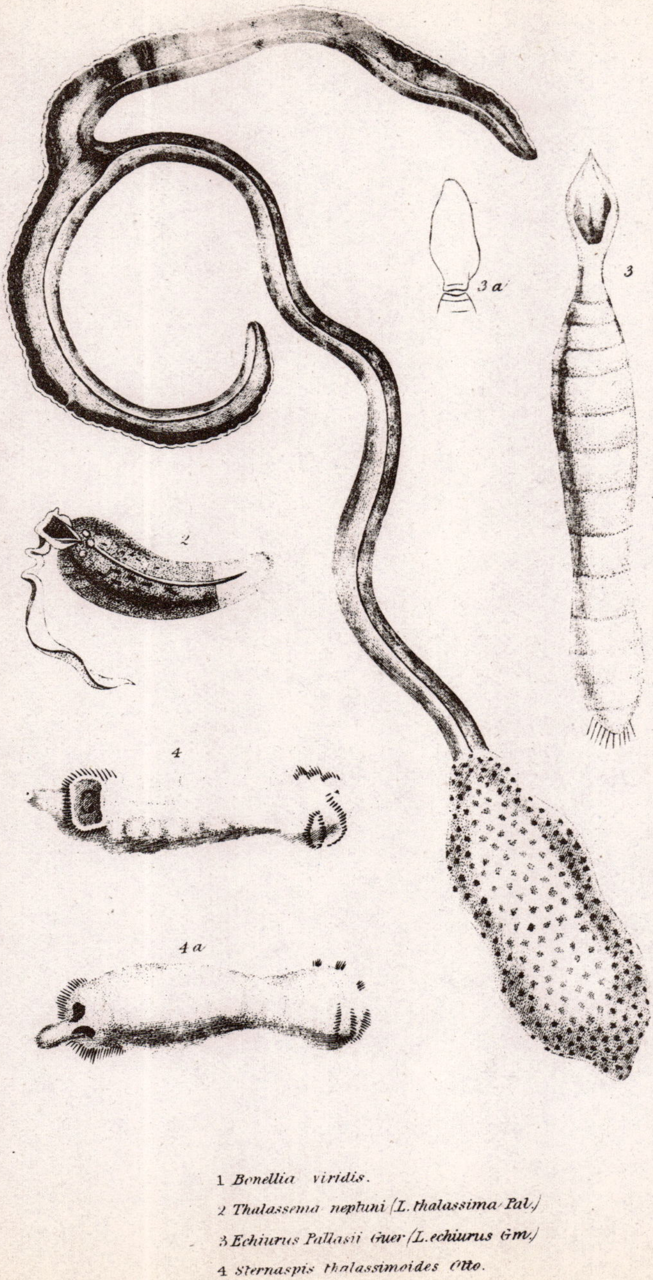

1 *Bonellia viridis.*

2 *Thalassema neptuni (L. thalassima Pal.)*

3 *Echiurus Pallasii Guer (L. echiurus Gm.)*

4 *Sternaspis thalassimoides Otto.*

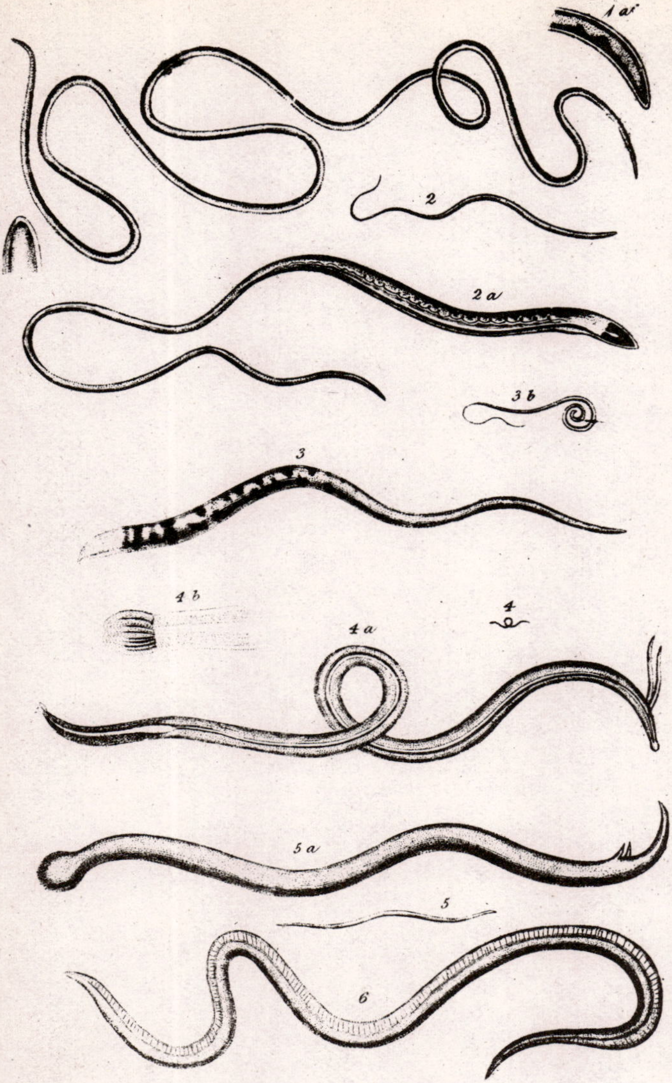

1 *Filaria medinensis.*

2 *Tricocephalus dispar.*

3 *Oxyurus curvula.*

4 *Cucullanus elegans.*

5 *Ophiostoma sphærocephala.*

6 *Asca lumbricoïdes.*

internal organs of generation consist in very long vessels, containing the semen, or the eggs, and having their issue at different points according to the genera.

FILARIA, *L.*,

Have the body elongated and slender, in the form of a thread, pierced in front with a round mouth. Externally they very much resemble the gordius. They are principally found in the cavities of animals, which have no external issue in the cellular substance, and even in the thickness of the muscles, and the parenchyma of the viscera. They are sometimes in parcels, and in innumerable quantities, enveloped in sorts of capsules. They are even found in insects, and in their larvæ. and in the visceral cavity of several of the mollusca.

The most celebrated species of this genus is,

Filaria Medinensis, Gm. Encyc. xxix. 3. (Vulg. *Guinea-Worm.*) Very common in warm climates, where it insinuates itself under the human skin, principally in the legs, and is sometimes developed there, to the length of ten feet and more, if we may credit the accounts of some writers. It can subsist there for several years, without causing any very lively sensations; but it also sometimes produces very dreadful pains and convulsions, according to the parts which it attacks. When it shows itself externally, it is seized, and drawn out very slowly, for fear of breaking it. It is as thick as the quill of a pigeon's feather. Its distinctive character is to have the end of the tail pointed and hooked.

TRICHOCEPHALUS,

Have the body round, thicker behind, and as slender as a thread in front. This slender part is terminated by a round mouth.

The most known is,

Trich. dispar. Rud. Gœtz. vi. 1—5. Encyc. xxxiii. 1—4.

One or two inches in length, of which the thick part occupies only one third. In the male this part is convoluted spirally, and a small penis is visible, which comes out near the tail. In the female it is straighter, and simply pierced at the extremity.

It is one of the most common worms in the large intestines of man, and multiplies beyond measure in certain disorders.

From the trichocephali have been distinguished,

Trichostomus, *Rus.* Capillaria, *Zeder.*,

The anterior part of which grows slender only by degrees; and the

Oxyuris, *Rud.*,

In which it is the posterior part of the body which is attenuated in the form of a thread.

A species is known belonging to the cœcum of the horse, *Oxyuris curvula*, Rud. Gœtz. vi. 8. Encyc. xxxiii. 5., from one to three inches in length.

Cucullanus,

Have the body round, more slender behind; the head blunt, invested with a sort of little hood, often striated; the mouth round.

These have hitherto been found only in fish. The most common is that of the perch. *C. lacustris*, Gm. Gœtz. ix. A. 3. Encyc. xxxi. 6., which also infests the pike, &c. It is viviparous, about an inch in length, of the thickness of a thread, and appears to be red, in consequence of the blood with which its intestine is usually filled.

Ophiostomus,

With the body of the preceding, are distinguished by a mouth cut cross-wise, and consequently provided as it were with two lips.

Some of them are found in the air-bladder of certain fishes, *Ophiost. cystidicola*, R. *Cystidicola,* Fisch. Monogr.

ASCARIS,

Have the body round, slender at both ends, and the mouth furnished with three fleshy papillæ, between which projects from time to time, a very short tube. It is one of the genera that is most numerous in species. They have been found in all sorts of animals. Those which have been dissected have exhibited a straight intestinal canal; and in the females, which form by far the greater number, an ovary with two branches has been observed, several times longer than the body, opening externally by a single oviduct, towards the interior fourth of the length of the animal. The males have but a single seminal tube, also much longer than the body, and which communicates with a penis sometimes double, which issues forth through the anus. The latter is pierced under the extremity of the tail.

M. Otto, and M. Cloquet, consider as the nervous system of these worms, two white filaments which extend, one along the back, and the other along the belly; two other thicker threads, extending, one on the right, the other on the left, are regarded by some as muscular, by others, as vascular, or even as tracheal.

Some have the head without lateral membranes.

The most known species,

Ascaris lumbricoïdes, L., is found without any sensible difference in man, in the horse, the ass, the zebra, the hemionus, the ox, and the hog. They have been seen more than fifteen inches in length. Its natural colour is white. It sometimes multiplies to excess, and may cause mortal maladies, especially in infants, to whom it occasions a variety of affections, particularly when it ascends into the stomach.

Other species have a small membrane on each side of the head. Such is,

Ascaris vermicularis, L. Gœtz. v. 1—6. Encyc. Meth. Vers. xxx. pl. x. 1. So common with infants, and in certain maladies of adults, in whom it produces an insupportable itching at the anus. It scarcely exceeds five lines in length, and is thicker in front.

STRONGYLUS, Müll.,

Have the body round, and the anus enveloped in the male, by a sort of pouch, variously configured; and from which issus a small filament, that appears to serve the purpose of generation. The female is destitute of these latter characters, which might sometimes cause her to be taken for an ascaris.

There are some of these strongyli which have ciliæ, or denticulations to the mouth. Such is

Strong. Equinus, Gm. *Str. armatus*, Rud. Müll. Zool. Dan. ii. xlii. Enc. meth. xxxvi. 7—15. Two inches long, with a hard spherical head, mouth furnished all round with small soft spines. The pouch of the male is divided into three leaflets. It is the most common of all the worms of the horse. It penetrates even into the arteries, where it occasions aneurisms. It is also found in the ass and the mule.

Others have round the mouth, only tubercles or papillæ. Such is particularly,

Strongylus gigas, Rud. *Ascaris visceralis*, et *Asc. renalis*, Gm. Rud., &c. The largest of the known intestinal worms; it is two or three feet long and upwards, and of the thickness of one's little finger. What is most singular is, that it is most frequently developed in one of the kidneys of divers animals;

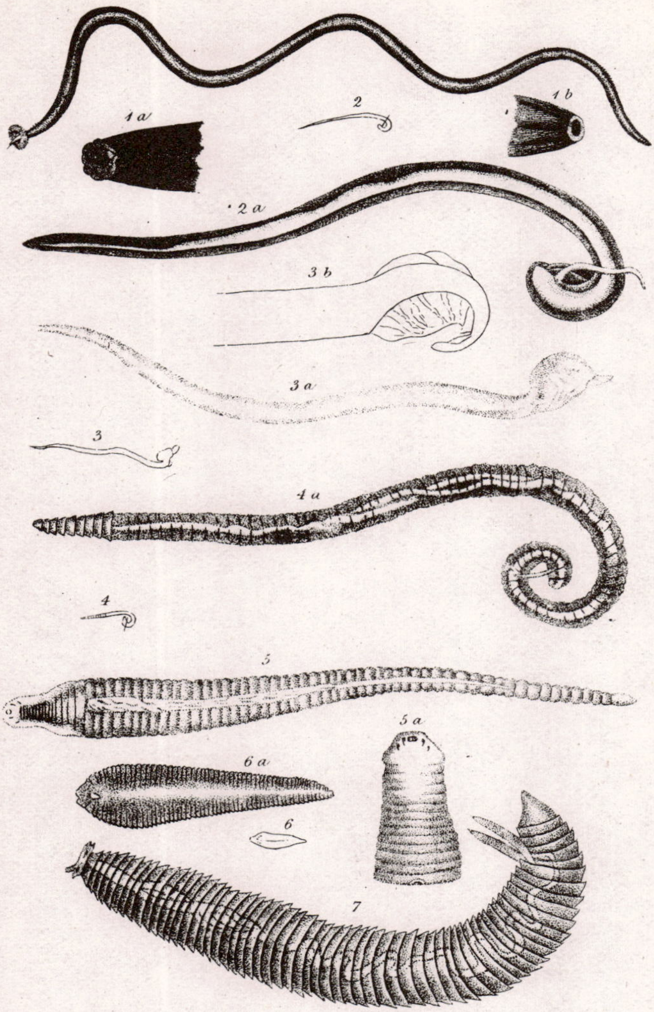

1 *Strongylus gigas.* 4 *Liorhynchus denticulatus.*

2 *Spiropterus storngulinus.* 5 *Linguatula (Pentastoma) tænioides.*

3 *Physalopterus clausus.* 6 *Linguatula.*

7 *Prionoderma ascaroides.*

such as the wolf, the dog, the marten, and even man, remaining there convoluted upon itself, causing this organ to swell, destroying the parenchyma, and, in all probability, occasioning the severest pains to the individual in which it has taken up its abode. These worms have been sometimes passed in urine, while they were still small. This species also sometimes inhabits other viscera. It is often of the finest red colour. It has six papillæ around the mouth, the intestine is straight and wrinkled transversely, the ovary simple and three or four times longer than the body, communicating externally by a hole a little behind the mouth, and as it would seem, its other extremity opening into the anus. A very fine white filament which extends along the belly, has been thought by M. Otto to be the nervous system.

From the ascarides and strongyli, have been lately distinguished

SPIROPTERA,

Whose body is terminated in a spiral, surrounded with two wings, from between which the penis issues.

It is said that a species of them is sometimes found in the human bladder.

There is one in the mole, *Sp. strumosa.* Nitsch., which passes itself into a ring which it pierces in the inmost coat of the stomach, and retains itself there by a small tubercle.

PHYSALOPTERA,

In which the posterior extremity has a bladder between two small wings, and a tubercle from which the penis issues.

SCELOROSTOMA. Blain.,

Which have at the mouth six small denticulated scales.

There is one in the horse, and one in the hog.

LIORHYNCUS, Rud.,

Which have the mouth in the form of a small proboscis.

LINGUATULA, PENTASTOMA, Rud.,

Have the body depressed and trenchant on the sides; and the transverse wrinkles are marked by strong and numerous crenulations. The skin is thin and weak; the head is broad and flatted; the mouth pierced underneath; and at each of its sides are two small longitudinal clefts, from which issue little hooks. The intestine is straight; the genital vessels long and contorted. Both have their issue at the posterior extremity. Near the mouth are two cœca, as in echinorhyncus. A white filament surrounds the mouth, and gives out two descending trunks, in which I think I have recognized some appearance of a nervous system.*

This genus connects the intestina cavitaria, with the parenchymata.

One is known. *Tænia lanceolé,* Chabert. *Polystoma tænioïdes.* Rud. Hist. ii. xii. 8—12.; *Pentastoma tænioïdes,* id. Syn. 123, which attains to nearly six inches in length. It remains in the frontal sinuses of the dog and the horse.

It is here that we should place

PRIONODERMA, Rud.,

Whose body and intestines are very similar; but the mouth is at the anterior extremity; simple, and armed with two small hooks.

But one is known, which attacks the silurus. *Cucullanus ascaroides,* Gœtz. pl. viii. f. 2, 3. Rud. Hist. xii.

I think that we should place at the end of the intestina of this order, but as a family sufficiently different, and which should be divided into several genera, when their economy shall be better understood,

LERNÆA, L.,

Whose body has pretty nearly the same internal and external organization as that of the intestina cavitaria; but is prolonged in front by a neck of corneous substances, at the end of which is a mouth variously armed, and surrounded or followed by productions of divers forms. This mouth and its appendages are insinuated into the skin of the gills of fishes, and fix the animal there. The lernææ are further distinguished by two cords, sometimes of moderate size, sometimes very long, or even very much folded, which hang from the two sides of their tail, and which may be their ovaries.

M. Surrirey has found in the cords of a lernæa, some eggs, which have appeared to him to contain an animal analogous to the crustacea, and very different from the lernæa itself. This fact, compared with what MM. Audouin, and Milne Edwards have observed, on the nicothoë of the lobster, has caused those naturalists to think that the lernææ may be, for the most part, crustacea, grown monstrous after they have been fixed. The males would remain free, and that, according to them, would explain why we never find any but females. But to establish this opinion, it would be necessary to find these males.

LERNÆA, proper,

Have an oblong body, a long and narrow neck, and sorts of horns around the head.

The most common is that which attacks the cod and other gadi. *Lernæa branchialis*, L. Encyc. vers. lxxvii. 2., from one to two inches in length; its mouth is surrounded with three ramous horns, which are, as well as the neck, of a deep brown. Its body, more enlarged, is bent like an S, and the two cords are contorted in a thousand ways. Its horns are rooted, as it were, in the gills of fish.

Another; *L. ocularis*, Cuv., is attached to the eyes of herrings and other fish. It has only short and simple horns, two larger, and two smaller. Its body is slender, its cords long and not folded.

There is one with small horns, unequal, and very numerous. *L. multicornis*, Cuv., on the gills of a serranus, belonging to the East Indies.

Another group,

PENNELLA, Oken.,

Has the head enlarged, furnished at the nape with two small horns, the neck corneous, the body long, wrinkled crosswise, and furnished behind with small filaments, disposed like the barbs of feathers. The two very long filaments originate at the commencement of this pennated part.

There is a species in the Mediterranean. *Pennella filosa, pennatula filosa*, Gm. Boccone mus. 286. Ellis. Trans. Phil. lxiii. xx. 15., seven or eight inches long, which penetrates into the flesh of the sword-fish, the tunny, &c. and torments them horribly.

A third group,

SPHYRIONS, Cuv.

Has the head enlarged on both sides like a hammer, some small hooks at the mouth, a slender neck, followed by a depressed and heart-formed body, which, beside the two long cords, supports on each side a thick fasciculus of setæ. *Chondracanthe lisse*, Quoy et Gaim.

A fourth,

ANCHOSELLA, (Cuv.),

Is fixed to the gills only by a single production, which proceeds from the under part of the body, and is directed backwards. *Lernæa adunca*, Strœm. Sondmœr. pl. i. f. 7, 8., common on many gadi.

A fifth,

BRACHIELLA, Cuv.,

Has two prominences, which form, as it were, two arms, and which are united in a single corneous part, by which the animal fixes itself to the gills. *Brachiella Thynni*, Cuv., &c.

A sixth,

CLAVELLA, Oken.,

Has none of these appendages, and is only fixed by the mouth. *Lernæa uncinata*, Müller, &c.

These last three groups have marked hooks to the mouth. Their cords are but little elongated. There are sometimes other appendages at the posterior part of the body.

After a fresh examination, I refer to the sequel of the Lernæa.

CHONDRACANTHUS, Laroch.,

Which also have hooks to the mouth; and on the sides of the body some appendages, very various in number and form; so much so, that in the course of time, it will be found necessary to establish many divisions of them.

Thus some have on each side two appendages, like arms, more or less prolonged. *Lernæa radiata*, Müll., &c.

Others have many pairs of them partly forked. *Lernæa cornuta*, Id., and many new species.

Or even still more subdivided. *Chondracanthus Zei*. *Laroche*.

There are some which have a slender neck, the body widened, and very unequally indented at the edges. *Lern. triglæ*, Blainv.

I still place at the sequel of this order, an animal which,

in some measure, approximates to it, but which may one day serve as a type for a new order. It forms a genus which I name

NEMERTES, Cuv.

It is a worm extremely soft and elongated, smooth, slender, flatted, and terminated at one extremity by a blunt point, pierced with a hole; widened, and broadly open at the opposite extremity, by which it fixes itself. Its intestine traverses the whole length of the body. Another canal, probably connected with generation, winds along its parietes, and finishes at a tubercle on the margin of the wide aperture. MM. Dorbigny, and de Blainville, who have seen this animal living, assure us that the wide aperture is the mouth.

The only known species, *Nemertes Borlasii*, Cuv. Borlase. Cornw. xxvi. 13., is more than four feet long. It remains sunk in the sand, and, it is said, attacks the anomiæ, which it sucks in their shell.

Near these nemertes should, probably, be placed

TUBULARIA, Renieri,

Equally large, and of a very elongated form; but which have a small mouth pierced under the anterior extremity.

OPHIOCEPHALUS, Quoy et Gaim.,

With the same forms, have the end of the muzzle cleft.

CEREBRATULA, Renieri,

Appear to differ only by having a shorter body.

THE SECOND ORDER OF INTESTINA.

PARENCHYMATA

Comprehends those whose body is filled with a cellular substance, or even with a continuous parenchyma, in which we observe at most, instead of all alimentary organs, some ramified canals which distribute the nutriment, and which, in the majority, derive their origin from suckers, visible externally. The ovaries are also enveloped in this parenchyma, or in this cellular substance. There is no abdominal cavity, nor intestine, properly so called, nor anus; and if we except some doubtful vestiges in the first family, nothing is distinguishable, which has the appearance of a nervous system.

This order may be divided into four families.

The first family,

Acanthocephala, *Rud.*,

Is attached to the intestines by a prominence, armed with recurved spines, which appears at the same time to serve as a proboscis. It only comprehends the genus

Echinorhyncus, *Gm.*,

Which has the body round, sometimes elongated, sometimes in the form of a sac, provided in front with a prominence like a proboscis, armed with small hooks, bent backwards, which can be protruded or retracted, by means of particular muscles.

We sometimes observe at its extremity, a papilla, or a pore, which may be an organ of absorption, but it is also certain that the animal, when plunged in water, swells in every part, and absorbs the liquid through its entire surface, where, it has been thought, a net-work of absorbent vessels was observable. In the interior, no other part is seen comparable to intestines, except two cœca, but little prolonged, attached to the base of its prominence. It is tubiform, and on each side a vessel extends over its entire length. M. de Blainville considers as a nervous system, a filament which stretches along the inferior face, but neither M. Rudolphi, nor M. Cloquet, will have this to be the case. Certain species have a distinct oviduct. In others, the eggs are extended in the cellular substance, or the parenchyma of the body. The males have a small bladder at the end of the tail, and internal vesiculæ seminales, very distinct. It is probable that they fecundate their eggs after they are laid.

These worms attach themselves to the intestines by means of their proboscis, and frequently pierce them. Accordingly individuals are to be found in the thickness of the tunics, and even in the abdomen, adhering to the intestines internally.

The largest species, *Echinorhyncus gigas*, Gm., Gœtze. 1—6. Encyc. xxxvii. 2—7, inhabits in abundance the intestines of the hog and wild boar, where the females sometimes arrive to the length of fifteen inches.

Certain species, besides the hooks of their proboscis, are armed with the like in some other part of their body.

HÆRUCA, *Gm.*,

Do not differ from echinorhyncus, except that their prominence is reduced to a single crown of spines, terminated by double hooks.

One is known which frequents the liver of rats, *Hæruca*

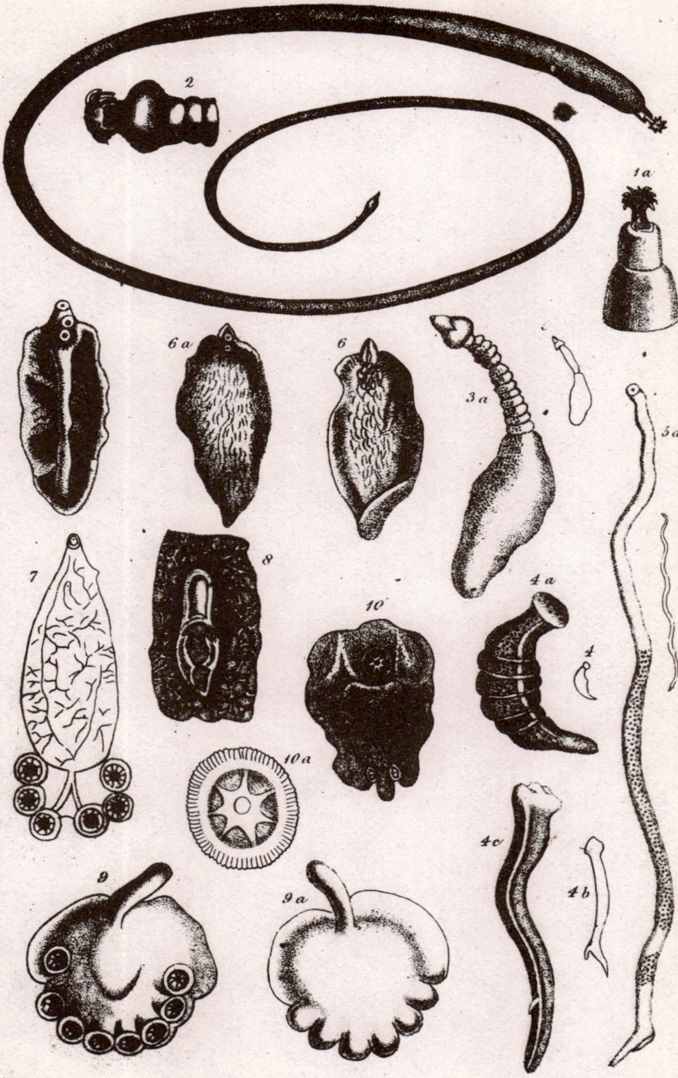

1 *Echinorhynchus gigas.*	6 *Distoma hepatica.*
2 *Hæruca.*	7 *Polystoma integerrimum.*
3 *Amphistoma longicollis.*	8 *Pol. pinguicola.*
4 *Caryophyllæus mutabilis.*	9 *Cyclocotyles Bellones.*
5 *Monostoma Bl.*	10 *Tristoma coccineum.*

muris, Gm., *Echinorh. hæruca*, Rud., Gœtz. ix. B. 12. Enc. Vers. xxxvi. 1.

The second family,

TREMADOTES, *Rud.*,

Comprehends those which have under the body, or at its extremities, some organs in the form of cuppers, by which they are attached to the viscera.

A single genus might be formed of them, to which might be given in common the name of

FASCIOLA, *L.*,

But which may be subdivided as follows, according to the number and position of the cuppers.

FESTUCARIA, *Schr.* MONOSTOMA, *Zeder.*,

Have but one cupper, sometimes at the anterior end, sometimes under this same end. They are found in several birds and fishes.

STRIGEA, *Abildg.* AMPHISTOMA, *Rud.*,

Have a cupper at each extremity. Some are found in several quadrupeds, birds, &c.

We should probably approximate to them

CARYOPHYLLÆUS, *Bl.*

In which he head is dilated, fringed, and has underneath a sucker furnished with two lips not very easily distinguished. Another similar sucker is sometimes perceivable under the tail.

But one is known, taken from divers fresh water fishes, and particularly common in the Bream.

DISTOMA, *Retz.* and *Zéder.*,

Have a sucker, or mouth, at the anterior extremity, and a

cupper a little farther behind, under the belly. The species are extremely numerous ; some are found even in the comb of the eye of some birds. But it also appears that some are not intestine, but inhabit at large the fresh and salt water.

The most celebrated is,

Fasciola hepatica, L., Schœf., Monogr., Copied., Encyc. Vers., pl. lxxx. i. ii., which is so common in the hepatic vessels of sheep, but is also found in those of many other ruminantia, of the hog, of the horse, and even of man. Its form is that of a small oval leaf, pointed behind, having in front a small contracted portion, at the end of which is the first sucker ; this opens into a sort of œsophagus, from which some canals proceed, ramifying through the entire body, and carrying thither the bile on which this animal is nourished. A little backwards is a small retractile tentaculum, which is the penis ; and immediately behind that is the second sucker. Some spermatic vessels, very much folded, fill the middle of the leaf. The ovary, which is found in all the individuals, is enchased in the intervals of the intestines, and the eggs issue through a convoluted canal, which ends at a small hole on the side of the penis. These animals couple reciprocally.

This worm multiplies greatly in sheep when they pasture in humid grounds, and occasions dropsy and death.

M. Rudolphi makes a division which he names ECHINOSTOMA, of the species which have in front a small enlargement, armed with hooks.

HOLOSTOMA, *Nitzsch.*,

Have a moiety of the body concave, and so disposed as to act altogether as a cupper. Their orifices otherwise appear similar enough to those of distoma.

Some of them are found in certain birds. There is one in the fox.

POLYSTOMA, *Zéder.*, (or more properly) HEXASTOMA,

Have the body depressed, smooth, and six cuppers ranged on a transverse line, under the posterior edge. Their mouth appears to be at the opposite extremity.

Some have been found in the urinary bladder of frogs, in the ovary of the human female, on the gills of some fishes, and in the nasal cavity of certain tortoise. (*Polyst. integerrimum*, Rud., &c.)

CYCLOCOTYLES, *Otto.*,

Have eight cuppers forming almost a complete circle under the hinder part of the body, which is broad, and supports in front a small proboscis.

But one species is known, very small, taken on the back of the Bellone, *Cyc. bellones*, Ott. Nat. ac Eur. xi. part ii. pl. xii. f. 2.

I also approximate to the fasciolæ, a subgenus which I name

TRISTOMA, *Cuv.*

Their body is a broad and flat disk, behind, at its inferior face, is a large cartilaginous sucker, which is attached to the body only by a short pedicle, and under its anterior edge are found two small ones, between which, a little behind, is the mouth. In the parenchyma of the body is a ramified circular vessel, the nature of which it is difficult to determine.

A species of an inch and more in breadth, of a lively red-colour, *Tristoma coccineum*, Cuv., is attached to the gills of many fishes of the Mediterranean, such as the sword-fish, &c.

One of the most extraordinary genera of this family is that of

Hectocotyles, *Cuv.*

Long worms, more thick and compressed, at the anterior extremity where the mouth is situated, whose inferior face is altogether furnished with suckers, ranged in pairs, and of a very considerable number, sixty or an hundred, and which support at the posterior extremity a sac filled with the convolutions of the oviduct.

The Mediterranean possesses a species four and five inches in length, with four hundred cuppers, which inhabits the octopus, and penetrates into its flesh. (*Hectocotyle octopodis*, Cuv., An. Sc. Nat. xviii. pl. xi.)

And another smaller, with seventy cuppers, which lives on the argonauta. (*H. argonautæ*, or *trichocephalus acetabularis*, Delle Chiaie. mem. part ii. pl. xvi. f. 1, 2.)

Perhaps it is here that should come

Aspidogaster, *Bœr.*,

Which has under the belly a lamina, hollowed with four ranges of small fossettes.

There is a very small one which is a parasite of the mussels. (*Asp. conchicola*, Bœr. An. Nat. Cur. xiii. part ii. pl. xxviii.)

I cannot avoid thinking that we should still approximate to Fasciola the greater part of those animals comprehended under the genus

Planaria, *Müll.*

Although they do not inhabit other animals, but merely the fresh and salt waters. In fact, their body is depressed, parenchymatous, without any distinct abdominal cavity; the alimentary orifice, placed under the middle of the body, or more behind, and dilating into a small proboscis, conducts, as in fasciola, into an intestine, whose numerous ramifications are hollowed in the thickness of the entire body. A vascular net-work occupies the sides ; behind the alimentary orifice is a double system of genital organs, and the sexual intercourse is reciprocal. Small black points are discoverable in the planariæ, which are probably eyes.

These animals are very voracious, and do not even spare their own species. They not only multiply in the ordinary way, but also very easily by division, and they sometimes even undergo spontaneous separation.

We have several of them in our fresh waters. (*Planaria lactea*, Zool., Dan., &c.)

Our coasts also abound with them, and especially with those of the largest size. (*Pl. aurantiaca*, Nob.)

There are some whose superficies appear to be hairy. (*Pl. brocchii*, Risso.)

Several have two tentacula in front. (*Pl. cornuta*, Müll.)

M. Dugés distinguishes from them

PROSTOMA,

Which have an orifice at the anterior extremity, and another at the posterior, and

DEROSTOMA,

In which the alimentary orifice is underneath, but nearer the anterior extremity.

To the first of these I approximate the PHŒNICURI, *Rud.*, or *Vertumnus Otto*, which have but one orifice at the anterior extremity.

But one species is known, *V. thethidicola*, Otto, Ac. Nat. Cur. xi. part ii. pl. xli. f. 2. A parasite of the *Thethys fimbria*, of a marked appearance, and often with a forked tail, which is the effect of laceration.

The third family of Parenchymatous intestina,

Tenioides,

Embraces those in which the head has two or four pores, or suckers, placed round its middle, which itself is sometimes marked by a pore, sometimes provided with a small proboscis, either naked or armed with spines. Sometimes there are four small proboscides thus armed.

Its most numerous genus is that of

Teniæ.

Their body elongated, often to an excessive degree, flat, composed of articulations, more or less marked, grows narrow in front, and generally supports there a square head, hollowed with four small suckers.

It has been thought that canals were perceptible, which proceed from these suckers, and extend along the margin of the articulations of the body. These last have each one or two pores, diversely placed according to the species, and which appear to be the orifices of the ovaries, which are themselves situated in the thickness of the articulations, when they sometimes assume a simple figure, and sometimes are divided into ramifications. The teniæ are in the number of the most cruel enemies of the animals in which they are developed, and which they appear to exhaust.

Some have no projecting part in the middle of the four suckers. Such is in man the

Tænia **lata**, Rud. *T. vulgaris.* Gm. Gœtz. xli. 5—9., whose articulations are broad and short, and have a double pore in the middle of each lateral face. It is very commonly of the length of twenty feet, and it has been seen even more than a hundred. The large ones are nearly an inch in breadth ; but the head and the anterior part are always very slender. It is extremely troublesome and tenacious. It is frequently found difficult of expulsion, by the most violent remedies.

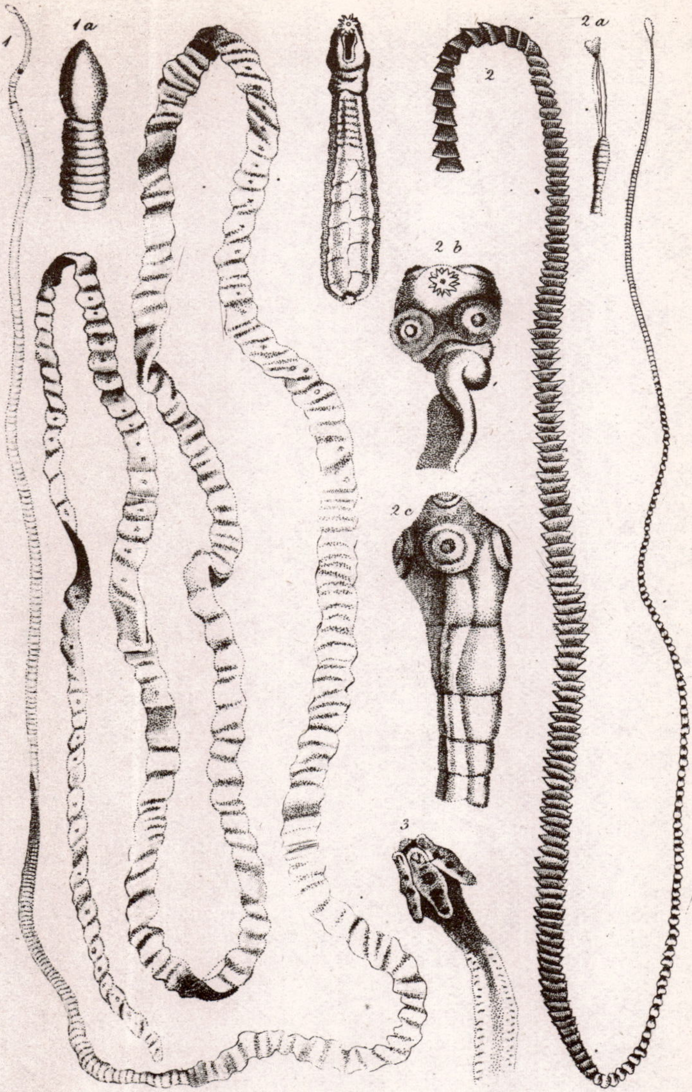

1 *Tænia lata.*

2 *Tæ. solium.*

3 *Head of Bothryocephalus coronatus.*

4 *Dibothryorhynchus lepidopii.*

Others have the prominence between the suckers armed with small points, disposed in radii. Such is again in man,

Tenia Solium, L. Gœtz. xxi. 1—7. Encyc. xl. 15—22. xli. 1—7, whose articulations, except the anterior, are more long than broad, and have the pore alternately at one of their margins. It is in general from four to ten feet in length, but some are found much larger. It is by no means the case that but one of these worms exists in an individual at one time, as is vulgarly supposed. Its detached articulations are called *cucurbitæ*. It is one of the most dangerous of the intestinal worms, and the most difficult to be expelled.

From these common tæniæ, in consequence of the form of their head, have been distinguished, the

TRICUSPIDARIA, Rud.,

Which M. Rudolphi now calls TRIANOPHORA, whose head divided, as it were, into two lips or two lobes, has, on each side, instead of suckers, two stings, with three points.

But one is known which inhabits divers fish, the pike, perch, &c. *Tenia nodulosa*, Gm. Gœtz. xxxiv. 5, 6. Encyc. xlix. 12—15.

BOTHRYOCEPHALUS, Rud.,

Whose head, instead of suckers, has only two longitudinal fossettes, placed opposite one to the other.

Some are found in various fish, and in some birds.

Among the Bothryocephali themselves, it is proper to distinguish

DIBOTHRYORHYNCUS, Blain.,

Which have at the summit two small proboscides, or tentacula, bristling with hooks.

But one is known, with a short body; lives in the lepidopus. Blainv. App. ad Brems. pl. ii. p. 8.,

FLORICEPS, Cuv.,

Which have four small proboscides, or tentacula, armed with recurved spines, by means of which they bury themselves in the viscera.

Certain species,

RHYNCOBOTHRIUM, Blain.,

Have the body long, articulated, and without bladder.

There is one sufficiently common in the rays; *Bothryocephalus corollatus*, Rud. ix. 12., some inches in length. Its head altogether resembles a flower.

Some others,

FLORICEPS,

Properly so called, have the body terminated by a bladder, into which it enters and is concealed.

TETRARHYNCUS, Rud.,

Appear to be only floriceps, reduced naturally to the head, and two articulations, instead of an elongated body, and several articulations.

One is found very commonly in the flesh of the tongue of the turbot, and of several other fishes. *Tetr. lingualis*, Cuv.

TENTACULARIA, Bosc.,

Differ only in the tentacula not being armed with spines.

Those have also been distinguished from the ordinary tæniæ, which, with a head similar to theirs, that is to say, with four suckers, have the body terminated behind by a

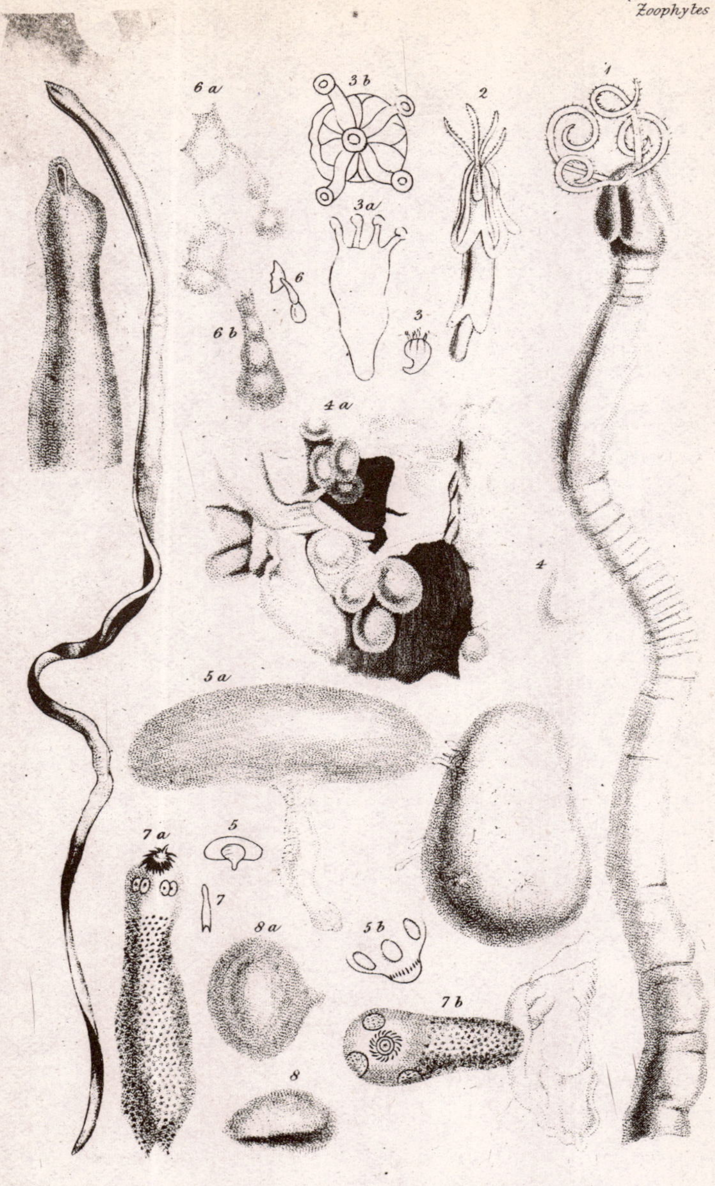

1 *Floriceps corollatus.* 5 *Cysticercus finna.*
2 *Tetrarhynchus lingualis.* 6 *Acrostoma amnii.*
3 *Tentacularia Boscii.* 7 *Cœnurus cerebralis*
4 *Cysticercus pisiformis.* 8 *Echinococcus.*
9 *Ligula simplicissima.*

bladder. Their articulations are not as distinct as in the preceding.

CYSTICERCUS, Rud. Vulgò. HYDATIDES,

Are those in which the bladder supports but a single body, and a single head. They are particularly developed in the membranes, and the cellular substance of animals.

There is one species which multiplies in a great number of quadrupeds, especially in the ruminantia. It is the *globular hydatid* (*Tænia ferarum ; T. caprina ; T. ovilla ; T. bovina ; T. apri ; T. globosa*, Gm.) Gœtz. xvii. A.B. Encyc. xxxix. 6—8.

Another is very common in hares and rabbits; *Tænia cordata ; T. pisiformis ; T. utricularis*, Gm. Gœtz. xviii. A.B. Encyc. xxxix. 6—8.

But the most celebrated is that which remains between the fibres of the muscles of hogs, and produces what is called in them measles. *Tænia cellulosa*, and *T. finna*, Gm. Blum. Abb. 4. Cah. pl. 39 It is small, and multiplies excessively in this disgusting malady, penetrating even into the heart, the eyes, &c. It appears that similar ones have been observed in some apes, and even in man. But it is said never to be found in the wild boar.

ACROSTOMA, Le Sauvage. Ann des Sc. Nat.,

Is very much akin to this genus. It lives in the amnios of cows.

CŒNURUS, Rud.,

Have several bodies and several heads, attached to the same vesicle.

One very celebrated species is known ; *Tænia cerebralis*, Gm. Gœtz. xx. A.B. Encyc. xl. 1—8., which is developed in the brain of sheep, destroys a part of its substance, and occasions a sort of paralysis, called the *staggers*, because it causes them to turn round involuntarily as if they were giddy. Some of these worms have also been seen in oxen, and other ruminants, in which they produce the same effects. The vesicle is sometimes as big as an egg. Its parietes are very slender, fibrous, and exhibit sensible contractions. The small worms are scarcely half a line in length, and enter the vesicle by contraction.

Here should, probably, come the genus,

Echinococcus, Rud.,

But I have not observed it, nor can I form a sufficiently distinct notion of it, for the purposes of classification.

Scolex, Müll.,

Have the body round, pointed behind, very contractile, and terminated in front by a sort of variable head; round which are two or four suckers, sometimes in the form of ears, or small tongues.

Only some very small ones are known, taken from some fish.

I have seen a large one; *Scol. gigas*, Cuv., which penetrates into the flesh of a species of bream, the *sparus raii*, L. the middle part of the body of which is inflated into a bladder, which in the living state contracts or enlarges alternately in its middle. It is the *gymnorynchus reptans*, Rud. Syn. 129.

The fourth family,

Cestoides,

Comprehends those in which no external suckers are observed.

But a single genus is known.

Ligula, Bloch.,

Are, of all the intestina, those which appear to be the most simply organized. Their body resembles a long riband. It is flat, obtuse in front, marked with a longitudinal stria, and finely striated cross-wise. No external organ is distinguishable, and in the interior, nothing is seen but eggs, variously distributed through the length of the parenchyma.

They live in the abdomen of some birds, and more particularly in that of several fresh-water fish, whose intestines they envelope and press to such a degree, as to cause them to perish. At certain periods, they even pierce their abdomen to get out.

There is one in the bream. *Lig. abdominalis*, Gm. *L. cingulum*, Rud. Gœtz. xvi. 4—6, which attains even to the length of five feet. These worms, in some parts of Italy, are considered as an agreeable food.

THE THIRD CLASS OF THE ZOOPHYTES.

Acalephæ, *Vulg.* Sea-nettles,

Comprehend zoophytes which swim in the waters of the sea, and in the organization of which some vessels are perceived which, in truth, are most frequently only productions of the intestines, hollowed in the parenchyma of the body.

THE FIRST ORDER OF ACALEPHÆ.

Acalephæ simplices,

Float and swim in the waters of the sea, by means of the contractions and dilatations of their body, although their substance is gelatinous, without any apparent fibres. The sorts of vessels which are seen in some of them, are hollowed in the gelatinous substance. They often come from the stomach in a visible manner, and do not give rise to any true circulation.

Medusa, L.,

Have a disk more or less convex above, similar to the head of a mushroom, and to which the name of *umbrella* has been given. Its contractions and dilatations concur in producing the movements of the animal. The edges of this umbrella, as well as the mouth, or the suckers, more or less prolonged into pedicles, which take its place, at the middle of the inferior face, are furnished with tentacula of very various forms and sizes. These different degrees of complication have given rise to very numerous divisions.

We shall give the general name of

Medusa (proper),

To those which have a mouth under the middle of the inferior surface, whether simply opening on the surface, or prolonged into a pedicle. And among the medusa proper

We may unite under the name of

Æquorea,

All those in which this mouth is simple and not prolonged or furnished with arms.

When they have no tentacula around the umbrella, they constitute the PHORCYNIA of Lamarck.

When the umbrella is furnished with tentacula all around, it characterizes the ÆQUOREA, more particularly so named, of Peron, one of the most numerous subgenera, especially in the seas of hot climates. *Medusa æquorea*, Gm. &c. &c.

Certain species are remarkable for laminæ, which furnish their inferior surface. Others, FOVEOLIA, Peron, are signalized by little fossettes, hollowed in the circumference of the umbrella. *Medusa mollicina*, Forsk., &c.

We may thus unite under the name of

PELAGIA,

Those in which the mouth is prolonged into a peduncle, or divided into arms. *Pelagia panopyra*, Peron.

In all these subgenera, there are no lateral cavities; but a much greater number of these medusæ, with simple mouths, have in the thickness of the umbrella four organs, formed of a plaited membrane, filled at certain periods with an opake substance, and which appear to be ovaries. They are most frequently lodged in as many cavities, open at the inferior face, or on the sides of the pedicle; and which, as I think, have been erroneously taken for mouths, because some little animals are occasionally caught in them. Some naturalists take them for organs of respiration; but it is more probable that this function is exercised on the margins of the umbrella. The tentacula, whether of the margin of the umbrella, or the circumference of the mouth, vary not only according to the species, but even according to age.

We unite under the name of

CYANEA, Cuv.

All the medusæ with central mouth, and four lateral ovaries.

The most extended, *Medusa aurita*, L. Müll. Zool. Dan.

lxxvi. and lxxvii., acquires with age four long arms. Its umbrella is finely ciliated all around. Some reddish vessels pass from the stomach to the circumference, subdividing in their course.

Another, *Med. chrysaora*, Cuv., has the edges furnished with long tentacula, and some fulvous or brown lines or spots, disposed in radii in its convexity. It is also very common, and varies much as to the spots.

We have given the general name of RHIZOSTOMA, to a portion of the great genus MEDUSA, comprehending the species which have no mouth open at the centre, and which appear to be nourished through the suction of the ramifications of their pedicle, or of their tentacula. They have four ovaries or more.

RHIZOSTOMA, (proper) Cuv.,

Are those which have in the middle a pedicle, more or less ramified, according to the species.

The vessels proceeding from the small ramifications of the pedicles, unite in a cavity of its base, from which branches proceed for all the parts of the umbrella.

The most common is the *Rhizostome bleu*. Cuv. Jour. de Phys. tom. xlix. p. 436. Reaum. Acad. des Sc. 1710. pl. xi. f. 27, 28. It is found every where on the sand of our coasts, when the sea retires, and its umbrella is sometimes nearly two feet in breadth. Its pedicle is divided into four pairs of arms, forked and denticulated almost *ad infinitum*, furnished each at the base with two corslets, equally denticulated. The umbrella has all around, in the thickness of its margins, a fine net-work of vessels.

According to the observations of Messrs. Audouin, and Milne Edwards, these medusæ live in society, or at least are always to be met with united in great numbers, and swimming in the same direction, the body being inclined obliquely.

Cephæa, Peron, are distinguished from the other Rhizostoma only by some filaments mixed with the denticulations of their pedicle. *Medusa Cephæa,* Forsk., &c.

The Cassiopææ have, properly speaking, no pedicle; their arms usually eight in number, sometimes branched, spring immediately from the inferior surface. *Med. Frondosa,* Pallas, &c.

Other species, without a central mouth, have none of these numerous ramifications to the pedicle, nor open cavities to lodge the ovaries. They may be united under the name of

Astoma.

Some, however, have still a large pedicle, furnished on each side with hairy filaments, which may serve as suckers. Lymnorea and Favonia, Per.

Others have not even these filaments, but a membrane in the form of a funnel at the end of the pedicle, and from the bottom of which the vessels appear to issue, which ascend into the pedicle, and spread out in the umbrella. *Geryonia,* properly so called; Peron. There is one in the Mediterranean; *Med. proboscidalis,* Forsk. xxxvi. 1.

This membrane is even wanting in others. Orythia, id. *Medusa minima,* Bart., &c.

There are some without any pedicle, but in which the under part appears furnished with small suckers along the passage of the vessels. Berenice, Peron. *Cuvieria eurisochroma,* Peron.

Finally, there are some in which no suckers are perceptible, the two surfaces being smooth, and without apparent organs. Eudora, Peron.

The Mediterranean possesses a species of these about the size of a five-franc piece, and to which the people give the name of that coin. *Eud. moneta,* N.

When these very simple animals assume more concavity, their

inferior surface becomes interior, and may perhaps be regarded as a true stomach. These are the CARYBDEA, Peron. Those in whose interior no traces of vessels are perceptible, do not properly differ from the *hydræ*, but in size. *Medusa marsupialis*, Gm.

It has been found necessary to separate from the medusæ, some genera united to them by Linnæus, on very slight grounds of relation, such as

BEROE, Müll.

They have an oval or globular body, furnished with projecting ribs, bristling with filaments or fringe, proceeding from one pole to the other, and in which we perceive vascular ramifications, and a sort of movement of the fluid. The mouth is at one extremity ; in those which have been examined, it conducts into a stomach which occupies the axis of the body, and on the sides of which are two organs probably analogous to those which we have called ovaries in the medusa.

Such is,

Medusa pileus, Gm. *(Globular Beroë)* Baster. i. iii. xiv. 6—7. Encyc. xc. 3, 4, with a spherical body, furnished with eight ribs ; two ciliated tentacula, susceptible of a great elongation, issue from its inferior extremity. It is very common in the north seas, and even in the channel on our coasts, and is considered as one of the aliments of the whale.

According to MM. Audouin and Milne Edwards there exists in the axis of these animals a cavity which goes from one pole to the other, and which communicates externally by means of an inferior aperture, which may be considered as the front mouth. In the superior third of this cavity is contained, and as it were suspended, a sort of straight and cylindrical intestinal tube, which has its external aperture immediately at the superior pole, and which supports on each side two granular cords (ovaries perhaps). The cavity is filled with a fluid in motion, which is seen to pass into two lateral tubes, that soon subdivide each into four branches, and terminate on the surface of the body, by opening into longitudinal canals. Through these canals the fluid is conducted into the ciliæ, which are constantly in motion, and appear to be respiratory organs. Finally, from the sides of each of the eight costal canals, spring an infinity of small vessels, or transverse sinuses, which establish a communication between them, and which sink into the surrounding parenchyma. On each side of the spheroid, and internally, we perceive two small masses which occupy each the bottom of a cavity or cul-de-sac, and give birth to two long contractile filaments, issuing through two circular apertures, situated towards the inferior third of the body. These filaments are subsequently divided into a great number of branches.

To this genus have been referred some more simple species, which merely resemble a sac, furnished with ciliated ribs, and open at the two ends. IDYA. Oken. *Berœ ovatus* Brug. or *Medusa infundibulum*, Gm.

There are some which are even destitute of ribs, and whose form represents that of a band without bottoms. DOLIOLUM. Otto. *D. Mediterraneum*, Otto.

CALLIANIRA, Peron, do not appear to differ from Beroe, except in having much more projecting ribs, united two by two, to form two sorts of wings. Their internal organization is not sufficiently known. *Callianira didiploptera*, Peron.

JANIRA, Oken., appear to be akin to the callianiræ; but on each side there are three large ciliated ribs, and two long filaments divided into branches. *Beroë hexagone*, Brug. Encyc. vers. pl. 90. f. 6.

ALCINOE, Rang., have the body cylindrical, open at one extremity, furnished on the other with two large wings, which by being folded over it, can envelope it altogether. Its cylindrical part is flanked with four projecting ribs, each terminated in a point, and has eight lines of ciliæ. *Alcinoe vermiculata*, Rang.

Ocyroe, id., have the same sort of body with four ranges of ciliæ, but without ribs; they have also similar wings, each furnished at the base with two ciliated points. *Ocyroë maculata*, id., &c.

It is also near Beroe that we should place

Cestum, Lesueur,

A very long gelatinous ribbon, one of the edges of which is furnished with a double range of ciliæ. The inferior one has some also, but smaller, and less numerous. It is at the middle of the inferior margin, that the mouth is situated; a large aperture which leads into a stomach, pierced through the breadth of the ribbon, and proceeding to a very small anus. From the extremity near the anus, proceed some vessels, which extend to the two extremities of the ribbon. At the sides of the mouth two sacs open, which are probably the ovaries. This animal may be compared to a callianira with two ribs, and whose wings were excessively prolonged.

The only known species,

Cestum veneris, Lesueur. Nouv. Bull. des Sc. June 1813. pl. v. f. 1., is from the Mediterranean. Its length, or rather its breadth, is more than five feet; its height, two inches. It is preserved entire with great difficulty.

The two following genera, which have also been united to the medusæ, might form a small family in this order, in consequence of the internal cartilage which supports the gelatinous substance of the body.

Porpita, *Lam.*,

Have this circular cartilage, and its surface marked with concentric striæ, crossing with radiating striæ. At the upper face it is clothed only with a slender membrane, which

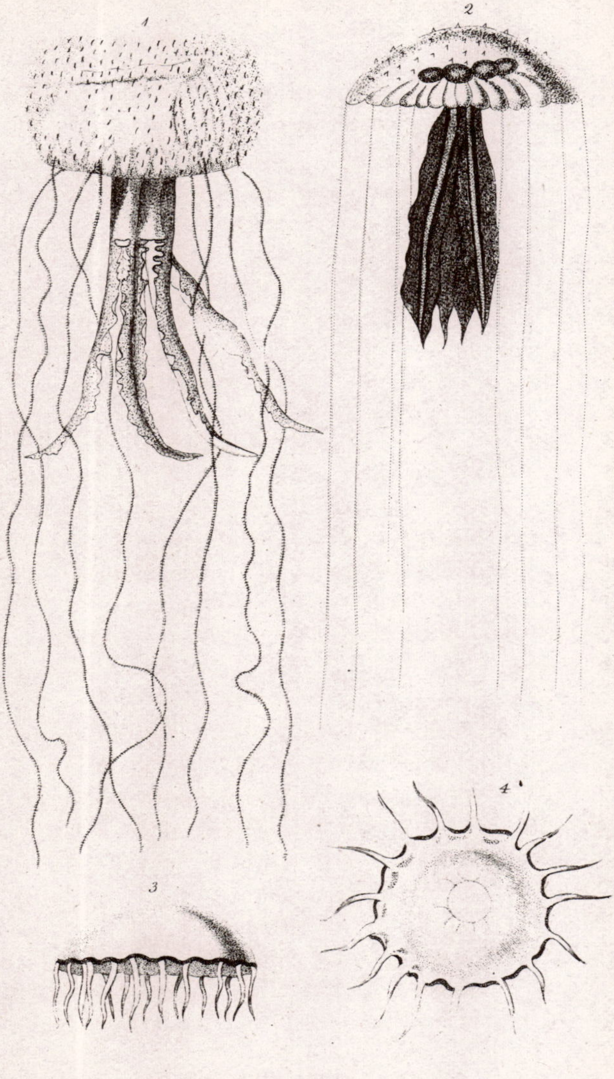

1 *Pelagia panopyra.*

2 *Cyanæa Labiche.*

3 *Equorea cyanogramma.*

1. *Rhizostoma Aldrovandi.*

2. *Cassiopea borbonica.*

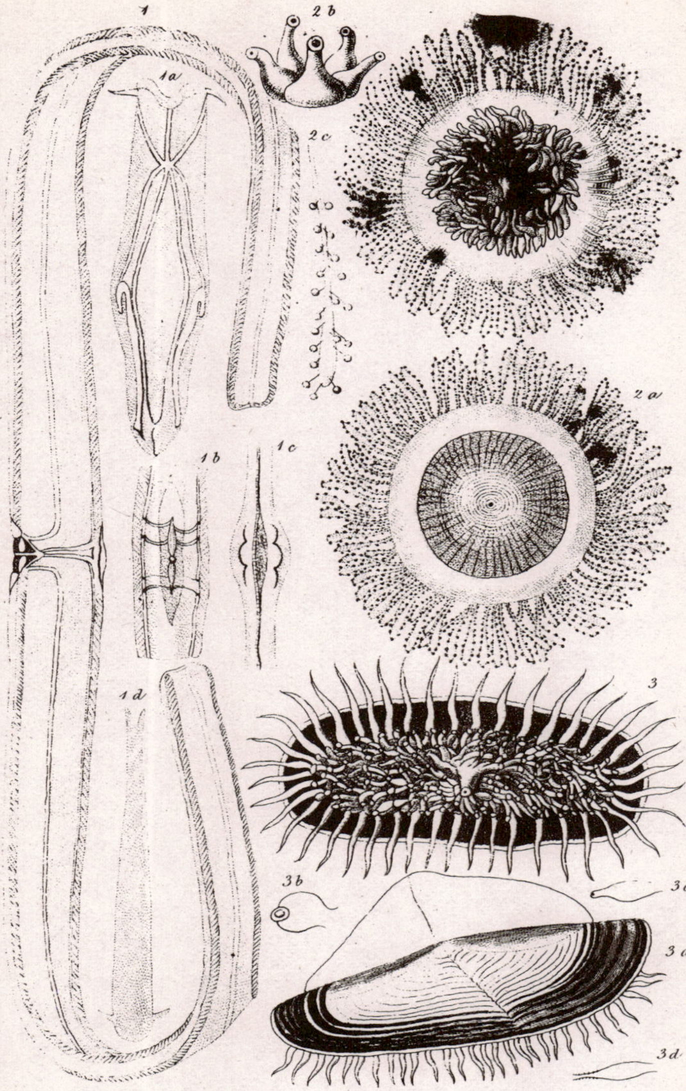

1 *Cestum veneris.*

2 *Porpita chrysocoma.*

3 *Velella limbosa.*

out-edges it. The inferior is furnished with a great number of tentacula, of which the external ones are longer, and furnished with little ciliæ, terminated each by a globule. They sometimes contain air; the middle ones are shorter, more simple and more fleshy. At the centre of all these tentacula is the mouth, in the form of a small projecting proboscis. It conducts to a simple stomach surrounded with a sort of glandulous substance.

But one species is known of a fine blue, belonging to the Mediterranean, and to warmer seas. (*Medusa umbella*, Müll.)

VELELLA,

Have, like the porpitæ, at the inferior face, a mouth in the form of a proboscis, surrounded with innumerable tentacula, of which the external ones are longer; but the latter are not ciliated, and what gives a more important character is, that the cartilage, which is oval, has on its upper face a vertical crest placed obliquely, and sufficiently raised. This cartilage is transparent, and has only concentric striæ.

One species is also known of the same colour, and living in the same seas, as the porpita. It is eaten fried. It is the *Medusa velella*, and *Holothuria spirans* of Gmelin.

THE SECOND ORDER OF ACALEPHÆ.

ACALEPHÆ HYDROSTATICÆ,

ARE recognized by one or more vesicles, usually filled with air, by means of which they are suspended in the water. Appendages singularly numerous, and varying in form, some of which probably serve as suckers, others perhaps as ovaries, and others, longer than the rest, as tentacula, unite themselves to these vesicular parts, to compose the whole apparent organization of these animals. There is no mouth that can properly be recognized as such.

PHYSALIA, *Lam.*,

Consist of a very large oblong vesicle, raised above into a projecting crest, oblique and wrinkled, and furnished underneath, towards one of its extremities, with a great number of cylindrical fleshy productions, which communicate with the vesicle, and terminate variously. Those towards the centre support groups, more or less numerous, of small filaments; those towards the sides are only bifurcated into two filaments, one of which is often very much prolonged. One of the extremities of the vesicle appears to have a very small orifice. But in the interior, we find, instead of intestine, only another vesicle, with more slender parietes, and which has some cœca extending partly into the cavities of the crest. For the rest, there is neither nervous, nor circulating, nor glandular system. The animal swims on the surface of the sea when it is calm, and employs its crest as a sail. It also has, in the living state, some very long filaments, more slender than the others, and sown, as it were, with pearls, or little drops. It is reported that their touch stings like that of the nettle.

There are some in all the seas of warm climates. (*Holothuria Physalis*, L., &c. &c. &c.)

PHYSSOPHORA, *Forsk.*,

Have sensible relations with the Physaliæ; but their bladder is much smaller in proportion, without a crest, often accompanied with lateral vesicles, and their various and numerous tentacula are suspended vertically under this vesicle, as a garland or a cluster of grapes.

1 *Berenice eurisochroma.*

2 *Geryonia probescidalis.*

3 *Cephea papuensis.*

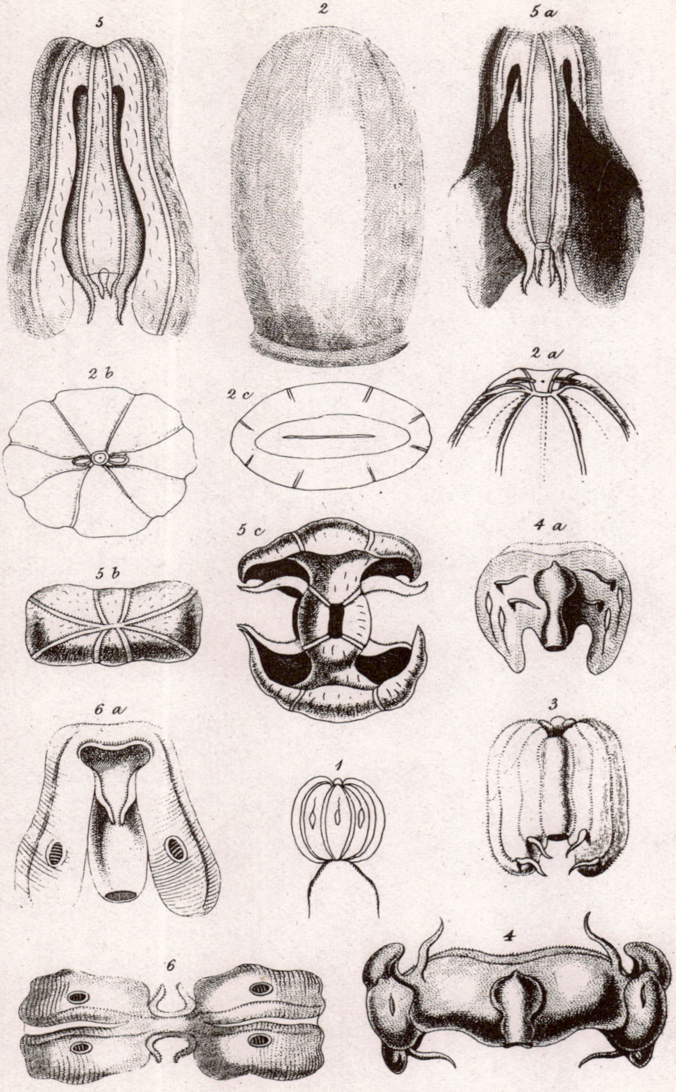

1 *Beroe pileus.*

2 *Ber (Idya) elongatus.*

3 *Beroe costata.*

4 *Callianira bucephalon.*

5 *Alcinoe vermiculata.*

6 *Ocyroe maculata.*

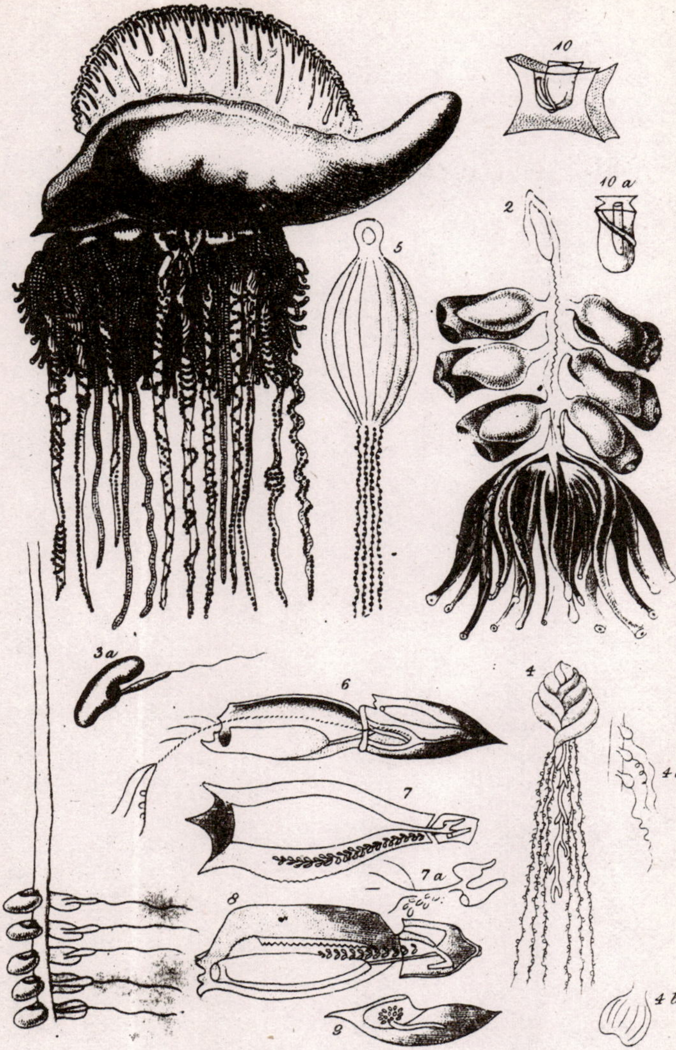

1 *Physalia atlantica.*	6 *Diphia dispar.*
2 *Physsophora disticha.*	7 *Calpe pentagona.* _ Quoy and Gaym.
3 *Phys. nicea.*	8 *Abyla trigona:* _ Jid.
4 *Hoppodius luteus.*	9 *Cymba sagitta.*
5 *Rhizophyza heliantha.*	10 *Cuboides vitreus.*

In

PHYSSOPHORA, (proper) *Peron*,

Between the superior vesicle and the tentacula, are found other vesicles, placed side by side, or one above the other, and of a form sometimes irregular, sometimes polyhedral, and forming by their union, prisms or cylinders; the tentacula, partly conical, partly cylindrical, partly formed of groups of filaments, or of globules, some in fine filiform, and susceptible of great elongation, form a cluster, or garland, at the inferior extremity. (*Physsophora hydrostatica*, Gm., &c. &c.)

HIPPOPUS, *Quoy* and *Gaive.*,

Have only lateral vesicles, almost semicircular, or in the form of a horse's hoof, crowded on two ranges, and thus forming a sort of ear, like that of certain grains, from which also hangs a garland, which traverses all those pieces. The contractions of those vesicles impress upon the whole a rapid movement.

CUPULITES,

Have their vesicles attached regularly on both sides of an axis, often very long.

RACEMIDES, *Cuv.*,

Have all their vesicles globular, small, furnished each with a small membrane, and united into an oval mass, which is moved by their combined contractions.

RHIZOPHYZA, *Peron*,

Have no lateral vesicles, but only a superior vesicle, and a long stem, along which the tentacula are suspended; some conical, others filiform. (*Physsophora filiformis*, Forsk.)

STEPHANOMIA, *Peron*,

Appear to be a third combination, in which the lateral vesicles, which, in the Physsophoræ proper, adhere to the top of the stem, above the tentacula, are prolonged over its length, and mingled with tentacula of various forms. (*Stephanomia amphitritis*, Peron.)

It is at the end of the hydrostatic acalephæ, that may be placed

DIPHYES, *Cuv.*

A very singular genus, in which two different individuals are always found together, one being emboxed in a cavity of the other, which arrangement, however, permits them to separate without destruction to individual life. They are gelatinous, transparent, and move pretty nearly like the medusæ. The one which receives the other, produces from its cavity, a chaplet which traverses a semi-canal of its companion, and appears to be composed of ovaries, of tentacula, and of suckers, like those of the preceding genera.

MM. Quoy and Gaymard, have established divisions in this genus, according to the forms and relative proportions of the two individuals.

Thus, in

DIPHYES, (proper.)

The two individuals are almost similar, pyramidal, with some points round their aperture, which is at the base of the pyramid.

In CALPE, the emboxed individual has still the pyramidal form, but the other is very small and square.

In ABYLE, the emboxed one is oblong, or oval; the other a little smaller, and in the form of a bell.

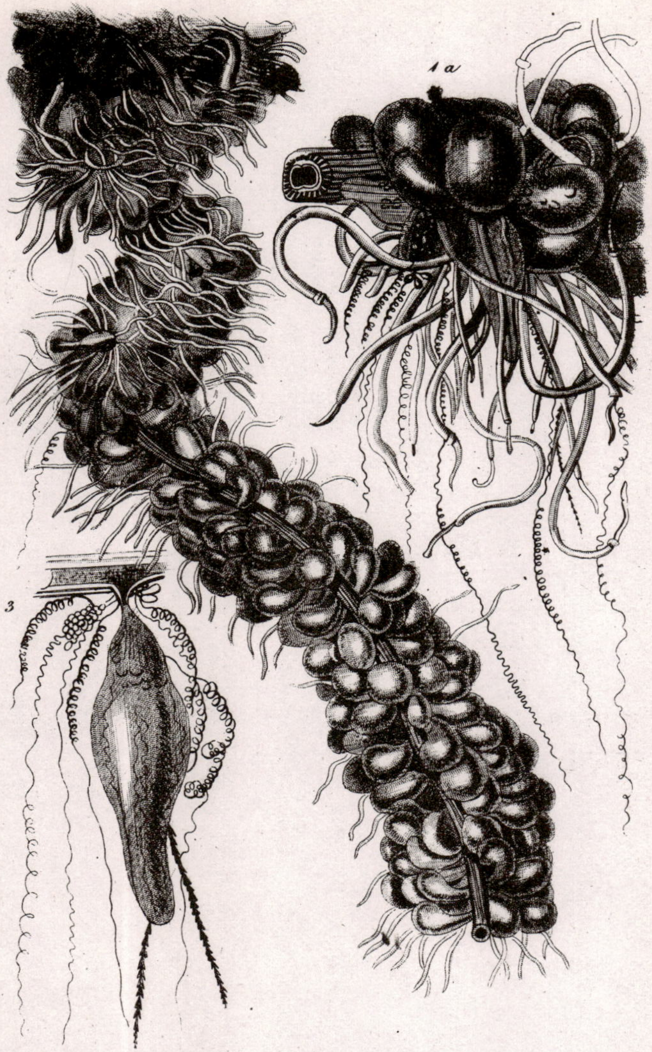

1 *Stephanomia uvaria.*

1a. *Part of same enlarged.*

3 *A separate sucker.*

In the CUBOIDES, it is the emboxed one which is small and bell-formed. The other is much larger and square.

In NAVICULA, the emboxed one is bell-formed; the other is equally large, but shaped like a shoe.

There are also several other combinations.

FOURTH CLASS OF THE ZOÖPHYTES.

THE POLYPI

HAVE been thus named, because the tentacula which surround their mouth give them some slight resemblance to the octopus, which the ancients called *polypus*. The form and number of these tentacula vary. The body is always cylindrical, or conical, often without any other viscera than its cavity, often also with a visible stomach, to which adhere intestines, or rather vessels, hollowed in the substance of the body, like those of the medusæ. In this case ovaries are also visible. The majority of these animals are susceptible of forming composite beings, by sprouting forth new individuals, like buds, nevertheless, they also propagate by eggs.

THE FIRST ORDER OF POLYPI.

POLYPI CARNOSI. (Vulgo. FIXED SEA NETTLES.)

COMPREHEND some fleshy animals, which have the habit of fixing themselves by their base, but several of which can also crawl upon this base or detach it altogether, and swim, or suffer themselves to be carried away by the motion of the waters. They are most frequently limited to the movement of expanding more or less the aperture of the mouth, which also serves the purposes of an anus. It is surrounded with tentacula more or less numerous, and leads into a stomach which terminates in a *cul-de-sac*. Between this interior sac and the external skin, is an organization tolerably complicated, but still rather obscure, consisting chiefly of vertical and fibrous leaflets, to which the ovaries adhere, similar to threads very much contorted. The intervals of these leaflets communicate with the interior of the tentacula, and it appears that the water can enter there and come out by little orifices around the mouth; at least the actinia sometimes ejaculates it in this way.

ACTINIA, *L.*

Their fleshy body, often adorned with lively colours, developing numerous tentacula, placed round the mouth, on several ranges like the petals of a double flower, has occasioned the name of *sea anemone* to be bestowed upon them. They are amazingly sensible to light, and open or close according as the day is more or less fine. When they retract their tentacula, the aperture from which these organs issue, contracts, and closes upon them like that of a purse.

Their power of reproduction is scarcely less than that of the armed polypi. They shoot forth again the parts which have been cut, and multiply by division. Their ordinary generation is viviparous. The little actiniæ pass from the ovary into the stomach and come out through the mouth. These zoophytes dilate their mouth considerably when they are hungry. They devour all sorts of animals, and especially crustacea, testacea, and small fishes, which they seize with their tentacula, and digest pretty quickly.

ACTINIA, (proper)

Are fixed by a broad and flat base. The species most common on our coasts are

Act. Senilis, L. Three inches broad, with a coriaceous, unequal, and orange-coloured envelope, and tentacula on two ranges, of a moderate length, and usually marked with a rose-coloured ring. It keeps principally in the sand, into which it instantly sinks back upon the slightest alarm.

Act. Equina, L. With a soft skin, finely striated; the colour usually of a fine purple, often spotted with green; smaller, the tentacula longer, and more numerous than the preceding. It covers all the rocks of our coasts of the channel, and ornaments them as though they bore the finest flowers.

Actinia Plumosa, Cuv. White, four inches and more in breadth; the edges of its mouth expand into lobes, all charged with innumerable small tentacula. There is an interior rank of larger ones.

Actinia Effœta. Rond. Lib. xvii. cap. xviii. Of a clear brown, striped longitudinally with whitish; of an elongated form, often more narrow towards the bottom; a smooth skin, and numerous tentacula. When it contracts itself, there often issue through the mouth some long filaments, which come from the ovaries. It attaches itself, in preference, on shells, and is extremely common in the Mediterranean.

THALASSIANTHUS, *Ruppel*, are actiniæ, with ramified tentacula. (*Thal. aster*. Ruppel.)

His DISCOSOMA, are some in which the tentacula are reduced to nearly nothing by their shortness. (*Discos. nummiforme*, Id.)

ZOANTHUS, *Cuv.*,

Have the same fleshy tissue, the same disposition of mouth,

and of tentacula, and an organization pretty nearly similar to that of the actiniæ. But they are united in a more or less considerable number, on a common base, sometimes in the form of a creeping stem. (*Hydra sociata*, Gm.) Sometimes in the form of a broad surface. (*Alcyonium mamillosum*, Ell.)

LUCERNARIA, Müll.

It would appear, should be approximated to the actiniæ; but their substance is softer; they fix themselves to fucus and other marine substances, by a slender pedicle; their upper part is dilated like a parasol; the mouth is in the middle. Numerous tentacula approximated in fasciculi, furnish the edges. Between the mouth and these same edges, are eight organs, in the form of cœca, which proceed from the stomach, and contain red and granulated matter.

In *Lucernaria quadricornis*, Müll., Zool., Dan. xxxi. 1—6, the edge is divided into four forked branches, each supporting two groups of tentacula. In the *L. auricula*, Ibid. clii. the eight groups are equally apportioned round an octagon edge.

SECOND ORDER OF POLYPI.

POLYPI GELATINOSI,

ARE not, like the preceding, invested with a hard envelope, neither do they produce in the interior of their structure an axis of ligneous, fleshy, or corneous substance. Their body is gelatinous, of a form more or less conical; its cavity serves as a stomach.

Hydra, *Linn.*

Present us the animals of this class reduced to their greatest possible simplicity. A small gelatinous trumpet-shaped body, whose edges are furnished with filaments, which serve as tentacula, constitutes their whole apparent organization. The microscope enables us to discover nothing in their substance but a transparent parenchyma, filled with grains, a little more opake. Nevertheless, they swim, they crawl, they even walk, by fixing alternately their two extremities, like the leeches, or geometrical caterpillars. They agitate their tentacula, and make use of them to seize their prey, which is visibly digested in the cavity of their body. They are sensible to light and seek after it. But the most marvellous property is that of constantly and indefinitely reproducing the parts which are taken from them, so that the individuals may be multiplied at will, by section. Their natural multiplication takes place by the young shooting forth from different parts of the body of the adult, and at first resembling branches of it.

Our dormant waters nourish five or six different species of them, which differ in colour, and the number and proportion of the tentacula.

The most celebrated, in consequence of the experiments on reproduction, to which it has first given occasion, is

Hydra viridis, Trembley, Pol. i. 1. Rœs. iii. lxxxviii. Encyc. lx. vi., which is, in fact, of a fine clear green colour. It is particularly found under the water-lentils.

Hydra fusca, Tremb. Pol. i. 3, 4. Rœs. iii. lxxxiv. Encyc. lxix., is more rare; of a grey colour; its body is not an inch long, and its arms are more than ten.

Corine, *Gœrtner*,

Have a fixed stem, terminated by an oval body, more consistent than that of the hydræ, open at the summit, and bristling on all its surface, with small tentacula. Some carry heir eggs at the bottom of this body. (*Tubularia coryna*, Gm. &c.)

CRISTATELLA,

Have in the mouth a double range of numerous tentacula, curved into a half-moon, forming a sort of plume of this figure, and attracting by their regular motion, the nutritive molecules. These mouths are supported upon short necks attached to a common gelatinous body, which moves like that of the hydræ. These animals are found in our dormant waters. To the naked eye they appear like small mouldy spots. (*Cristatella mucedo*, Cuv.)

VORTICELLA,

Have a fixed stem, often branched, and very much divided, each branch of which terminates by a body, in the form of a trumpet, or a bell. We behold issue from the aperture, in two opposite groups, some filaments, which maintain a continual motion, and attract the nutritive molecules. The species are numerous in our seas, and the majority too small to be well distinguished but by the microscope. They form bushes, arbuscula, plumes, and take other shapes, all very agreeable.

PEDICELLARIA

Are found between the spines of the echini, and are considered by divers authors, as organs of these animals. Nevertheless, it is more probable that they are polypi, which choose these parts as their asylum. A long slender stem terminates by a cornet, furnished at its extremity with tentacula, sometimes in the form of filaments, sometimes in that of leaves.

THIRD ORDER OF POLYPI.

POLYPIFEROUS POLYPI

Form this numerous series of species, which for a long time were regarded as marine plants, and whose individuals are, in fact, united in great numbers, to form composite animals, for the most part fixed like vegetables, whether they form a stem or simple expansions, by means of the solid supports which line them at the interior. The individual animals, more or less analogous to the actiniæ, or hydræ, are all connected together by a common body, and receive their nutriment in common; so that what is eaten by one, profits the general body, and all the other polypi. They have also a community of volition; at least, this is certain with respect to the species not fixed, such as the pennatula, which are seen to swim by the contractions of their stems, and by the continued movements of their polypi.

The name of *polyparia* has been given to the common parts of these composite animals. They are always formed by deposition, and by strata, like the ivory of the teeth; but sometimes they are at the surface, sometimes in the interior of the composite animal. These different positions have given rise to the establishment of the families of this order.

The first family,

POLYPI TUBIFERI,

Inhabit tubes, the common gelatinous body of which traverses the axis, like the pith of a tree, and which are open either at the summit, or at the sides, to allow the polypi to pass.

Their most simple polypi appear principally to resemble the hydræ, and the cristatellæ.

TUBIPORA, *L.*,

Have simple tubes, of a stony substance, each containing a polypus. These tubes are parallel, and united together from space to space, by transverse laminæ, which has caused them to be compared to the tubes of an organ.

The most known species, *Tubipora musica*, L. Seb. iii. cx. 89, is of a fine red. Its polypi are green, and have the form of the hydræ. It abounds in the Indian Archipelago.

It would seem, that it is to the tubipora that we should approximate some fossil polyparia, likewise composed of simple tubes, such as the CATENIPORA, Lam., in which the tubes are arranged upon lines, intercepting empty meshes; *Tubipora catenulata*, Gm. The FAVOSITES, id. *(Corallium Gothlandicum)* composed of hexagonal tubes, crowded closely together, &c.

TUBULARIA, *L.*,

Have simple or branched tubes, of a corneous substance, from the extremities of which the polypi issue forth, and show themselves.

The polypi of the fresh-water tubulariæ,—PLUMATELLA, Bosc. appear very much akin to the crystatellæ, in the disposition of their tentacula.

We have some which creep over the plants of our dormant waters. (*Tubularia campanulata*, Rœs.)

TUBULARIA MARINA,

Have polypi with two ranges of tentacula; the exterior one being developed in radii, the interior raised in the style of a tuft.

Our coasts produce one, *Tub. indivisa,* Lam. Ellis. Corall. xvi. c., with simple tubes, two or three inches high. Similar to blade of straw.

TIBIANA, *Lamour.,*

Have zig-zag tubes, which give out from each angle, a small open branch. (*Tibiana fasciculata, Lam.*)

CORNULARIA, *Lam.,*

Have small conical tubes, from each of which issues a polypus, with eight denticulated arms, like those of the alcyones, gorgones, &c. (*Tubularia cornucopia.*)

ANGUINARIA, *Lam.,*

Have small cylindrical tubes, adhering to a creeping stem, each of which is opened laterally near its extremity, for the passage of a polypus. (*Sertularia anguina,* Ellis.)

CAMPANULARIA, *Lam.,*

Have the ends of the branches through which the polypi pass, expanded into the form of bells.

Lamouroux distinguishes them into CLYTIA, which have climbing stems. (*Sertularia verticillata,* Ellis, &c.)

And into LAOMEDEA, which have not climbing stems, and in which the bells are smaller, and with shorter branches. (*Sert. dichotoma,* Gm., &c.)

SERTULARIA, *L.,*

Have a corneous stem, sometimes simple, sometimes branched, and in its sides some cells of very various forms, occupied by polypi, all attached to a gelatinous stem, which traverses the axis after the manner of the pith of a tree. These zoophytes have the appearance of little plants, equally delicate and agreeable to view. Their propagation goes on by eggs, or gemmæ, which are developed in some cellules, larger than the others, and of different forms.

The various directions of their cellules, have given rise to their distribution into several subdivisions. Thus when the little cells are ranged on one side only over the branches, they are then the AGLAOPHENIA of M. Lamouroux, which M. de Lamarck names PLUMULARIA. (*Sert. Myriophyllum*, Gm., &c.)

When they are assembled in certain places, like the small tubes of an organ, they constitute AMATIA, Lamour., or SERIALARIA, Lam. (*Sertularia lendigera*, Ellis.)

We may distinguish from them the species in which the cellules thus disposed, surround the stem with a spiral figure.

When the cellules are placed around the stem, in horizontal rings, these are the ANTENNULARIA, Lam., which M. Lamouroux had named CALLIANYRA, and has since changed into NEMERTESIA. (*Sertularia antennina*, Gm. &c.)

Thus the name of SERTULARIA proper, remains to those species only in which the cells are on both sides of the stem, whether opposed, *Sert. abietina*, Gm., &c. or alternate. *Sertularia operculata*, Ell., &c. M. Lamouroux distinguishes the first under the name of DYNAMENES.

The extreme smallness of the cellules has also caused him to establish the genus THŒA, (*Sertularia hœlicina*, Gm. &c.)

The second family is that of

POLYPI CELLULOSI,

In which each polype is adherent in a corneous or calcareous cell with their parietes, and does not communicate with the others but by an external tunic very much attenuated, or by fine pores, which traverse the parietes of the cellules. These polypi have a general resemblance to the hydræ.

CELLULARIA, L.,

Have cellules disposed so as to form branching stems in the manner of the sertulariæ, but without any tube of communication in the axis. Their substance besides, is more calcareous.

Lamouroux distinguishes among them the

CRISIA,

Whose cellules in two ranks, usually alternate, open on the same face. (*Sertularia eburnea*, Gm., &c.)

ACAMARCHIS,

Which, with the same disposition, have a vesicle at each aperture. (*Sertularia neritina*, Gm.)

LORICULA,

In which each articulation is composed of two cells, back to back, whose opposite orifices are towards the top, which is widened. (*Sertularia loricata*, Ell.)

EUCRATIA,

In which each articulation has but a single cellule, with an oblique aperture. (*Sertularia chelata*, Gm., &c.)

We may approximate to them

ELECTRA, *Lamouroux*,

In which each articulation is composed of several cells, disposed like a ring. (*Flustra verticillata*, Gm.)

We should separate from them those which have cylindrical articulations, empty in the interior, and hollowed at all their surface with cellules, disposed like a quincunx. They conduct to the flustræ, and perhaps to the corallinæ. I name them SALICORNARIA, (*Cellularia salicornia*, Ellis, &c.)

Flustra, *L.*,

Present a great number of cellules, united like the combs of bees, and sometimes covering divers bodies, sometimes forming leaves or stems, of which one side only is furnished with cellules in certain species, and both in others. Their substance is more or less corneous. (*Flustra foliacea*, Gm. &c. &c.)

Cellepora, *Fabri.*,

Present accumulations of little cellules, or calcareous vesicles, crowded one against the other, and each pierced with a small hole. (*Cellepora hyalina*, Gm., &c.)

Tubulipora, *Lam.*,

Are accumulations of little tubes, the entrance of which is as broad as or broader than the bottom. (*Millepora tubulosa*, Gm.)

Some bodies exist in the sea, pretty similar to polyparia, in their substance and general form, but in which no polypi have as yet been observed. Their nature is therefore doubtful; and some great naturalists, such as Pallas, and others, have considered them to be plants. Nevertheless, there are several who regard them as polyparia, with polypi, and with cells exceedingly small. If this conjecture be true, they must belong to the present order. Those among them whose interior is filled with corneous filaments present, however, some analogy with the Ceratophytes.

Corallina, *L.*,

Have articulated stems, supported on a sort of roots, divided into branches equally articulated, at the surface of which no pores are visible, and in which it has not been possible to perceive any polypi.

They are divided as follows :

CORALLINA proper,

Have their articulations calcareous, of an homogeneous appearance, without any perceptible bark.

The bottom of the sea is entirely covered on certain shores, with the *Corallina officinalis*, L. Ell. Corall. xxiv. a. A. b. B., the articulations of which are in a reversed oval, the small branches disposed like pinnated leaves, and themselves supporting other branches similarly disposed. It is white, reddish, or greenish. It was formerly employed in pharmacy, because of its calcareous substance.

Lamouroux again distinguishes from them, but rather slightly,

AMPHIROE.

The articulations of which are elongated. (*Corallina rigens*, Sol. et Ell., &c.)

JANIA,

Which merely have the branches more slight, and the articulations less cretaceous. (*Corallina rubens*, Ellis, &c.)

CYMOPOLIA,

In which the articulations are separated one from the other, by corneous intervals. Their surface exhibits more marked pores.

M. de Lamarck had already separated,

PENICILLUS, Lam. *Nesea* Lamouroux,

Which have a simple stem, composed internally of corneous fibres, woven together, and as it were, *felted ;* encrusted with a calcareous plaster, and terminated by a bundle of articulated branches, analogous to those of the ordinary corallines. (*Corall. penicillus*, &c.)

HALYMEDE, *Lamouroux*,

Have stems articulated and divided like the corallines; but the substance of their articulations, which are very broad, is penetrated in the interior by corneous filaments, which are easily deprived of their calcareous crust, by acids. (*Corallina tuna*, Sol. et Ell.)

FLABELLARIA, *Lamarck*,

Have no distinct articulations, but form large foliaceous expansions, composed like the articulations of the halymedes, and the stem of the penicillæ, of corneous filaments, encrusted with a calcareous envelope. (*Corallina conglutinata*, Sol. et Ell., &c.)

GALAXAURA, *Lamouroux*,

Have stems divided by dichotomy, but their branches are hollow. (*Corall. obtusata*, Sol. et Ell., &c.)

LIAGORA, *Id.*,

Have hollow stems, divided by dichotomy, but without articulations. (*Corall. marginata*, Sol. et Ell., &c.)

It is, perhaps, at the end of the corallines that we should place

ANADIOMENE, *Lamouroux*,

Vulgarly known under the name of *Corsican moss*, and so usefully employed as a vermifuge. It is composed of articulations regularly disposed in branches, of a substance a little corneous, covered with a gelatinous coat. (*Anadiomene flabellata*, Lamour.)

Among those productions without apparent polypi, which by conjecture are referred to the polyparia, there are few more singular than

Acetabulum, *Lam.*

A slender and hollow stem, supports a round and thin plate, like a parasol, striated in radii, crenulated at the edge, and having at the centre a small smooth disk surrounded with pores. No polypi are perceptible in it. The radii of its disk are hollow, and contain greenish grains, which has caused it to be regarded as a plant by Carolini.

There is one of them in the Mediterranean. (*Tubularia acetabulum*, Gm. Donat. Adri. iii. Tournef. Inst. cccxviii.)

Polyphysa, *Lam.*,

Have, like the preceding, a slender and hollow stem, but which supports at its summit, a packet of small closed vesicles, instead of a disk formed of tubes. (*Pol. aspergillum*, Lamour.)

The third family,

Polypi corticati,

Comprehends the genera in which the polypi are all attached by a common, thick, fleshy, or gelatinous substance, in the cavities of which they are received and which envelopes an axis of variable form and substance. The polypi of those which have been observed, are a little more complex than the preceding, and approach more to the actiniæ. In their interior a stomach is distinguishable, from which proceed eight intestines, two of which are prolonged into the common mass, and the others terminating sooner, appear to fill the place of ovaries.

They are subdivided into four tribes.

The first is that of

Ceratophyta,

In which the interior axis has the appearance of wood or horn, and is fixed. Two genera are known, both very numerous.

ANTIPATHES, Lin. Vulgò. *Black coral,*

Have the substance of their axis branched, and of a ligneous appearance, enveloped in a bark so soft, that it is destroyed after death. Accordingly, in cabinets they resemble branches of dry wood. (*Ant. spiralis*, Sol. et Ell.)

GORGONIA, *L.*,

Have, on the contrary, this ligneous or corneous substance of their axis, enveloped in a bark, the flesh of which is so penetrated with calcareous grains, that it dries up on the axis, and often preserves its colours in a state of great beauty and vivacity. It is dissoluble in acids. The polypi of several species have been observed. They have each eight denticulated arms, a stomach, &c. like those of the coral, and the alcyone. (*Gorgonia pinnata*, Gm., &c.)

M. Lamouroux distinguishes from these,

PLEXAURUS,

Whose thick bark, with cellules not projecting, makes but little effervescence in acids. (*G. Crassa*, Gm., &c.)

EUNICEA,

Whose bark, organised like that of Plexaurus, has projecting nipples, from which its polypi issue forth. (*Gorgonia antipathes,* Seba, &c.)

MURICEA,

Whose bark, moderately thick, has projecting nipples, covered with imbricated and bristling scales. (*M. spicifera,* Lam.)

PRIMNOA,

Whose elongated nipples are imbricated, hanging one upon the other. (*Gorg. Reseda*, Gm., &c.)

The second tribe,

LITHOPHYTA,

Has the interior axis of a stony substance and fixed.

ISIS, *L.*,

Have the axis branched, and without impressions or cellules, hollowed at its surface. The animal bark which envelopes it, is mixed with calcareous grains, as in Gorgonia.

CORALIUM, *Lam.*,

Has its axis without articulations, and only striated at its surface.

It is to this subgenus that belongs

The *coral of commerce*. *Isis nobilis*, L. Esp. i. vii., celebrated for the fine red colour of its stony axis, and the fine polish of which it is susceptible, which renders it particularly adapted for the composition of various trinkets. Its fishery is very productive in several parts of the Mediterranean. Its bark is cretaceous and reddish. Its polypi, as in many other genera, have eight denticulated arms.

MELITE, *Lam.*,

Have the stony substance of their axis interrupted by inflated knots, of a substance similar to cork. (*Isis ocracea*, Espern.)

ISIS, properly so called. *Lam.*,

Have it interrupted by strangulations, the substance of which resembles horn. Their bark, thick and soft, falls more easily than in the preceding. (*Isis hippuris*, L., &c.)

M. Lamouroux yet distinguishes from the Isis, properly so called,

MOPSEA,

The bark of which is thinner, and more adherent. (*Isis dichotoma*, Seba, &c.)

MADREPORA, *Lin.*,

Have their stony part sometimes branched, sometimes in rounded masses, or in extended laminæ, or in leaflets; but always furnished with lamellæ, which are united concentrically in points where they represent stars, or which end at lines, more or less serpentine. In the living state, this stony part is covered with a living bark, soft and gelatinous, and bristling with rosettes of tentacula, which are the polypi, or rather the actiniæ, for they have generally several circles of tentacula, and the stony laminæ of the stars correspond, in some respects, to the membranaceous laminæ of the body of the actiniæ. The bark and the polypi contract upon the slightest touch.

The varieties of their general form, and of the figures which result from the combination of their lamellæ, have given rise to several subdivisions, many of which, however, enter one into the other. They cannot be definitively established, until the relations of the polypi, with these dispositions, shall be known.

When there is but a single circular star, or in an elongated line, with very numerous laminæ, these are the FUNGIA, Lamarck. *Mad. fungites,* L. Their animal truly represents a single actinia, with large and numerous tentacula, and whose mouth corresponds to the sunken part, where all the laminæ end.

We find among the fossils, stony polyparia, with a single star, which appear to have been free from all adherence. These are TURBINOLIA, Lamarck, Madr. turbinata, L. CY-

CLOLITHES, *Mad. porpita*, Linn., and TURBINOLOPSES, Lamour. *Turbin. ocracea*, Lamour.

When the madrepore is branched, and there are no stars, but at the end of each branch, it is a CARYOPHYLLIA, Lam. The branches are striated; at each star is a mouth surrounded with many tentacula. (*Madr. cyathus*, Sol. et Ell., &c.)

OCULINA, *Lam.*,

Have small lateral branches, very short, which gives them the appearance of having stars along the branches, as at the end. (*Madr. virginea*, L., &c.)

MADREPORA (properly so called) *Lam.*,

Have all their surface bristling with little stars, with salient edges.

His POCILLOPORA have in the surface small sunken stars, and pores in the internals. (*Madr. damicornis*, Esper.)

In SERIALOPORA, the little stars are ranged in lineal series. (*Madr. seriata*, Pall.)

ASTREA,

Have a broad surface, most frequently convex, hollowed with crowded stars, each of which has a polype furnished with numerous arms; but on a single range, at the centre of which is the mouth. (*Madr. radiata*, Sol. et Ell., &c.)

When the surface is plane, or in broad laminæ, sown with stars on one side only, the animals are named EXPLANARIA. (*Madr. cinerascens*, Sol. et Ell.)

The PORITES are in some sort ramous astreæ. (*Madr. porites*, Sol. et Ell.)

When this surface is hollowed with elongated lines, as it were with valleys, separated by hillocks, furrowed cross-wise, the animals are MEANDRINA, *Lam.*

In each valley, from space to space, open some mouths, and the tentacula, instead of forming rosettes around these mouths, form a range along the sides of each valley. Some species have none at all, but the margin of each mouth is merely festooned. (*Madr. labyrinthica*, Sol. et Ell., &c.)

If the hillocks which separate these vales, are raised in leaflets, or crests furrowed on both sides, the species are PAVONIA. There are mouths in the bottom of the valleys, and usually without tentacula. (*Madr. agaricites.*)

When these hillocks are elevated in cones, as if they were salient stars, M. Fischer names them HYDNOPHORA, M. Lamarck, MONTICULARIA. They should be distinguished according as their polypi are at the summit of the salient part, as in the oculinæ, or in the bottom of the concave parts, as in the meandrinæ. (*Madr. exesa.*, Sol. et Ell.)

AGARICINA,

Are composed of laminæ, hollowed on one side only, by valleys which are themselves furrowed. (*Madr. cucullata*, Sol. et Ell.)

It has been thought that we might approximate to the madrepores in general, certain polypariæ ; SARCINULA, *Lam.*, formed of cylinders, whose section forms a star, in consequence of the projecting lamina, which traverse the interior, *Madr. organum*, Lin. ; when there is a solid axis in the middle of the laminæ, these are STYLINÆ. These polypariæ may also, perhaps, be related to the tubipora.

MILLEPORA, *L.*,

Have their stony part of very various forms, and its surface only hollowed with small holes, or pores, or even without any apparent holes.

M. Lamarck distinguishes

DISTICHOPORA,

In which the pores, more marked, are ranged on both sides of the branches. (*Millepora violacea*, Pall.)

Among those in which the pores are equally divided, we distinguish

MILLEPORA, (proper) *Lam.*,

Solid, diversely branched. (*Mill. alcicornis*, Pall.)

When their pores are not apparent, as sometimes happens, they are named NULLIPORA. (*Millepora informis*, L., &c.)

ESCHARA, *Lam.*,

Which have expansions flatted like leaflets. (*Millepora foliacea*, Ell., &c.)

RETEPORA, *Id.*,

Which are escharæ, pierced with meshes. (*Millepora cellulosa*, Ell.)

ADEONA, *Lamouroux*,

Which are escharæ supported on an articulated stem. There are some entire, and some pierced with meshes. (*Ad. grisea*, Lamour.)

The third tribe,

POLYPARIA NANTIA,

Whose axis is stony, but not fixed.

PENNATULA, *L.*,

Have the body common, free from all adherence, of a regular and constant form, and able to move by the contractions of its fleshy part, and also by the combined action of its polypi. This body is fleshy, susceptible of being contracted, or dilated, in its divers parts, by means of fibrous layers which enter into its composition. Its axis encloses a simple stony stem. The polypi have generally eight denticulated arms. Most of the species shed a lively phosphoric light.

Whatever may be the general form of the pennatulæ, they have always one of their extremities without polypi. This has been compared to the tubular part of the feathers of birds.

PENNATULA, (properly so called) *Cuv.*,

Which, having given their name to the whole genus, have themselves derived it from their resemblance to a quill. The part without polypi is cylindrical and terminated in a blunt point. The other part is furnished on each side with wings, or barbs, more or less long, and broad, supported by spines, or stiff setæ, which spring from their interior, and bristle one of their edges, without, however, being articulated with the petrous stem of the axis. It is from the centre of these barbs that the polypi issue forth.

The Ocean and the Mediterranean both produce

Pennat. rubra, et *Penn. phosphorea*, Gm., Albinus Annot. Acad, i. vi. 3—4. Which has the stem between the barbs very rough behind, except on a line which traverses its length.

We find more particularly in the Mediterranean,

Pennatula grisea, Gm., Albinus, Annot. Acad. 1, 2. larger, with broader barbs, more spiny, and a smooth stem.

VIRGULARIA, *Lam.*,

Differ from the pennatulæ, only because their wings, much shorter in proportion to their total length, are deprived of spines. (*Pennatula mirabilis*, Müll., Zool., Dan. xi., very different from the true *Pen. mirabilis* of Linnæus.)

These wings sometimes represent only simple transverse ranges of tubercles. (*Pennat. juncea*, Pall. et Gm.)

SCIRPEARIA, *Cuv.*,

Have the body very long and very slender, and the polypi isolated, ranged alternately along both sides. (*Pennatula mirabilis*, Linnæus.)

PAVONARIA, *Cuv.*,

Have also the body elongated and slender, but have polypi only on one side, and they are crowded there in the shape of a quincunx. (*Pennat. antennina*, Bohatsch.)

RENILLA, *Lam.*,

Have the body short, and instead of the part, which in the proper pennatula is furnished with barbs, a broad kidney-formed disk, supporting the polypi on one of its faces. (*Pen. reniformis*, Ellis.)

VERETILLUM, *Cuv.*,

Have a cylindrical body, simple, and without branches, furnished with polypi in a part of its length. Their bone is usually small, and the polypi large. We can trace more easily than in any other composite zoophyte, the prolongations of their intestine in the common stem.

We have one in the Mediterranean, *pennatula cynomorium*, Pall., Misc., Zool. xiii. 1—4; *alcyonium epipetrum*, Gm., Rap. Ac. Nat. Cuv. xiv. p. 2. xxxviii. 1. Often more than a foot in length, thicker than one's thumb, remarkable for the brilliancy of the light which it sheds. Finally,

OMBELLULARIA, *Cuv.*,

Have a very long stem, supported by a bone of the same length, and terminated at the summit only by a branch of polypi. (*Pennatula encrinus*, Ellis.)

We find in the sea, and among the fossils, some small

petrous bodies pierced with pores, which have been thought to approximate to the millepora. If they were, in fact, enveloped with a bark, containing polypi, they would be moveable polyparia, and should rather be approximated to the pennatulæ. Such are

The OVULITES, *Lin.*, in the form of eggs, hollow internally, and often pierced at both ends; the LUNULITES, orbicular, convex, striated, porous on one side, and concave on the other; the ORBULITES, orbicular, flat, or concave, porous on both sides, or at the edges. If the DACTYLOPORA is free, as M. de Lamarck thinks, it would also appertain to this division. It is a hollow ovoïd, open at both ends, with two envelopes, both pierced with meshes, like the retepora.

In the fourth tribe, the animal bark encloses only a fleshy substance, without either osseous or corneous axis.

ALCYONIUM, *Linn.*,

Have, like the pennatulæ, polypi, with eight denticulated arms, and intestines stretching into the common mass of the ovaries; but this mass is not sustained by an osseous axis. It is always fixed to the body, and when it is elevated into trunks or branches, we find nothing in its interior but a gelatinous substance, traversed by several canals, surrounded by fibrous membranes. The bark is harder, and hollowed with cellules, into which the polypi retire more or less completely.

We have in abundance in our seas the

Alcyonium digitatum, Ell., Coral. xxxii., which is divided into thick, short branches; *Alc. exos.*, which has more slender branches of a fine red, &c.

Linnæus, and his successors, have united, on rather slender grounds, to the alcyonia, divers marine bodies of various tissue, but always without visible polypi. Such are

THETHYA, *Lam.*

The interior of which is altogether bristled with long siliceous spirals, which are united on a central nucleus equally siliceous. Their crust presents, as in the sponges, two orders of holes; the one closed by a sort of trellis-work, must be for the entrance of water; the other, gaping, destined for its exit.

We also place at the sequel of the Alcyonia,

SPONGIA, *L.*

The sponges are marine fibrous bodies, which exhibit nothing perceptibly, but a sort of thin, fine, gelatine, which dries up, and scarcely leaves any trace, and in which, as yet, no polypi, or any other moveable part, have been observed. It has been said that the living sponges undergo a sort of tremor, or contraction, when they are touched; that the pores with which their superficies is pierced, palpitate in some degree; but the existence of such movements has been contested by Mr. Grant, and other writers.

The sponges assume innumerable forms, each according to its species, as those of shrubs, trumpets, vases, tubes, globes, fans.

Every one is acquainted with the *common sponge, spongia officinalis,* which is found in great brown masses, formed of very fine fibres, flexible, elastic, and pierced with a great number of pores, and small irregular conduits opening one into the other.

FIFTH AND LAST CLASS OF THE ZOOPHYTES, AND OF THE ENTIRE ANIMAL KINGDOM.

INFUSORIA.

It is customary to place at the end of the animal kingdom, some beings so small that they escape the naked eye, and which remained undistinguished, until the microscope had, in some sort, revealed to us another world. The majority of them present a gelatinous body of the most extreme simplicity, and this, undoubtedly, is their proper place ; but there have also been left among the infusoria, some animals much more complicated in appearance, and which resemble them only in their minuteness, and the habitat in which they are usually found.

Of these we shall make our first order, still, however, insisting on the doubts which yet remain relative to their organization.

FIRST ORDER OF INFUSORIA.

ROTIFERA,

Are distinguished, as we have just said, by a greater degree of complication. Their body is oval and gelatinous. A mouth, a stomach, an intestine, and an anus near the mouth, are distinguishable. Behind, the body is most frequently terminated by *a tail,* variously constructed, and in front it supports a singular organ, diversely lobated, with denticulated edges, the denticulations of which execute a successive vibration, which might cause the belief that this organ consisted in one or several denticulated and revolving wheels. One or two prominences on the neck, have even appeared to some observers to support eyes. This revolving organ does not serve to conduct the aliments towards the mouth; it might be suspected that it has some relations with respiration.

FURCULARIA, *Lam.*,

Have the body without armature, the tail composed of articulations, which enter one into the other, and terminate by two filaments.

It is upon one of them, the *Furcularia,* or *Rotifère des toits,* that Spallanzani has made his famous experiments, as to their resurrection. Covered with dust in the gutters, it dries up, but yet, after several weeks will resume life and motion, if it be moistened with a little water.

TRICHOCERCA, *Lam.*,

Do not appear to me to differ from the Furculariæ, but in a little less development of their vibratile organs. (*Trichoda paxillum,* Müll., &c.)

VAGINICOLA, *Lam.*,

Appear to be only trichocercæ, enveloped in a transparent case; but in this instance, there is reason to fear, that some optical illusion has taken place. (*Trich. innata.* Müll.)

TUBICOLARIA, *Lam.*,

Do not differ from the furculariæ, except that they live in

small tubes, which they construct with foreign molecules, but which make no part of their body, like those of the polyparia. Their rotatory organ, nevertheless, appears out of the tube, pretty nearly after the manner of the head of the polypi.

We have one common enough on the confervæ of our marshes. *Vorticella tetrapetala*, Blumenb., Dutrochet. Ann. Mus. xix. xviii. 1—10, whose rotatory organ is divided into four lobes.

BRACHIONUS, *Müll.*,

With rotatory organs, and a tail pretty nearly similar to those of the furculariæ, carry a sort of membranous, or scaly buckler, which covers their back, like that of certain monoculi.

SECOND ORDER.

INFUSORIA HOMOGENEA,

Whose body exhibits neither viscera, nor other complications, and often does not even present any appearance of a mouth.

The first tribe

Comprehends those which, with a gelatinous body, more or less contractile in its various parts, still presents as external organs, some ciliæ more or less strong.

They are named URCEOLARIA, *Lam.*, when they have the form of a trumpet, from which the ciliæ issue, as in the polypi called *Vorticellæ ;* TRICHODÆ, when with a flat body, these ciliæ are at one extremity ; LEUCOPHRA, when they surround the whole body; KERONA, when there are some large ones, representing a sort of horns; HIMANTOPUS, when these pretended horns are elongated into a kind of filament.

The second tribe

Presents those which have no visible external organs, unless we consider a tail as such.

CERCARIA, *Müll.*,

Have, in fact, their oval body terminated by a filament. To this genus belong, among others, the animalculæ, which are observed in the semen of different animals, and on which so many fantastic hypotheses have been founded.

When this filament is forked, as sometimes happens, M. de Lamarck names these animals TURCOCERCA.

VIBRIO, *Müll.*,

Have the body slender, and round, like the little end of a thread.

It is to this genus that belong the pretended eels in glue and vinegar. *Vib. glutinis et aceti.* These last often distinguishable by the naked eyé. It is asserted that they change skin, that they have sexes, produce living young in summer, and eggs in autumn. The frost does not destroy them. The first appear in the glue or paste of diluted flour.

ENCHELIS, Müll.,

Have the body oblong, softer, less defined than that of vibrio.

CYCLIDIUM have it flat and oval; PARAMECIUM, flat and oblong; KOLPODA, flat and sinuous; GONIUM, flat and angular; BURSARIA, hollow like a sac.

The most singular of all are

PROTEUS, *L.*

No determined form can be assigned to them; their figure

changes every moment, and successively assumes all kinds of limits, sometimes rounded, and compact, sometimes divided and subdivided into strips in the most fantastic manner. (*Proteus diffluens*, Rœsel.)

MONAS, *Müll.*,

Resemble, in the microscope, little horns, which move with great quickness, although without any apparent organ of motion.

VOLVOX,

Have a small globular body revolving on itself, and often enclosing smaller globules, which, doubtless, serve to propagate the race.

SUPPLEMENT

ON THE

FOURTH AND LAST GRAND DIVISION

OF

ANIMALS.

THE ZOOPHYTES.

UNDER this complex denomination, which signifies *animal plants*, we comprehend those animals, which, not being symmetrical, are not capable of being divided into two similar sides, situated at the right and left of a secant plane, or imaginary line, which should pass through the length of the animal, all the parts of which might be referred to this plane. But, on the contrary, in the animals in question, the parts are disposed, in a manner more or less regular, round a common point, taken as a centre, or the axis of the body. This has caused them sometimes to be compared to flowers, of which all the parts have also this disposition. From this, Pallas has derived the denomination of *centrina*, which he has given to one division of these animals, which has been changed to radiata, radiated animals, and *actinozoaria*, by some writers.

This last type of the animal kingdom, we do not find designated under a collective name, except by the ancient and the most modern zoologists.

Thus Aristotle, who appears, nevertheless, to have known some species of the principal classes which constitute it, has never employed the word zoophytes as a collective name or otherwise, although, in relation to the sponges, he has said, that they partake more of plants than of animals, and that it may be doubted whether they are animals or vegetables. But the compound word zoophytes, is not to be found in his works, though some authors have erroneously attributed it to him.

It does not appear that he was acquainted with the animals which are designated at present under the name of *holothuria.* He certainly employs this denomination, the etymology of which is unknown; but he applies it to beings which have not the faculty of locomotion, though they are not attached any where, which leads us to presume, that by this word he indicated the actiniæ, which, nevertheless, were designated by him under the name of *acalephos,* and which, in fact, he classes among the animals which partake at once of the animal and plant.

Aristotle, on the contrary, was perfectly well acquainted with the echini and the asteriæ, which he designates, the first under the name of sea-urchins, the second, under that of sea-stars. But he has made them animals of his division of testacea; an approximation which we shall find to have been admitted, even to the end of the last century. He distinguishes several species very well, such as *spatangus brissus,* the sea urchins proper, and a smaller species. But still it does not seem that his distinction is established on characters sufficient to enable us to recognize at the present day, in any very certain manner, the animals of which he intended to speak.

As for the asteriæ or sea-stars, which in one passage he enumerates among the equivocal beings between the animal and plant, while in another he ranges them with the testacea, the little that he says is very incomplete, and not very easily to be understood.

The medusæ also appear to have been known to Aristotle, but he confounded them with the actiniæ, properly so called, under the common denomination of sea-nettles, *acalephe* and *knide*, which signify nettle. These, again, are beings whose nature is equivocal between the plant and the animal. In fact, he tells us, it is the property of an animal to move itself, to direct itself towards its food, to feel what it encounters, and to use in its defence the firm and hard parts of its body. But to have an organization extremely simple, to attach itself with facility to rocks, and to have a mouth, but no apparent orifice for the issue of the excrements; all this is more of the nature of the plant. In other places of his works, Aristotle enters into some details of organization and manners, respecting his acalephai. Among other matters, he tells us that there are some which remain fixed upon rocks, and other submerged bodies, and others which are detached from them; an observation which has led several authors to think that he was speaking of actiniæ and medusæ; but this opinion is not free from doubt.

Of all the rest of the animals which constitute the zoophytes of modern zoologists, it does not appear that Aristotle was acquainted with any except the sponges, respecting which he gives us some details of tolerable extent.

As to the animals which he calls *polypi*, it is well known, that they are not the same as those so called in the present day. We have already seen in our observations on the mollusca, that they are the octopi, respecting which he has left us some valuable remarks.

It is not certain whether his *Pneumon* be a medusa, as some authors assure us, and not a testaceum. As to his tethys, it is evident that they are our ascidiæ.

Pliny, as might be well imagined, has not added much to what Aristotle has said, concerning the zoophytes. He has confined himself to translating the Greek names of urchins, sea-stars, nettles, and sponges, by those of *echini*, *stellæ marinæ*, *urticæ marinæ*, and *spongiæ*, without adding any thing to the little said by Aristotle. Neither has he, any more than the Greek philosopher, made use of the term zoophytes, though he has most certainly declared that these beings were neither plants nor animals, but something of an intermediate nature.

Nor has Ælian employed this denomination of zoophytes, or animal plants; and if we do find in different parts of his collection, the names of urchins, sea-stars, marine lungs, it is only in relation to some peculiarities, altogether insignificant, and even completely erroneous.

Oppian, in his poem on fishing, has introduced nothing more respecting these animals, than the authors which preceded him.

Sextus Empiricus appears to have really been the author who first employed the term zoophytes; but it does not seem that he did so for the purpose of indicating the beings which Aristotle regarded as intermediate between animals and vegetables. He tells us that they are beings which are found in the roads, and are produced by fire.

Isidore de Seville, and much later, Albertus Magnus, employed this expression for the true zoophytes; but these writers have added nothing to the observations of the ancients on the natural history of the animals. The first translators of Aristotle, Budæus, and Theodore Gaza, also employed it, and since their time, it has been generally adopted.

Wotton, in the very remarkable work which he published upon animals, also employs the same word for the same beings. In fact, his zoophytes comprehend the tethys, the holothuriæ, the star-fish, the sea-lungs, the sea-nettles, and the sponges. In the same author the expression *purgamenta maris,* is to be found for a division of beings, with the relations of which he was not acquainted.

From that period, all the naturalists, at the revival of letters, employed the classic denomination of zoophytes; but there was always some uncertainty in the application which they made of the names left by the ancients, to the objects which they had under their immediate eye. Moreover, they ranged among the zoophytes, animals of classes altogether different, which they designated by names derived from some rude resemblance with terrestrial beings. Thus Belon placed there the anatifæ, or pollicipes, with the sponges; the holothuriæ, and the tethys, which he appears to have known very little, and confounded together, although his tethys is evidently an ascidia. On the contrary, he ranged the sea-nettles, a denomination which he reserved for the actiniæ, among the mollusca, in the same manner as he treated of the urchins, and sea-stars, among the testacea, specifying them, however, in a tolerably complete manner.

Rondelet, a short time after, in adopting the same divisions, made pretty nearly the same confusion; but he began to make known, not only some new species, but some animals of genera altogether new. He applied, in a definite manner, the denomination of holothuria, to the animals which we know at the present day under this name. Nevertheless, he placed one species among the sea-nettles, and on the other hand, referred to holothuria, a species of firola. He clearly distinguished the tethyes, which are the ascidiæ of the present day; and he applied, in a definitive manner, the name of free sea-nettles to the medusæ, and that of fixed sea-nettles to the actiniæ, supporting this distinction by recognizable figures.

These different improvements were inserted in his great Dictionary, by Conrad Gesner, published for the first time in 1604. In fact, he gave a synoptic table of the species cf sea-nettles, divided as Rondelet had divided them. The urchins and sea-stars are united among the testacea; but the escharæ and pinnatulæ constitute the marine zoophytes. Some other species were also indicated, and figured by this author. He likewise perfectly perceived the order of the gradation of organization in this last division from the sponges which approximate the closest to the plants, up to the conchs which are preceded by the univalve shells.

Aldrovandus exhibits, perhaps, better than Gesner, the state of zoophytology, because his compilation is methodical. In that work, we find these beings forming the last division of the whole animal kingdom, and composed of the actinia, under the name of fixed sea-nettles; of the medusæ, under that of free sea-nettles; of the alcyones, under that of marine lungs, and *malum granatum;* of the holothuriæ; of the ascidliæ (named tethyes); of the pinnatulæ *(pinnæ marinæ);* of the lobulariæ, under the name of *manus marinæ*, and probably of the encrusting species.

The echini are definitively placed among the testacea; but by a very remarkable singularity, the asteriæ are placed at the end of the division of the insects. There is no division for the *purgamenta maris.*

Here terminates the first part of the history of this science, in which we find the denomination of zoophyte generally adopted, with the notion that the beings ranged in this division, were intermediate between animals and vegetables; but still it contained as yet but the smallest number of the beings which modern zoologists have subsequently referred to it.

Towards the middle of the age in which the work of Aldrovandus had made known the state of natural history in general, appeared one of the most interesting works respecting the natural history of the zoophytes, a work which commences the long series of those for which we are indebted to the naturalists of Italy on the same subject. This was the Natural History of Ferrante Imperato, of Naples. Besides a number of observations on the living animals which have been since ranged among the zoophytes, (some, perhaps, erroneously) we find there, on the corals, the madrepores, the tubipores, &c., the bases of the opinion generally adopted as to the truly animal nature of these organized bodies; but before the truth of this opinion had been recognized, they were successively placed in the other two kingdoms.

The ancients, who had a very imperfect knowledge of corals—the genius of Aristotle having left them nothing on the subject—determined, after the mere consideration of the external form, to make vegetables of them, from whence the names of *lithophyton* and *lithodendron*, under which they were known for a long time after Dioscorides. Before him, we find them designated by the denominations of *coralium*, *curalium*, and, finally, of *corallium*, (the etymology of which is uncertain), in Theophrastus, Pliny, and Ovid.

At the revival of learning, the numerous commentators on Dioscorides went little farther than himself. It therefore appears that Imperato was the first who had a glimpse of the gradual passage from the corals to the tubulariæ, and to the madrepores, and who recognized in these last, the animal character grow more and more pronounced, to such a degree that he compares them to the velellæ. It is also in this original author, that we find for the first time, the terms pore, madrepore, millepore, retepore, tubipore, as well as those of fungite, astreolite, porpite, &c., which have since been applied to those determinate forms, which we call genera. We also find in him the denominations of alcyone, (already employed by Dioscorides), of coralline, of sertulariæ, and many others, which have been adopted as designatory of genera by modern zoologists.

These germs remained, however, for a long time buried, insomuch, that these organized bodies, whose existence Imperato had marked by good figures, and by particular denominations, were regarded during the whole course of the 17th century, as appertaining to the mineral kingdom, by some writers, or to the vegetable by others. Nevertheless, these different authors, (deceived as they were respecting the nature of corals, which they divided into lithophytes and keralophytes, according as their solid part, the only part which was known, was calcareous or corneous), did not the less augment the number of species, and divide them into genera, which they endeavoured to characterize in a more precise manner. Thus it was that the corallines, and sertulariæ, which they placed among the mosses, the escharæ, the alcyones, and even the pinnatulæ, of which they made fuci, were successively and very clearly established into distinct genera.

From this time, we remark, however, many authors already who, like Boccone and Lluid, suspected the animal nature of some of these productions. Thus the former, although he would have it that the coral was a stone, and not a plant, maintained, erroneously, perhaps, that the *alcyonium asbestinum* was a hive of animals, and the latter was of opinion that the *tubularia indivisa* should be regarded as a zoophyte.

These different facts coincided with the point of time at which the classification of plants began to take as its bases, the character furnished by the flowers. It was natural, therefore, that Marsigli, probably awakened to the subject by the apothecaries of Marseilles, who considered some flowers as coral, described as such the polypi, which he had observed in the alcyone palmata, in the true coral, and in the antipathes. Thus the opinion of the botanists, who claimed all the corals, all the polyparia, as belonging to the vegetable kingdom, appeared to be confirmed, and the true nature of these beings was still unknown for some time, although some chemists had made the observation that the principles which entered into their composition were much more animal than vegetable, and Marsigli himself had remarked that the flowers of the coral disappeared, when it was put into fresh water, or withdrawn from the water altogether. Accordingly, the moment was at hand when they should definitively pass into the kingdom to which they really belong, although, even in 1700, Tournefort published a memoir to distinguish marine from maritime plants, and in which he employs the manner in which he supposes the madrepores to grow, to establish his opinion on the germination and vegetation of stones. Reaumur himself, in 1727, published a memoir to explain how stony bodies can vegetate, by supposing, that in the coral, for example, it was the bark alone which vegetated, and which formed a stem, by depositing the red grains with which it is filled. Rumph, who had occasion to examine a great number of living corals in the Indian Archipelago, where they are spread in profusion, having established a particular division for the zoophytes, was perhaps the first who demonstrated the animal nature of these pretended plants, in many species. But it was only in 1727, that Reaumur reported to the Academy of Sciences, the celebrated discovery made by Peysonnel, in the Mediterranean, of the animality of the lithophytes, proving that what Marsigli had described and figured as the flowers of coral, were real aggregated animals, altogether analogous to the actiniæ, and by no means what he had himself described as the flowers of marine plants in the memoir just mentioned, and before; consequently, that we must regard the madrepores, millepores, and in general all the lithophytes, or aggregated testæ, as the habitations of these animals.

This important discovery, to which Peysonnel was undoubtedly conducted by the observations of Marsigli, was not, however, immediately adopted; and Reaumur himself, in the memoir in which he reported it, endeavoured to contest its evidence, and was even afraid to name its author. But he was obliged to admit it, when Trembley, in a letter addressed to him in the month of December, 1740, had published all the singularities of the natural history of a little animal, known in the fresh waters of Europe, and which, already signalized by an anonymous writer, in the transactions of our own Royal Society, had been forgotten for more than ten years. We find, in fact, in the fresh water polypus, named hydra by Linnæus, the naked type of the coral animals.

Shaw, the traveller, in his Voyage to Barbary, vainly proposed to regard as simple nutritive radicles, the undulating filaments which he saw come forth from the stelliform impressions of the *madrepora ramea,* and from some other aggregated living madrepores. The discovery of Peysonnel acquired all the corroboration which it deserved, especially from the coasting voyages of Jussieu. These voyages were undertaken, one in the Channel, and the other in the Atlantic, for the express purpose of verifying and extending this discovery, by applying it to a greater number of beings, which was done for the tubulariæ and other genera. The name of polypus appears to have been first employed by Bernard de Jussieu, to designate the little animals, which, inhabiting pretended marine plants, are provided on the head or body with horns, *(tentacula)* which serve them as hands or feet, to take their aliment or to walk.

Reaumur, from this time thoroughly convinced, fully adopted the views of Peysonnel, thus confirmed by Jussieu and Guettard. He created the name of *polypier (polyparium)* since generally adopted, perhaps without sufficient discrimination, to express the solid part, of whatsoever nature it might be, on which those little animals live, to which, with Jussieu, he gave the general denomination of polypi. This name he had given to those discovered by Trembley, because their tentacula or horns appeared to him to be analogous to the arms of the sea-animal which the ancients had named *polypous.* Thus definitively returned into the animal kingdom an entire and an extremely numerous class of beings, which from the intimate nature of their union, had been so long considered as vegetables, and which, considered in part, had been recognized as animals approximating to the actiniæ, and consequently entering into the grand division of the zoophytes.

Notwithstanding all these circumstances which we have now detailed, and more which it might prove tedious to enumerate, Linnæus, who in the first editions of the *Systema Naturæ*, had imitated Ray, in placing the lithophytes, in the vegetable kingdom, still preserved some doubts. In a dissertation on the corals of the Baltic, after enumerating successively the objections to the opinions of those who have maintained that these bodies were minerals, vegetables, or animals, he professes that he considers it a difficult matter which opinion to choose. Some time after, however, he was convinced of the truth, and in the sixth edition of his immortal work, admitted them into the animal kingdom.

Thus at this second period of the science, the zoophytes were definitively ranged in the animal kingdom, by systematic writers. But they were still very far from being grouped and united in a suitable manner, as we shall see in our examination of the particular researches of some writers on the subject.

One of the first works which tended to bring this science to perfection, was published by Vitali Donati, on the Adriatic

Sea, in which he has described the animals of a considerable number of polyparia, already figured by Imperato.

It was also at the same period, that the more or less flexible polyparia, known under the names of Sertulariæ, &c. were much better distributed, owing to the remarkable labours of Ellis upon the corallines—labours which have served as the basis of every thing valuable which has been subsequently performed respecting the genera of these animals. This writer, however, has not been eminently happy in the methodical distribution of the numerous species which he has examined. He has united almost all of them under the common denomination of corallines, as Ray had done, who regarded them as plants.

Notwithstanding these new researches of Ellis, which seemed irrefragably to confirm the discovery of Peysonnel, some authors, and especially Hill, Targione, and Baster, were still inclined to oppose the system—so difficult is the progress of truth. But Ellis refuted these objections so completely in the Philosophical Transactions, that Baster himself gave way and adopted his opinion.

While the division of the zoophytes was augmenting in number and consistency, by the approximation of newly discovered beings, or of such as had been for a long time removed from it, the groups which had been anciently admitted into it, became more extended and better known by the particular labours of zoologists and travellers. Thus Link, in 1755, published a monograph of the asteriæ, which still forms the basis of all that has been done on the distribution of the species of this very remarkable family. Others followed in the same track, and by their attention to particular subdivisions, illustrated and extended our knowledge of these animals.

About the same time the observations of Trembley or

animals of pretty small dimensions, led to the study of animals still smaller, to which the name of *microscopic* has been given. Leuwenhoëk and Hartsoëker led the way, but were ably followed by Hill, Rœsel, and others, who considerably augmented the number of discoveries in this kind. The difficulty of observation, and the want of proper principles to guide most of the observers, prevented those animals from being sufficiently known, to justify systematic writers in assembling them in a single group, or even uniting them to the zoophytes. All subsequent zoologists left them in the same state, just as if the degree of size was of necessity in relation with the degree of organization.

The first systematic author in which we find the microscopic animals arranged, appears to be Hill. As, however, he did not admit the system of subdivision founded upon Aristotle, and has no class under the name of zoophytes, it is not easy to analyze his labours. The animals to which the most recent zoologists give this name, are placed by this author in sections exceedingly remote from each other. The infusoria, under the name of *animalculæ,* are placed altogether at the commencement of the animal kingdom. It would be equally superfluous and disagreeable to follow him through his groups, and to repeat his learned, but very cacophonous appellations. It is sufficient to remark, that in his system, or no system, the medusæ, actiniæ, hydræ, and asteriæ, are in the same section with the naked mollusca, and that the setigerous annelida are between the insects properly so called, and the amphibia, vertebrated animals.

The work of Pallas on the zoophytes, may be considered as one of the most classic, and the best executed which we possess in zoology. He does not, however, by any means treat of all the animals now comprehended under the name of zoophytes; nor, indeed, are scarcely any of the animals which the ancients regarded as intermediate between vegetables and animals, to be found among the zoophytes of Pallas; but he has all those with which they were not acquainted, or which they believed to appertain to the mineral kingdom, that is, their *Corallia*.

SUPPLEMENT

ON THE

FIRST CLASS OF ZOOPHYTES.

THE ECHINODERMATA.

THIS term was invented by Klein, and applied by him with sufficient exactitude, as he only comprehended under this name the *echini* of Linnæus, (*echinites,* Lam.) all of which have the skin covered with hard prickles, varying somewhat in form, which has caused them to be compared to hedge-hogs. But Bruguières, by uniting under this name the asteriæ to the echini, has certainly not proved so correct in nomenclature, inasmuch as none of the latter present any traces of prickles on the surface of the skin. M. Lamarck, and the authors who have followed him, have rendered this name still more equivocal by including under it the holothuriæ, and even the sipunculi, which may be considered almost as genuine worms. M. de Blainville has restrained this name to the three orders, holothuriæ, echinites, and asterias; but is yet of opinion, that the name of *placyrodermata,* would better distinguish them, as indicating the principal character which they present, which is the existence of a great number of tentacular and respiratory suckers, by means of which the animal performs the function of locomotion.

The ASTERIÆ, or STAR-FISH, whose species are greatly multiplied in all known seas, are remarkable for their stellated form. Their body is invested with a coriaceous substance, bristling with tubercles or spines, or covered with scales. In some species it has the figure of a flatted pentagon. In others, the angles of the disk extend in lobes, or are elongated into radii, most frequently five in number, sometimes more numerous, and simple, or divided into ramifications.

The mouth, armed with five teeth, of a calcareous substance, is placed at the central part of the animal underneath. No tentaculum is remarked there. It leads into a membranaceous sac, of small extent, which serves as a stomach, and from which the excrements are rejected, and issue through the same aperture. O. Fabricius, and Bosc, are, indeed, of opinion, that the excrements are filtrated through an osseous tubercle, which is observable in the back of the asteriæ, a little on one side. But this opinion cannot be well founded, inasmuch as this tubercle is not existent in all the species of the genus. It seems to be generally wanting in the ophiuri. Besides, we are not acquainted, as it would seem, with any other animal, in which it is necessary that the excrements should pass through a filter, to issue forth from the sac, or alimentary canal. Some short, cylindrical, retractile tentacula, very numerous, ranged in pairs underneath, or on the sides of the divisions of the animal, are developed on the instant, in which it is desirous to hook itself, or to walk. We see on its upper part, but only while it is in the water, an infinity of small conical tubes. The mouth is a sort of respiratory organ, which constitutes with the stomach, and the ovary divided into as many pairs of branches as there are arms, all the apparent organs of the asteriæ. Some osseous articulated stems form the skeleton of each branch. A net-work of the same nature strengthens the rest of the envelope.

The asteriæ are very voracious; they walk with difficulty,

and very slowly. When they swim, they are seen to present their body obliquely to the action of the water, and to agitate their radii slightly. When they want to descend, they suspend all motion, and allow themselves to fall perpendicularly to the bottom. When they have succeeded in getting a shell, they suck out the animal from it through the aperture. They possess a wonderful power of reproduction.

The text is sufficiently explanatory of the organic differences which distinguish the genera.

ENCRINUS was a name first proposed by Ellis, to designate a very singular animal, whose place is still uncertain enough in the natural series, so much so that while many naturalists, with Guettard and Ellis, place it with the family of asteriæ, others with Linnæus, range it with the polyparia, near the Isis. Ellis, in 1761, read to the Royal Society, a very interesting memoir on this same genus, having made his observations on an individual which came from the coast of Barbadoes. This genus does not appear even yet to have been examined with all the attention it deserves, for it seems indubitable that some of its species, no longer in our present seas, must have been very common there anciently, since nothing is more multiplied in certain calcareous strata, than these fossil remains, known under the names of entrochi, encrinites, &c. Our limits will not permit us to enlarge on the characters of this genus, for which we must refer to the text of Cuvier.

These animals, very probably, live in the bottom of the sea, at considerable depths; but it is not even known whether they are fixed there, which, however, appears very likely to be the case. It is to chance alone that the discovery has been owing of three or four individuals only, which exist in the European collections, and which come from the American seas.

Under the denomination of ECHINUS, Linnæus, and the

zoologists of his school, comprehended all the animals, more or less orbicular, whose cretaceous envelope bristles with a number of spines, of a form more or less variable, and constantly calcareous. This has caused these animals to be compared to hedge-hogs, and accordingly, they are popularly termed *sea hedge-hogs*, or *urchins*, and sometimes sea chestnuts. But at the present day, among modern zoologists, since the researches of M. de Lamarck, this name is reserved for a certain number of species, for those which really better merit the title, in consequence of the long spines with which they are armed.

The exterior envelope, which determines the form of an echinus, (which is described in the text) can be compared to nothing which exists in other animals. In the greatest part of its extent, it is formed by two membranes, one external, and thicker, the other internal, and so thin, that the name of pellicle perfectly suits it, and between which there is a tolerably thick and solid testa, completely calcareous, and composed of a great number of small polygonal pieces, evidently immovable, but not cemented, at least, during the growth of the animal. In the neighbourhood of the mouth and of the anus, the skin is not thus solidified, accordingly, it is sensibly thicker, and much more resistant.

The testa of the echini is entirely calcareous, almost without any mucilaginous or animal part, and is fibrous perpendicularly at the surfaces, which proves that the mode of growth, though taking place upon the edges, nevertheless differs much from the same operation in the shell of the mollusca.

The pieces which constitute the testa of an echinus may be divided into coronal, and terminal. The coronal are those which by their union form the most important part, the most extended, and which circumscribe the body in its circumference; and the terminal are those which surround the buccal orifice, and the anal orifice, and which fill the two apertures more or less considerable, which are left below or above, by the assemblage of the coronal part.

The coronal pieces are subdivided into ten groups or series, which radiate from one orifice to the other, a little like the ribs of melons, and which form areas alternately full and perforate, equal or unequal. The name of *ambulacraria,* is given to the series which are perforate, and that of *anambulacraria,* to those which are not.

The anambulacraria are themselves constantly formed of two series of pieces, more or less hexagonal, and usually transverse, which are united at one extremity in the middle of the anambulacrarium, and at the other, but less angularly, with the ambulacraria. Each piece is raised at its external surface, with a variable number of mammillæ, more or less projecting, well-rounded, polished at their summit, and widened at the base, without any trace of perforation.

The ambulacraria, sometimes more narrow than the others, are, nevertheless, likewise formed of two series of polygonal pieces, united angularly together in the middle line of the ambulacrarium, and externally with the pieces of the anambulacraria. They are also raised with mammillæ, more or less salient; but, besides, they are pierced at their external side by pores, variable in number, and in disposition, for each species, but which always traverse the testa from one part to the other. This is what constitutes the ambulacra, properly so called.

The breadth of the anambulacraria is generally greater at the middle than at the two extremities; but this is not the case with the ambulacraria. They are always larger towards the mouth, and the last presents at the interior a sort of apophysis or lamina, pierced with a hole in its middle, and which affords an attachment for the motor muscles of the teeth. These are named *auriculæ.*

The skin which surrounds the mouth is scarcely rough;

there may, however, be remarked on it, some pairs of subcircular scales, a little concave, and which are exactly placed two by two, in the direction of the ray, which should go into the interstice of the teeth; each is pierced with an orifice.

Around the anus the coronal pieces fill almost completely the spaces which are left by the areas. They are, like these latter, ten in number, alternately great and small. All of them are usually granular, and pierced with a hole, much broader, however, in the large than in the small ones, which correspond to the ambulacraria; the large ones correspond to the anambulacraria.

The holes with which the pieces of the ambulacraria are pierced, give passage to small tentacular cuppers, proceeding from the interior lamina of the skin, perhaps, from the hollow respiratory laminæ in their entire length, and terminated at their extremity by a small swelling susceptible of being dilated into a cupper, or into a disk denticulated at its circumference. These organs are remarkable for the great contractility which they possess, and can re-enter completely into the interior, something like the horns of snails, or be considerably elongated at the exterior.

Another portion of the locomotive apparatus of the echini, is that from which their name is taken, though more frequently these organs merit the name of little clubs, or tubercles, rather than that of bristles. What they offer in common, is the having at their base, a small spherical concave head, with a circular pad above. Their length, their form, and their thickness, are extremely variable, and generally in relation with those of the mammillæ of the testa. Their structure is equally peculiar. When broken they have a vitreous sort of brilliancy. Their external surface is almost always finely striated, and they are composed of concentric strata, each of which is formed of a great number of irradiated fibres.

These organs, articulated like a knee on the mammillæ of

the testa, are put in motion in every direction by the external lamina of the cutaneous envelope, which attaches itself to the circumference of the pad of their base, and which appears stronger, and more evidently muscular at the spines of the base of the echinus. In desiccation, it is impossible to perceive any distinct muscular fibres, and sometimes even any muscles, properly so called.

These animals are all aquatic and marine. They, nevertheless, constantly live on the sea-shore, in rocky and sandy places. They are very rarely found abandoned by the tide. It would seem, that if too much advanced to recede, they possess the faculty of sinking more or less deeply into the sand. In this case, it is easy to recognize the place where they are, by the existence of a small hole in the form of a funnel, which is remarked at the surface of the sand. They sink and recede considerably less when the weather is fine, than when it is tempestuous.

The echinus, in locomotion, which is never very quick, makes use of its tentacular suckers, and its prickles, and especially of the inferior ones. But it appears, that this cannot take place, except on a resistant soil. In the first case, it elongates as far as possible (and it is astonishing the extent to which it can be done) a certain number of the suckers, which are in the direction in which it wishes to go. It attaches these strongly to some solid body, causing a vacuum by means of the cuppers which terminate them, after which it contracts them, and thus draws its body towards this point. By thus reiterating the same manœuvre, the echinus may, without doubt, advance with some degree of rapidity. In the second case, when it employs its prickles, it extends those on the side where it wishes to move, to the utmost possible degree. Then it lowers them, pushing itself on with those on the opposite side; and as it has some in all directions, it is evident that it can walk in all ways. In general, its progress is performed by turning, although the animal, nevertheless, finishes by arriving at the end which it desired to attain.

The echini are said to be eminently carnivorous. It is even admitted that they feed upon crustacea and bivalves, but this is probably rather a conjecture deduced from the strength of their jaws, than a fact ascertained by observation. M. de Blainville has opened many, either taken alive, or preserved in spirits of wine, and never found any thing but sand in their stomach. M. Bosc, however, was a witness of the mode in which an echinus got possession of a crustaceum; and it appears, that as soon as the latter was caught by some of the tentacular suckers, it was speedily masticated, and swallowed.

We are yet ignorant of the mode of reproduction in the echini. We only know that it is in spring that they deposit their spawn, which appears to contain an almost countless number of eggs; and it is probable, that it is rejected in a mass all at once. But it does not appear that any naturalist has witnessed this.

Genuine echini are known in all the quarters of the globe. The largest and most numerous, however, belong to the seas of warmer latitudes.

The HOLOTHURIÆ were placed by Linnæus and Bruguières among their molluscous worms; and subsequently ranged by Pallas, Cuvier, and Lamarck, who better studied their relations near the asteriæ and echini. We shall not repeat the characters of this genus here.

Many authors, such as Hill, Brown, and Baster, have given the name of actinia to this genus. Linnæus, for some time, gave it that of *priapus*. Gœrtner preferred the denomination of *hydra*, and confounded these animals with the actiniæ proper. This was imitated by Bohatsch. Pallas returned to the name of actinia, which he divided into two sections; one comprehending the holothuriæ, and the other the true actiniæ.

The holothuriæ, which the ancients usually comprehended under the vague name of *purgamenta maris*, or of *pudenda marina*, are found, as it would seem, in every sea, the deepest parts of which they most particularly inhabit, even as far as three hundred feet and more. It is even to this cause that the peculiarity seems owing, of these animals vomiting, as it were, their intestinal canal, in consequence of the great difference of pressure. They more particularly remain in the oozy bottom, and in the anfractuosities of the rocks, where they fix themselves by means of sorts of cuppers, or papilli, from tentacula, with which certain parts of their bodies are provided. They can, by means of these organs, draw themselves along over the submarine bodies, and thus change place. But it appears, that they can do so likewise, either by alternate flexions of the body, after the manner of worms, or even by filling their body with water, and shooting it forcibly through the anus, so as to be able to swim, assisting themselves with their tentacula. They feed on pretty strong marine animals. It is thought that they are hermaphrodites, and that they reproduce by internal gemmulæ, like the actiniæ, whence it has been said, that they are viviparous. Fabricius, in fact, tells us that he found in the anal part of one individual a young holothuria, swimming freely.

The organization of the holothuriæ had been studied by many anatomists, and among others, by M. Bohatsh, and by Vahl, but in an incomplete kind of way. M. Tiedmann, in his dissertation, which gained him the prize proposed by the Academy of Sciences in Paris, has added very considerably to our acquaintance with this subject.

The species of this genus appear to be rather numerous, and especially so in the seas of cold countries. But they have been in general too ill described to be characterized. Some small generic groups have been introduced by some naturalists. Thus M. de Lamarck has separated the species whose tentacula are formed by small denticulated and pedicled disks to form his genus fistularia. M. Oken had already extended these divisions much further, since he has established four genera according to the consideration of the disposition of the retractile tubes, and the form of the body. As these animals are extremely altered in appearance, when they have been kept for some time in alcohol, it becomes rather difficult to characterize the species. They seem to form a passage from the radiata to the worms, and perhaps even to the mollusca; for, in fact, we find some species which have altogether the form of the thalassema, and others which exhibit at the first glance, a resemblance to some dorides.

SUPPLEMENT

THE ENTOZOA.

THE reader will doubtless have remarked that the allocation of these animals among the zoophytes, or radiata, will admit of some controversy, inasmuch as they are neither plant-like nor radiated, and appear to have much more analogy with the annelides, which they would seem more properly to follow in the natural series.

The study of the worms, which are entirely apodal, and which have no trace of appendages serving for locomotion, was pursued but little until the middle of the last century, and more especially that of such species as live continually in the interior of other animals. Attention to them seems to have been chiefly originated by a prize proposed by the Royal Academy of Sciences in Copenhagen.

The ancients have scarcely noticed the round and flat worms, and that solely in a medical point of view, confining themselves to those which infest the human species. They have further limited their labours in this way to mere nomenclature, without giving any external description.

The few authors who occupied themselves with the history of the distinction of animals, previously to the revival of literature, have added but little to our acquaintance with this subject. From Isidore de Seville, Albertus Magnus, Belon, and Rondelet, we derive nothing in addition to what the ancients have left us; and Gesner and Aldrovandus, who compiled with so much patience, and frequently with so much sagacity, all that had been said by their predecessors, one in an alphabetical, the other in a systematic form, have collected nothing new respecting the worms, and more particularly respecting the intestinal ones.

It was only, in fact, towards the end of the seventeenth century, that the science of helminthology may properly be said to have had its birth, and this birth took place in Italy, the mother of almost all modern art, science, and literature. To that country belong the names of Redi, Malpighi, and Vallisnieri.

The popular name of *worm*, under which these animals have been included, sufficiently indicates that their body is almost always cylindrical, more or less elongated, attenuated at the two extremities, and of a diameter infinitely less than the length. In some the body is more perfectly cylindrical, in others sacciform. Even a certain number resemble bladders, as the hydatids; or very depressed laminæ, as the fasciolæ, &c. Whatever be its form, it is always perfectly symmetrical, as in all other binary animals, and most frequently the dorsal face may be distinguished from the ventral, by a little more convexity in the former than the latter.

It is but seldom that we can trace in the body of these animals the distinction of head, neck, belly and tail. Nevertheless, it sometimes happens that the anterior extremity is enlarged, and is well distinguished from the rest of the body, and then the animal is provided with a cephalic enlargement, as in tænia, &c. But most usually the body, attenuated in front, swells out by degrees, and again becomes attenuated at the posterior extremity.

In many of these animals articulations may be observed, but they are sometimes extremely indistinct. In other cases the articulations are so much separated, that they seem to form a sort of chain, which breaks with the greatest facility.

Never, or scarcely ever, are any appendages observable on each side. The worms, in fact, are easy to be recognized by this great simplicity in the external form.

Their organization, too, in general, is but little complicated, and often, in consequence of their smallness and the transparence of their tissues, it may be observed through the teguments. It is particularly when they are young, and living, that this transparence is almost perfect.

The external envelope, or skin, is almost always confounded with the muscular substratum, which serves for locomotion.

The dermis is consequently not distinct; it is always very soft, of a nature almost mucous, and we cannot distinguish above it, any other of the parts of the skin, such as they have been analyzed in this envelope, in superior animals. Thus, neither papilla, nor epidermis can be recognized. The vascular net-work alone is sometimes tolerably developed, and we may remark there a pigmentum sufficiently marked, at least in some species; but all that are parasites in the intestinal cavities, or parenchyma of animals, are constantly white, unless they should be filled with some colouring matter, which may be perceived, in consequence of the transparence of their exterior envelope.

In a very small number indeed have any indications been observed of special organs of sensation.

The apparatus of locomotion consists solely in the muscular stratum which doubles the skin, and the intestine itself.

We cannot even say that these animals have true muscles. The muscular stratum is merely divided into eight longitudinal bands, by the dorsal, ventral, and lateral lines, which are themselves composed of interrupted fibres. In the vesicular species, where the body terminates by a bladder, the muscular fibres radiate all around its parietes.

As there are no appendages, it is evident that there can be no muscles for the purpose of moving them, though some indistinct fasciculi may be observed in some of the exterior species.

The apparatus of nutrition is also very simple in the entozoa.

As for that of digestion, or the intestinal canal, it is generally extended from one extremity to the other of the body, at least when it is complete. At other times it is irregular or ramified, and, finally, in a certain number of species it is vascular, or even sometimes completely imperceptible, or even nullified.

When complete, it extends almost directly from one extremity to the other, but never without being free or distinct, and consequently without any serous or peritoneal membrane. It is truly comprehended in the cellular parenchyma, which constitutes the mass of the body.

The mouth, almost terminal, is almost constantly very small and circular.

It is never armed with teeth, properly so called, whether calcareous or corneous. Sometimes certain projections, or swellings, of the muscular contractile tissue may be found, the edges of which bristle with remarkably fine denticulations.

As there is no real armature of the anterior orifice of the intestinal canal, we may easily conceive that there never exists in these animals any distinct buccal mass.

Neither are there any salivary glands.

We may say, that in general, in the extent of the intestinal canal, there is no distinction of œsophagus, of stomach, properly so called, of small intestine, and of rectum; such at least is the case in all the truly intestinal worms.

No organs have been met with that could be regarded as a liver, and still less as a pancreas. If the first does exist, which appears by no means probable, it is contained in the parietes themselves of the intestines.

The termination of the intestinal canal, or anus, is always medial, and yet approaching to be terminal.

This is the first sort of digestive apparatus found in the entozoa. In the second sort, there is still a terminal mouth, in the form of a sucker, and without armature; but the intestinal canal, which originates from it, after having been prolonged towards the middle of the body, is lost immediately in vasculariform ramifications, which proceed into all its parenchyma, and then there is no trace of anus.

Finally, in the third sort, the digestive system does not commence by a single medial orifice, or by a true mouth. There is even no true intestinal canal, but two or more anterior and lateral pores or orifices spring from the vessels which are united to two lateral trunks; these are prolonged throughout the entire length of the body, anastomosing together, and ramifying, without doubt, into the parenchyma of the animal. This is the organization of the tæniæ, &c.

In the ligula, which are the most simple of all binary animals, we can no longer recognize any trace of an intestine, of whatever sort it may be, and consequently no circulatory system.

A special respiratory apparatus exists in none of the animals of this class. The organ, therefore, of respiration is essentially limited to the skin, nor is there, in fact, any true respiration, except in certain families, where something like circulation may exist.

The blood, or recrementitial fluid, has been but little studied.

The apparatus of decomposition consists solely of the organ of generation; for no other glands exist in any genus of this group. There are several distinct sorts of it, since it is sometimes composed of female and male parts separated on different individuals, and sometimes of such parts on the same individual; at other times of female parts alone, and in fine, we sometimes find scarcely any indications whatsoever, of this apparatus.

The existence of a nervous system in these animals is very doubtful, at all events in those which may be properly termed entozoa. No one has been able to recognize any thing of the sort in the tæniæ, or neighbouring genera.

The organization of the product of generation in both, in the exterior and interior worms, has not been sufficiently studied. Such eggs as have been discovered are oval, and very regularly formed. They all appear to be in reality eggs, with a distinct envelope containing grains. But this was the most that could be distinguished, even with the microscope.

The physiology of worms presents nothing which can really be considered as peculiar to themselves.

Their general sensibility appears to be considerable. In fact, at the slightest contact of a solid body, or even of a liquid of a heat or nature different from that in which they are plunged, they torment themselves, and twist about in all directions. The tæniæ do this almost as much as the others, but only in a less extended portion of their body at once. A special sensibility they do not seem at all to possess, nor any perception of bodies, whether by savour or odour, and still less by the luminous rays which they send forth.

The contractility of worms is evidently very great, and that almost in all their tissues, so that they have considerable power of changing form and dimensions, as may be particularly observed in some of them, which are almost in a continual state of agitation. Their locomotion is, therefore, sometimes tolerably active.

They seem, in general, to possess but little tenacity of life. Some, indeed, such as tæniæ, have been observed to live many hours after having been extracted from the intestinal canal of an animal which had been dead for several days. The ascarides and distomæ may be preserved alive for many days. Rudolphi even speaks of one species of ascaris, the *A. spiculigera,* which was resuscitated, after being extracted from the intestine of a bird, which had been for twelve days in very strong spirits of wine.

We are in possession of no proof that they can reproduce any individual part which may have been lost or taken away.

We may say, in a general manner, that all these animals comprehended under the name of worms, or entozoa, live constantly in a fluid, and never, or at least very rarely in the atmospheric air. This fluid may be either living, or at least constituting part of a living body. In this case, they are more peculiarly entitled to the appellation of intestinal worms, or entozoa; and, in fact, the great majority of the worms are intestinal.

There is no tissue, nor any constituent parts of living bodies in which some of these worms have not been found. They are most frequently at the surface of some portion of the exterior envelope, which has re-entered, to form either the intestinal canal, the lungs, or organs of respiration, or the genito-urinary apparatus. But at other times they are found in the tissue itself of these parts, in the parenchyma, as in the brain, the vessels, the muscular system, &c.

Moreover, there is scarcely any animal in which some of these intestinal worms have not been found, at least if we speak of all the classes of the vertebrata; for among the invertebrata, they seem to be more rare. But as they have been as yet so little sought for in these latter tribes, it is impossible to assure ourselves that they are so few as has been generally asserted and believed.

It has been remarked that no intestinal worms have as yet

been discovered in the guinea pigs, though they have very frequently been sought after in those animals.

Among vertebrated animals, the entozoa are in general more common and more numerous in the aquatic species than in the others; more so in the females than in the males; more so in the young than in the old; and, finally, more so in weak individuals than in vigorous subjects.

It has been for a long time admitted in a general way, that each intestinal worm more peculiarly infests some one animal, and even a determinate part of that animal; but this opinion can by no means be supported in any positive manner. In fact, it is certain that the ascaris lumbricoïdes, for example, is found in the human species, in the horse, the pig, and some other animals. It is equally proved that the same species of tænia is to be found in the cat and in the dog. Nor are we less assured that the bothryocephalus punctatus is to be met with in the barbel, the turbot, and several other species of pleuronectes.

Nevertheless, there are some species which are evidently peculiar to some particular species of mammifera, such as the human tænia and the bothryocephalus.

Locomotion in these animals is but seldom, even slightly extended, and a certain number of them do not possess the faculty at all, but remain fixed in the parts in which they live. These, then, are strictly parasites, altogether fixed, as are the echinorhynchi, and the majority of the tænioïdes. There are only, in such cases, some partial movements, or a sort of undulation in those parts of their bodies which are not adherent. It is not the same with several other species, such as the ascarides, and even the porocephali, in which there is a true general locomotion, or total transport of the body, in the parts which they inhabit, for it is certain that the ascarides lumbricoïdes, whose normal position is in the small intestine towards its origin, do sometimes ascend into the stomach, or descend into the large intestines. They have even been known to penetrate into the peritoneal cavity, after having perforated the intestine, or taken advantage of some previous perforation which had been made there.

The nutriment of the worms is generally animal, and in a fluid state, as may be supposed from the formation of their buccal orifice. Among the external worms it would seem that there are some exceptions to this rule; but with such we have but little to do in the present place.

The mode in which this nutriment is taken is very simple, since in general it is drawn in by the successive action of the parietes of the intestine, performing the office of a sucking pump. This is quite obvious in the medicinal leeches, which are so analogous in structure to the animals of which we are now more particularly treating, and the like probably also takes place with the ascarides. But in these last, nutrition must also take place by the action of the skin, plunged in the mucous or chyleous matter, which lines or fills the intestinal canal which they inhabit. As for the porocephali, the echinocephali, and particularly the bothryocephali, and ligulæ, it is clearly evident that they can have no other mode of nutrition, the intestinal canal being no longer existent in them, or being reduced to a merely vascular state. All the nutriment is taken by the action of the suckers of the oral extremity, or by those of each ring as in the tæniæ, the bothryocephali, and lastly, by the pores of the skin alone, as must necessarily take place in the ligulæ.

In those in which the sexes are distinct, as in all the ascarides, it is certain that the males are always obviously less common than the females, and, most undoubtedly, must be considerably less numerous. There are even some species in which it is very rare to find any, as in the ascarides, properly so called. There are more in the strongyli.

The mode in which the intercourse of generation takes place, is but little known. M. Bremser has observed it in the strongyli. The pouch or sac of the male embraces the vulva of the female.

The result of generation after this intercourse, or even when no intercourse takes place, as is the case with all the monogamous species, is, as we have already said above, often innumerable eggs, as in the tæniæ, for example. These eggs do not adhere to the mother. They are formed in the meshes of the cellular tissue which constitute the ovary, and are rejected through a determinate orifice, in most cases, or through a simple rupture, as in the tæniæ.

Be this, however, as it may, they fall into the substance in which they are to be developed, without receiving any particular arrangement or disposition from the mother; with the hirudines, which we have so often said are very analogous animals, if, indeed, they ought not to be placed in the same class, the case is different. Their eggs are united in small packets, by a general mucous envelope, and are arranged by the mother in determined circumstances. Such appears never to be the case with the entozoa.

We know nothing, or scarcely any thing respecting the mode in which their eggs are developed. M. Rudolphi thinks that their development is rapid. Nevertheless, it appears, that in the bothryocephalus, so common in the *pleuronectes maximus*, the young subjects bear no resemblance to the mother. They at first exhibit no trace of articulation. The cephalic enlargement constitutes almost alone the whole of the body of the animal, which is terminated rather abruptly, by a small caudiform elongation. In the next degree of development, the body increases in size. The articulations become sensible by simple lateral denticulations. They are as yet few in number and very long. By degrees they cut and wrinkle, and other articulations become defined. This circumstance, at the same time that the body is continually shooting out, finishes by giving to the latter a very variable length, but which sometimes exceeds two feet. We may then observe some of these articulations, and more frequently the posterior ones, fill by degrees with eggs, of somewhat a different bulk, but of the same oval form, and which after they issue forth, in their turn produce young bothryocephali.

There are two interesting questions respecting the development of the intestinal worms; first, can such and such species only, be developed in the body of such and such animals, and may not this development be sometimes continued in another animal?

The successful experiment of Pallas, in introducing the eggs of the *tænia elliptica*, into the abdominal cavity of a dog, is a reply to the first question.

As for the second, it appears certain that the ligula, when it is found in fishes, never presents itself in the adult state, or with developed ovaries, which is quite the reverse of what is observed when it is taken in an aquatic bird. M. Rudolphi, therefore, supposes, that born in the fish, it only acquires its full development in passing into the birds, and that this is owing to the heat of their body. He brings forward, moreover, in support of this opinion, the case of a bothryocephalus, which in the imperfect state (*B. solidus*,) is found in a species of gasterosteus, and in the adult state (*B. nodosus*,) in aquatic birds only. He further explains by this, the curious fact, that in the northern countries of Germany, and in Denmark, where this little fish is common, the aquatic birds are infested with the *B. nodosus;* while in Southern Germany, where this gasterosteus does not exist, the aquatic birds have not this bothryocephalus.

We shall particularize only one species of this disgusting

class of animals; namely, the *F. medinensis*, belonging to the filariæ, commonly called the guinea-worm.

This species, the most celebrated of all, is very long. The edges of the mouth are inflated, and the point of the tail is inflected. It is of the size of a small cord, and of the same diameter almost throughout. Its head is provided with a sort of sucker, formed by the inflation of the lip which surrounds the mouth, the orifice of which is very small. The tail is terminated by a sort of inflected hook. Its colour is the same as that of most of the worms that live in the interior of animals; that is, a dirty white, passing to yellow in alcohol. In length, it appears to vary considerably. Kœmpfer speaks of a foot, of a cubit, and more. Grundler describes the one that he has seen, as being three feet and a half, Rhenish measure. Kemsemuller says it frequently exceeds two ells. Gallandat gives it eight or nine feet; and in fine, Fermin carries its length to eight or nine ells, which appear, indeed, to be somewhat improbable. Be all this, however, as it may, this worm appears hitherto to have been found on the human species alone, in the cellular tissue of different parts, and especially in that of the legs, towards the malleoli. It also appears to be endemic in the burning regions of the old and of the new continent. The names of *medina-worm*, and *guinea-worm*, have been given to it, from the places where at first it was observed.

There are among observers, great dissensions respecting the origin of this worm. Some think that it is exterior, that it is a true *gordius*, and that it insinuates itself into the skin of such persons as walk barefooted; that it deposits its eggs there, grows, and is developed there, and produces by its presence such painful symptoms, that it has received the denomination of *furia infernalis*. Proofs have, in fact, been given that it may exist in this manner for a considerable length of time. Other authors maintain that it is a worm altogether interior, and allege, in support of their opinion, that it has never been found out of the body of man; that it is entirely similar to the other species, and especially to the filaria of the simia; and that it is more probable, that it is born in the interior of the parts; that it may exist there for months, nay, entire years, without producing any sensible accidents; and that it is only when it approaches the skin, when it pierces it, that these accidents may become serious enough to produce intense pain, or other alarming symptoms. This is pretty nearly the opinion of M. Rudolphi, and it appears the most probable. Nevertheless, some persons, more versed in the art of surgery than in zoology, and influenced, no doubt, by the existence of a sort of inflammatory tumour, which the presence of the worm produces at the skin, have ventured, in these latter times, some doubts concerning its real existence, thinking it might be nothing but the cellular tissue itself, struck with death, which thus moulded itself, as it were, into a worm, in traversing the thickness of the skin. M. Delorme, in a letter inserted in the eighty-seventh volume of the Journal of Physical Science, has shown by facts, how erroneous is this opinion; he has confirmed all that was known concerning the symptoms, and even the treatment of the affection which follows the appearance of the worm at the skin. The symptoms are a tumour with redness, and most violent pain. Soon a little orifice appears, through which the worm puts forth a small part of its body. The treatment consists in seizing this part and rolling it with much caution round a small stick, which is turned very gently every day, for fear of breaking the body of the animal, which would render the extraction more difficult; besides that, the presence of the remaining part which would putrify, might occasion accidents still more fatal. The observation has been made, that the people who walk barefooted, like the negroes, are more frequently affected by it than others; and that it is towards the malleoli that the affection takes place. This is difficult enough to explain on the hypothesis that these animals issue forth from the splanchnic cavities; for it does not very clearly appear why very nearly all of them should come out at the same place. The medina worm may yet become the subject of very interesting observations.

SUPPLEMENT

ON

THE ACALEPHÆ.

In our supplementary observations on this class, we must chiefly confine ourselves to the *Acalephæ simplices*, which are mostly comprised in the great genus Medusa of Linnæus, and to which M. de Lamarck has given the name of *Medusariæ*.

These animals are extremely numerous in all seas, but more particularly in those of warm climates, have been remarked at all times by the inhabitants of the sea shore, and by all authors of natural history, from Aristotle down, though they are scarcely of any utility to the human race. But the singular property which they possess of being luminous to a great degree in darkness, or obscurity, and that of producing a painful sensation similar to the sting of nettles, when any of them are touched, must have occasioned them to have been observed early. Accordingly all maritime people have particular denominations to designate them. These names almost always indicate one of these two properties, such as *knide, acalaphe, urtica marina, sea-nettle*, &c.

These acalephæ have a regular form, very circular, hemispherical, more or less convex above, and concave underneath, with a simple, rounded, medial orifice, usually very large, surrounded, or not, with appendages of variable form. The union, in a greater or less portion of their edge, of these buccal appendages, constitutes a common pedicle, the cross-wise attachment of which divides the orifice into four parts.

The hemispherical, and principal portion of the body, is termed, as we see in the text, *umbrella*. The buccal appendages are called arms. The part composed by their union is called pedicle.

The umbrella, always regularly circular, is sometimes very much depressed above, as it is underneath. At other times it is subcylindrical from its great elevation. It is rarely globular, finally, and most frequently it is nearly hemispherical. Its edges, or the line of junction of the convex with the concave part, are sometimes entirely smooth, rarely raised into angles, a little salient, or sub-lobate, or tuberculous. Most frequently they are furnished with tentacular filaments, more or less elongated, and which have been called *tentacula*. We remark, also, in a certain number of species, in different points of the circumference of the umbrella, similar organs, at regular intervals, the use of which is unknown. They are designated by the name of auricles. The aperture of the middle of the concave face is sometimes very great, round, or squared. It is sessile, or at the extremity of a sort of labial elongation, in the form of a proboscis, or funnel, more or less elongated. In the circumference of this aperture, whether sessile or not, are often remarked appendages or arms, sometimes rather long, of a fixed number, and which are divided or ramified in all their extent, or at their extremity only. Between these divisions are sometimes seen some organs which Pallas and Peron have compared to the cotyledons of vegetables. These appendages are often attached to the circumference of the sessile mouth, and sometimes more or less high on the proboscis. But it sometimes happens that they are also united in a more or less considerable portion of their extent. From this results a pedicle, sometimes very thick, which appears to divide the mouth into four parts.

These acalephæ, which vary considerably in bulk, since, though there are some truly microscopic, there are others which attain to several feet in diameter, and weigh fifty pounds, are of all animals those which have the least of solid substance. They are composed, as it were, of a sort of jelly, more or less consistent, perfectly transparent, which, in consequence of the loss of life, resolves itself into a limpid salt water, leaving as a residue but some grains of membranaceous parts equally transparent.

The tissue of the medusæ is not then really homogenous, though it appears to be so. Their skin, or envelope, is nevertheless of an extraordinary thinness, not distinct. This may be considered but as the boundary of their tissue a little condensed. Observed with a microscope, M. Gaëde perceived that it was furnished with small grains, each of which appeared to be composed of grains still smaller. Might this be the source of the viscous matter, which transudes from all parts of the body, and which Peron says that he has observed in individuals put into sea-water frequently renewed, to enable them to preserve all their vital activity, and which is so abundant, that the thirtieth portion of water is as much charged as the first. This does not seem probable. It might, perhaps, rather be believed that this is the origin of the eminently phosphorescent substance, which Spallanzani has remarked in certain parts of the body of the luminous medusæ, and which possesses properties different from those of the liquor which issues from a wound. The latter has the taste of salt water, and the other causes a painful sensation, to that degree that having touched it with his tongue, Spallanzani felt a burning impression which lasted more than a day; a drop having by chance fallen upon his eye, the pain was still more intense. The caustic quality of this humour is not, however, always in accordance with the phosphorescent property, since there are species which are not luminous, but which, nevertheless, produce a sting.

These animals are in general perfectly colourless, and resemble the purest and most transparent rock-crystal. There are some, however, which have coloured parts, reddish, fine ultra-marine blue, greenish, &c.

The apparatus of sensation in the medusa appears to be limited to the skin. The name of tentacula, which has been given to the filaments, more or less elongated, which border the umbrella, the use assigned to them, as well as the brachial appendages in certain species, might cause it to be suspected that these organs enjoy a more exquisite sense of touch. But there is nothing in the organization of these parts to confirm this suspicion, and it is not even certain that these organs serve for the uses which are attributed to them.

No trace of nervous system has ever been observed in these animals, nor is it probable that any exists.

If, after investigating the organization of the medusæ, we turn our attention to the study of their functions, we shall still find many other phenomena equally worthy of observation.

Their general sensibility appears to be extremely obtuse; and, perhaps, it is the same with the special sensibility of the marginal and buccal tentacula, whose power of contractility, however, appears to be very great. The medusæ do not seem to feel the hand that seizes them.

Their locomotion, which is very slow, and denotes a very feeble degree of muscular energy, nevertheless, appears to have no cessation, since, being of a specific gravity more considerable than that of the water in which they are immersed, these animals, so soft, that it is probable they could not rest upon a solid ground, are obliged to move incessantly to sustain themselves in the fluid which they inhabit. Accordingly, they are in a continual motion of systole, and diastole. Spallanzani, who has observed them carefully in their motions, says that those by which they change place, are executed by the approximation of the edges of the umbrella, so that its diameter diminishes in a very considerable manner. By this a certain quantity of water contained in the stomachs, and in the cavity of the umbrella, is expelled with greater or less force, and the body is projected in an inverse direction. Returned by the cessation of the muscular force to its first state of development, it contracts itself afresh, and makes a new step. If the body is perpendicular to the horizon, this succession of contraction and dilatation causes it to ascend; if it is more or less oblique, it advances more or less horizontally. To descend, it is sufficient for the animal to cease its movements; its weight alone draws it down. It is never in an inverted position, with the convexity of the umbella underneath. Neither the tentacula nor the arms appear to be employed in these movements of transportation, at least the latter, according to Spallanzani, are always extended following the body. Some ingenious experiments related by this observer, prove that it is only the muscles of the marginal zone of the umbrella, which cause it totally to contract, since, on removing them, the remainder of the umbrella undergoes no change, while the removed zone continues its movements of systole and diastole. In spite of this almost continual action of the locomotive faculty, the medusæ do not appear to be able to overcome the smallest current, but are constantly carried away by it.

According to all observers, the medusæ feed on little animals, on mollusca, worms, crustacea, and even fish, which they attract towards their mouth, by means of the appendages with which it is armed. Spallanzani has supposed this, because he saw a small fish which was attached to one of the appendages of an individual which he had just caught. M. Gaëde says positively that he found in the stomach of the medusæ which he dissected, some small fishes and nereids. M. de Chamisso and Eysenhardt, in their memoirs on these animals, assure us that they have found several times in the ventricles some heads and remnants of fish, as it were digested. Several other naturalists, who have had copious opportunities of observation, aver the same thing; and M. de Blainville has found some small fishes in the equorea, and even in the rhizostoma. But he queries whether these little animals were seized by the medusæ for the purpose of nourishment, or that they came there accidentally. The last opinion is that of M. Cuvier, at least regarding the rhizostoma, which appear to him to derive their nourishment through species of suckers, as we have already stated.

We have been hitherto in ignorance, and, probably, shall long remain so, concerning the duration of life in the medusæ, as well as the history of their development. It is probable that they are rejected by the mother in a perfect state, and differ from her only in size. It is known that they are larger in spring and summer, that is, at the time when their ovaries are distended by the eggs which they contain, and that in the other part of the year they are smaller. It is also known, that the appendages acquire with age a development and a complication, which they did not at first possess.

We find some species of these animals in all the seas of cold climates, as well as in those of warm, and more especially far out at sea. Each, according to the observations of MM. Peron and Lesueur, appears to be confined to determinate portions of the globe, where the individuals are united in innumerable troops, and sometimes form many square leagues in extent. If they appear and disappear sometimes at determined periods, that, doubtless, depends upon the regular winds and currents which carry them away and bring them back. They are sometimes thrown in great quantity upon the shores of our climate, where endeavours have been made to turn them to some advantage. It has been attempted, but without much success, to extract ammonia from them. They have been more beneficially employed in the way of manure upon arable land.

All the medusæ in a state of death and putrefaction, appear to be phosphorescent; but there is only a small number which appear to be so in the living state. We are indebted to Spallanzani for a great number of curious experiments on this subject. He first endeavoured to discover what the parts are which more particularly possess this singular property. And he found that they were, 1st. the great tentacula, or arms. 2nd. the muscular zone of the umbrella; and 3rd. the stomachal cavity. The rest of the umbrella only shines by transmitted light. He then occupied himself in observing to what cause the phosphorescence is owing, and he became convinced that it was to a peculiar glutinous humour, which issues from the surface of the three parts just mentioned. This, however, as we have said farther back, is altogether different from that which issues from the body, and even from these same parts, when they are cut. It is very corrosive, and its application on the hand, and on the tongue more especially, occasions a lively sensation of pain. Expressed into different liquids, as into salt water, but particularly into fresh water warm, or milk, it communicates to them a phosphoric light. A single medusa thus expressed into twenty-seven ounces of cow's milk, rendered it so resplendent, that one might have read the characters of a letter by it, at the distance of three feet. At the end of eleven hours, it retained some degree of light; when it had lost this altogether, it was renewed by stirring it, and, finally, when this means no longer produced any effect, it was again obtained by heat, care being taken that the heat was not too strong. The dead medusa also still possessed for a considerable time the phosphorescent property, and it was renewed by pouring fresh water upon it, even some time after it had ceased to shine. On the living animal, it is stronger in the state of contraction than in that of dilatation, which is easily explained, because it is the part which is particularly contractile, which exhales the phosphoric humour. The light may be suspended for more than half an hour, which depends on the cessation of the oscillations, and nevertheless the phosphorescence continues, although to a degree much less intense, in the dead animal, even to putrefaction. The phosphorescence is increased by giving a commotion to the parts of the animal, or even by rubbing it with the hand. When it is living, it communicates to the water in which it is plunged, its phosphoric property, but half as much again in fresh water as in salt.

Certain of these animals possess another property which is more hurtful, which is that of producing a very sharp pain, when they touch any part of our skin, which has occasioned them to be called *sea-nettles.* Dicquemare, who has made experiments in reference to this subject on himself, with the *cyanea cœrulea*, relates the effects of them in these terms; "The pain is pretty nearly similar to that which is felt on touching a bunch of nettles; but it is stronger, and endures about half an hour. In the last moments reiterated stings are felt, but proportionally more faint. There appears a considerable redness in all the part which has been touched, and swellings of the same colour, which have a white point in the middle. After the end of some days, when the pain is gone by, the heat of the bed will cause the blisters of the skin to re-appear." This effect appears owing to a caustic humour, which issues from the skin of the medusa. Is it different from that which produces the phosphorescence?

This appears probable, since, as we have observed before, the species noticed by Spallanzani, which was eminently phosphorescent, produced no effect of urtication. The species which possess this property in the living state, have it also when dead. Certain others possess it in so small a degree, that it becomes sensible only in the softest parts of the skin.

Attempts have been made to ascertain whether the medusæ are susceptible of a reproduction of the parts which have been removed from them; but such does not appear to be the case.

The medusæ serve as food to several other animals. The actiniæ seize these acalephæ on their passage, and draw them by degrees into their stomach. The whales also destroy an immense quantity of them; but it appears that these are species or individuals of an exceeding smallness, with which the waters of the sea, inhabited by these great animals, are filled, and that they are there with many other animals of different types, but which are likewise almost microscopic.

The PHYSALIÆ, which constitute the type of the second division of acalepha, are a very singular race of animals, noticed for a long time by sailors, who give them the names of *galleys*, *frigates*, or even *ships of war*, in consequence of the elegant manner in which they seem to sail on the surface of the waters. They have received the name of *physaliæ*, or *sea-bladders*, in consequence of their resemblance to a bladder, or even that of *sea-nettles*, because it appears that they produce the same effect upon the skin as the medusæ. Some recent writers have thought proper to place them among the mollusca; and certain it is, that they present little or nothing of the radiated arrangement; but as the Baron observes, the total absence of internal and complicated organs, of which he has satisfied himself in many large individuals, will not allow us to admit of the notion that the physalia may be one of the mollusca.

M. Tilesius writes thus concerning the physaliæ: all the physaliæ consist of a long bladder inflated with air, floating on the water, having upwards a sort of comb, which answers as a sail, and underneath some long tentacula, which constitute at once the mouth and helm. Although to examine these is difficult enough, because they burn more strongly than nettles, when they are touched, he has been enabled to distinguish three different kinds at least, on separate individuals.

M. Tilesius has made most observations on the sting produced by touching these animals. He ascertained that the burning sensation which is felt, when one has touched more or less strongly the tentacula of a living physalia, and which is more intense than that produced by nettles, is owing, not to a mucous matter which covers them, as he had supposed at first, but to some little hairs of a rose colour, which the mucosity introduces into the pores of the skin. In fact, one day when he was severely burned by handling too much the tentacula of a physalia, after having tried without success to calm the pain, with vinegar, solution of salt-petre, salt, sulphuric or ammoniac acid, he only achieved the point by frequent lotions of soap and water on the affected parts, having previously carefully taken out the little hairs with a tweezers. We must, nevertheless, believe that the mucous matter itself also possesses this burning property; for the same observer found, that on washing himself in a porcelain vase, in which a physalia had been preserved, and which had not been sufficiently cleaned, his lips, nose and cheeks, were severely burned.

The physaliæ live in the waters of the sea, at tolerably great distances from the shore; except when they are driven thither by currents, or by the wind. Observers not having seen them, except at the surface, it has been generally admitted that they are always there, the bladder being partly out of the water, and the tentacula, &c. more or less deep in the sea. M. Tilesius adds that these animals sound with their tentacula all the bodies which may be near them under water, that their suckers are applied upon wood, stone, and even upon glass, and porcelain, and deposit a mucous matter there which communicates to these bodies the same burning property possessed by the tentacula themselves.

SUPPLEMENT

ON

THE POLYPI.

In the fourth class of this great division, the first and most extraordinary animals we meet, are the ACTINIÆ, vulgarly known by the name of *sea-anemones*, or *fixed sea-nettles*. Their body is fleshy, very contractile, usually remaining fixed upon its base, but able, however, to change place, either by crawling on this same base, or walking upon this tentacula.

When the body of the actinia is the most contracted, it represents a hemisphere, with a small aperture at its summit. Such is the position which these animals retain, when they are stranded, or when the sea is troubled and the sky overcast. But when they are hungry, or the weather is fine, they expand and blossom. The small aperture then becomes as broad as the base, and the body represents a short cylinder. The mouth is at the centre of the superior base of the cylinder, and the whole circumference is furnished with several ranges of tentacula, which represent extremely well in their fine colours and arrangement, the petals of certain double flowers. The mouth conducts into the stomach, which is a sac wrinkled internally, but without any other issue, at least visible, except the mouth. The interval between the stomach, and the external skin of the body, is filled with very singular viscera, which have not been yet sufficiently developed to enable us to communicate a very clear notion of them. Some membranaceous and vertical laminæ proceed from one of its surfaces to the other, like the radii of a circle. They are not of equal height. Their superior edge is furnished with very complicated fringes. There are also found in the interior part of the body, some long, tortuous, and very slender intestines. We do not find in the body of these animals, any thing which can be regarded as nerves or blood-vessels.

Through the experiments of Reaumur, of Baster, and more especially of Dicquemare, the history of these animals has become extremely curious. Their multiplication takes place in two ways; the first is natural, and consists in a spontaneous dilaceration of a portion of the ligaments of the base, which is performed by the constriction of this part. Then may be observed, escaping by this dilaceration, one or several small portions of the animal, which in a little time become new actiniæ of the same species as that of which they had composed a part. Dicquemare obtained an artificial multiplication, at least equally singular. He separated some parcels from the base of an actinia, and beheld them almost immediately form so many new animals.

In the month of May, 1772, Dicquemare cut off all the tentacula from an actinia, and in a short time after they rebudded. He cut them afresh on the 30th of July following, and they were reproduced in the same way in less than a month. An actinia cleft through the middle of the body, appeared, after the end of some months, as completely organized as before it was mutilated. These experiments, and many others, the detail of which might prove tedious, demonstrate in a positive manner the faculty possessed by the actiniæ of reorganization in their destroyed parts.

Their food consists of small mollusca, shrimps, little crabs, and medusæ. They seize them with their tentacula, keep them in the interior of their body for ten or twelve hours, and subsequently void through the same aperture the solid parts which they have been unable to digest. In the sea-water, they can support very long fasts. It has been observed already that the more hungry they are, the more they dilate their aperture; it sometimes even happens that they turn their stomach altogether, and render it convex, instead of concave, as it was before. They perish directly in fresh water. They are found attached by the base. When the animal is desirous of changing place, it possesses several methods of exercising this movement. It either slides slowly upon its pedicle, or, detaching its base altogether, it swells itself with water, and then becoming almost as light as the volume of water which it displaces, the least agitation is sufficient to impel it further. When it wishes to fix itself, the body contracts, the water escapes, it proceeds to the bottom, and its base becomes glued to the surface of the first body within its reach. Though no organ of vision is observable in these animals, a strong light evidently incommodes them. They appear strongly affected by it. Dicquemare has observed that those actiniæ from which certain parts were abstracted, were more sensible to the glare of light than they had been before they were mutilated. Severe cold does not annoy them. They may be enclosed in a piece of ice, left there all night, and on the following day they shall be found alive. Placed under a pneumatic machine, they support the effects of the vacuum without inflating themselves, and without appearing in the least degree exhausted when the air is restored to them. Dicquemare has made some further observations which may prove useful to mariners. He perceived that all the changes of weather were announced by some extraordinary movements in the actiniæ which he brought up;

and he arrived at this result, that the movements of these animals were in accordance with those of the barometer.

The actiniæ possess no maleficent quality, and are eaten in several countries.

Dr. Spix has given some very curious details respecting the organization of these animals, which he observed and dissected upon the coasts of the channel.

In considering the gelatinous polypi, we shall confine our remarks chiefly to the HYDRA.

The hydræ are animals exceedingly simple, which we can scarcely compare to any thing but filaments of small thickness, fixed at one of their extremities, by means of a sort of sucker, and provided on the other with a crown of cirrhi, or tentacula, more attenuated than the finest hairs, to the number of ten or more, and possessed of extreme contractility. This disposition of the tentacula and even their uses, caused these little animals to be compared to the polypi of the ancients, now denominated *octopus*, and occasioned Reaumur to give them the name of polypi. The structure of the different parts of the body of the hydræ is throughout completely uniform. In fact, we discover in them, even with the assistance of the microscope, nothing but a sort of parenchyma, formed of globules, and cellular tissue, and which is capable of contraction, particularly in the tentacula, to such a degree as to disappear almost completely. Accordingly, the general sensibility of these animalculæ is exquisite, so that they can feel or perceive the light, and distinguish it from the shade; not that we would be understood to say with some writers, that they actually possess the power of vision, through the medium of the general envelope; but they are in the predicament of plants, which direct themselves towards the light, the effects of which they experience, without otherwise perceiving the bodies which transmit it. The hydræ have no traces of the organs of sensation, except that of touch, which is perfect in the tentacula, with which the mouth is armed. All the parts of their tissue are capable of contraction, but without any possibility of our perceiving in them distinct muscular fibres. The hydræ are, however, capable of changing place altogether, and they do so after the manner of the geometrical caterpillars, and of some leeches. Most frequently, however, they remain fixed by the posterior extremity, and extend, more or less, their body and tentacula in the different directions necessary for seizing their prey. They feed upon very small aquatic insects, such as monoculi, or naïades. They attract them towards themselves by almost continual movements of the tentacula, enlace them in their numerous folds, agglutinate them by means of some secretion, or some mode of suction, and finally direct them towards the aperture of the mouth, which is in the centre of the circle formed by those tentaculæ. This mouth, which is capable of being dilated into a sort of calyx, communicates into the stomach, which is hollowed in the parenchyma itself, of the body of the little animal, without any distinct parietes, except the skin at the exterior surface. Accordingly, the similitude of the external and internal paries is so great, that Trembley, in one of his most curious experiments, has proved that the little animal may be turned inside out, almost like the finger of a glove, and that digestion and absorption can be as well performed by the external as the internal side. This sort of stomach has no posterior orifice, and when the prey, whether digestible or not, has remained there for some time, it is rejected, in the latter case entire, and in the former such parts are expelled, as were not susceptible of digestion. Hence it appears, that there is no choice respecting the bodies which the little animal introduces into its stomach, and that the stomach alone discriminates by its own action upon them, whether they are suitable or not. These animals, moreover, can support a very long fast, probably, because they absorb directly from the medium in which they live.

The reproduction of the hydræ is still more simple, if it be certain that it takes place by buds in all parts of the external surface of the body. During the summer, it is said, that a little germ is observed to project from some part of this surface, which enlarges by degrees, and assumes the figure of the mother. Very soon from its free extremity we see likewise the tentacula sprout forth; and at the end of a longer or shorter time, which depends a little on circumstances, more or less favourable, the young hydra, which, while it was attached to the mother, sought and attracted its prey like her, and fed upon it, ends by being detached from her, and proceeds to fix itself upon some submerged body, where it reproduces in its turn in the same manner. Sometimes it even reproduces while attached to the mother; insomuch that even eighteen of these animals have been reckoned thus united.

The reproductive faculty of the hydræ, carried to the extent observed by Trembley, must tend to make us believe that this power is extended to all parts of the body. In fact, after the very delicate experiments, but authenticated beyond all doubt, of a philosopher so entirely worthy of credit as Trembley, it is evident that not only the various parts of the body, cut longitudinally or transversely, can reproduce the parts that are deficient, and thus form so many complete animals, often in two days only; but even sometimes a portion of tentaculum can be developed, and produce a perfect hydra, which Roësel assures us that he has observed.

But it is not only by gemmation, or by artificial or spontaneous scissure, that the hydræ can be reproduced. Jussieu, Trembley, Roësel, and Pallas himself, have observed, that towards autumn, they eject from their parenchyma some eggs, which fall, are preserved during the winter, and are not developed until spring. It is also said, that the individuals which have their origin in this mode of reproduction, are always smaller than those which have come by gemmation. It has been questioned, however, whether these are in reality eggs.

Thus the history of the hydræ, in considering it as well authenticated, brings to general physiology some considerations of the greatest importance, since it exhibits to us an organized body, composed of a homogeneous tissue, consequently without distinction or separation of organs, not even of skin and muscular fibres, endowed with a very great sensibility, which permits it to feel the light, extremely contractile in all its parts, which can seize small animals of greater resisting solidity than itself, introduce them into its stomach, and digest them; and that not only with the surface which is habitually digestive, but also with the external surface, artificially become internal. It displays to us an animal, that can graft itself upon another, or by a true continuity of substance, can form, by the union of several individuals, a complex animal, with many heads, thus realizing in nature, the hydra of the fable; considerations which would lead us to believe that the hydræ are very inferior to the polypi of the madrepores, and to those of the pennatulæ, &c. which are evidently much more complicated in their organization, and that consequently they ought to form the last link of the chain of radiated animals.

Hydræ are to be found in all fresh waters, and in those of the sea, but more particularly, as it would appear, in the first, when they are dormant, providing that be pure. It is during summer that we must look for them, on all bodies submerged in such waters; for, during winter, it seems that they contract themselves and sink into the mud. They may be easily procured by taking a certain quantity of water-lentils, and putting them into a vessel full of water. After a certain term of repose, the hydræ are observed to begin to move and agitate their tentacula, as in their ordinary position. Trembley, who thus preserved and studied them for several consecutive years, used to feed them with daphniæ, and other small aquatic animals.

The VORTICELLA of *Müller*, is a genus concerning which some controversy has existed, as to whether it should be placed with the polypi or the infusoria. The common characters of the animals composing it are, that they are naked, contractile, and provided with rotatory organs. But they present enormous differences, some being binary animals, appendiculated, and very complex, others appearing radiated and very simple.

The organization of the true vorticellæ is much more simple than that of the species retrenched from them by M. de Lamarck. In fact, the vorticellæ exactly resemble a flower of the lily of the valley, supported upon a long filament. This filament is cylindrical, and a little enlarged at its terminal part. The body itself is formed like a small purse or monopetalous flower, having its edges widened, and provided with groups of short and very fine ciliæ, opposed laterally, which should make the vorticellæ binary animals. When these little beings are in their state of complete developement, they are attached to some submerged bodies by the extremity of the pedicle, this part being extremely stretched, as well as the body, at the anterior of which the two fasciculi of ciliæ are agitated with great rapidity. From this action results a sort of double vortex, or whirlpool, which separates, or drives out, the extremely fine molecules which are found in the ambient fluid, and which is regarded as serving to direct the prey towards the buccal cavity. On the least shock, these little animals contract themselves quickly, undulating their pedicle, which being fixed, serves them as a point of rest.

At other times they are observed to detach themselves spontaneously, and swim, drawing after them their pedicle extended in a right line. Finally, they sometimes fix themselves by the enlarged part of their body, and appear to move by means of their appendages. This is chiefly observed in individuals whose pedicle is short, or even nothing, which is the case with the *urceolariæ*.

It appears that there are vorticellæ attached one upon the other, so that they seem to constitute composite animals.

The vorticellæ, properly so called, multiply by natural sections, the body dividing by little and little in the middle, in such a manner, however, that the pedicle remains to a single individual. The promptitude with which this scission takes place is in proportion with the state of the external temperature, so that in fine weather, the multiplication of these animals goes on with wonderful rapidity.

On the approach of winter, they produce oviform germs or buds, which are preserved in the water during the whole of this season, and are developed in spring.

The vorticellæ particularly live in fresh and stagnant waters, fixed upon all the bodies which are to be found there.

Müller has observed that the mode of generation in the species called *V. racemosa*, is altogether peculiar. An adult individual fixes itself upon some body. Then from its own body, or at the base, germinate eight similar bodies, which, in a few hours, are raised upon their own proper pedicles. In a short time, each of these new bodies gives birth to eight others, which provided in their turn with their pedicles, go on *seriatim* to propagate in the same manner. During this time the pedicles of the first and second order cross like the branches of a vegetable. As to the pedicle of the mother, and which supports all the others, it preserves the same length.

Previously to taking a short review of the principal groups of the CORALLIFEROUS POLYPI, &c., it may not be unnecessary to say a few words in general on the bodies more or less solid, with which these animals are united, and which are now generally termed by naturalists *Polyparia*. This term, of very general, perhaps too general an extension, may be thus defined. A polyparium is a fixed envelope, more or less solid, calcareous, or corneous, in which a polypus resides, and which is the evident result of a transudation from its body of an excretion through certain pores of its skin, of matters sufficiently composite to form, by their approximation, a concrete body more or less solid, and altogether inorganic. This definition, however, is strictly applicable only to the madrepores and escharæ of Linnæus, and is not equally so to the other divisions. Therefore, if we persist in generalizing this name, it will be necessary to define it as a solid, calcareous, or corneous, the residuum of one or several polypi, without any attention to its mode of formation, or the manner in which the polypi are placed there, and then the fibrous mass of a true alcyon, the fleshy mass of a pennatula, the corneo-calcareous lamina of an eschara, the plant-like stems of cellaria and sertularia, the calcareous, arborescent, frondescent masses of the madrepores, will be equally polyparia. We may even add to these the beautiful tufts of the corallina, on the supposition that they support polypi, which, however, is by no means clearly made out.

In considering the nature of polyparia, we find them to be of several sorts, according as they are calcareous, or stony, corneous, fibrous, corticiferous, gluey, or fleshy.

The CORALLINES form a genus of organized bodies, on the nature of which, although it is very common in all the seas of Europe, and has for a long time been employed in therapeutics, authors are by no means agreed, some regarding it as appertaining to the vegetable, and some to the animal kingdom; but our limits forbid us entering into this question, or adding to the statement of the text, however brief, any thing beyond the observations of de Blainville, who says, that notwithstanding all the pains he bestowed in observing corallines in the shade, or in the sun, or in the small holes of rocks filled with water some time after the sea had retired, with a very strong microscope, he was never able to discern the least trace of animals, or even of filaments which might issue from them. If, after having viewed the exterior of a coralline, we come to study the internal structure, we shall not find, as some authors say, that it is a fibrous corneous axis, surrounded by a calcareous crust, but on the contrary, that it is a sort of cellular tissue, in the meshes of which the calcareous matter is deposited; and in fact, when we put a coralline into a weak acid, it is softened absolutely like a bone, without being diminished in volume, without assuming another form, or even changing colour. All this causes M. de Blainville to doubt that the true corallines can be formed by distinct polypi. But he puts the query—is the coralline really a vegetable? On this point he is by no means assured, although all the Italians, who have been the principal observers of these sorts of bodies, appear to be perfectly convinced that it is.

Every one knows that under the name of CORAL (*Corallium*) is commonly understood a sort of arbusculum, more or less branched, stony, calcareous, sometimes of a fine red colour, sometimes more or less roseate, or even altogether white. It has been employed from time immemorial in the manufacture of toys, and other objects of ornament, and it gives rise to a fishery and trade of great extent in different parts of the Mediterranean.

The polypi which inhabit the cellules of the surface of the coral, are very soft, altogether white, and not very transparent. Their body, or belly, is cylindrical, and entirely concealed in the cellule, to which, doubtless, it is adherent by its extremity, which appears to be continued with the vessels of the fleshy and common envelope. This at least is the opinion of M. de Blainville, who founds it on a supposed analogy with the pennatulæ, which he has dissected. Donati, however, tells us expressly that it is entirely detached and separate. Be this as it may, the body of the polypus is terminated by eight appendages, disposed in radii round the mouth, or an aperture which Donati informs us is made by a shell, a little widened at its root, with a large aperture at the summit, and hollowed by eight broad furrows, between each of which arises a sort of ridge. Between two of these elevations, is placed one of the tentacular appendages, which are thus upon the same plane. They are all perfectly equal, conical, a little compressed, and provided on each side with appendages, or barbles, regularly decreasing from the base to the extremity. Donati adds, that he has seen at the inferior part of the body of some polypi, some little hydatiform bodies, rather round, extremely small, soft, transparent and yellowish. He thinks, with reason, that these are the eggs, or reproductive corpuscula.

The coral lives in the Mediterranean sea only at considerable depths, though rather variable. The coral seems to require at least ten years to render its growth complete. Propagation takes place thus: the eggs, or reproductive corpuscula, rejected through the mouth of the animal, or rather, perhaps, through the orifices which are at its margin, fall upon some body, and adhere there by their soft and gelatinous nature. They extend a little, and begin to grow, particularly at the part in contact, which enlarges, and moulds itself on the submarine body. From the middle of this sort of coral-drop, rises a tubercle, in which are evidently seen an interior cavity, and eight wrinkles, or plates at its superior part, but without aperture. The polypus, in its interior, is as yet but in the state of fœtus, but it attains its growth successively. All the parts become developed, and it is then that the capsule opens, to permit it to put forth its tentacula, to seize its nutriment, and perhaps to respire. The growth of the central part then becomes more rapid. It deposits calcareous matter in the middle. It sprouts more and more, and is developed by means of new polypi in indeterminate points, so that one might say that the polyparium is almost totally independent of the polypus, and that its hard part, or axis, is always softer towards the extremities of the branches, than at any other place.

Chemical analysis has proved that the coral, or at least its axis, is entirely composed of carbonate of lime, for it completely dissolves in nitric acid.

Imperato was the first writer who made use of the term MADREPORE *(Madrepora)* which, in its application by subsequent naturalists, has undergone several modifications. Without troubling our readers with an account of these, it will be sufficient to say that the name is now reserved to certain lamelliferous tree-like polyparia, the surface of which is bristled with projecting cellules. Their classification is unfortunately based only on a knowledge of the polyparium, or cretaceous mass produced by the animals, for of the latter, very little that is satisfactory has been ascertained. Imperato was the first who suspected that the madrepores of Linnæus belonged to animals. Rumph, who had occasion to observe a great number in the Indian seas, confirmed this opinion, but he observed in them nothing but a sort of animal jelly covering the polyparium. Finally, Peyssonel removed all doubts upon this subject, and to Donati and Cavolini, we are most indebted for any details respecting the species.

Of the madrepores thus defined, none are known in the seas of Europe, and hitherto they have been met with only in those of South America and India. Fixed by their base at considerable depths, they elevate more or less their foliaceous expansions. We are totally ignorant of their mode of growth, multiplication, and death. We merely know that the polyparium, which is entirely calcareous, is of a closer tissue, near its base, and that, on the contrary, the extremities of the ramifications are always more porous. The inferior cellules are always more effaced, (the reverse is the case with the upper) and the extremity of the branches is often terminated by an infundibuliform excavation tolerably deep.

It is said that the formation of the numerous reefs in the South Seas, the Indian Ocean, and the Red Sea, is owing to the very rapid growth of the madrepores, and particularly of that species called *muricata*. It is certain that most of the islands in those seas rest on a calcareous soil, entirely composed of stony polyparia, and that their highest mountains are thus composed. But it may be difficult to prove that the madrepores are the species which are found there in the greatest number. On this subject we are deficient in positive observations.

PENNATULA is a genus of true zoophytes, established by Linnæus for a set of animals extremely singular, whose form in the most common species resembles that of a quill, from which its name is derived. These animals are composed of a common part, or stem, most generally containing in its exterior a long calcareous stick, and a certain number of polypi disposed in rather a fixed manner upon a part of the stem, or on some appendages which are added to it, and which constitute the barbs of the quill. To these species modern zoologists have reserved the name of pennatula.

Without following naturalists in a very detailed description of these animals, we may say in brief, that a pennatula is a body of a determinate form, binary, symmetrical, composed of a muscular contractile tissue, most frequently supported by a solid calcareous part, produced by a particular membrane, and of an areolar tissue, as it were spongy, susceptible of a sort of erection by the introduction of an aqueous fluid; a body with which is in organic communication a considerable number of little animals, each having a buccal orifice, surrounded with a rank of pennated tentacula, and an oviferous sac, developed in the tissue itself of the pennatula.

The physiological phenomena which the pennatula presents is extremely interesting, since it exhibits the example of a truly composite animal, that is, one in which animals, more or less in number, really perfect as far as comports with the grade of organization to which they belong, form part of a common, living, contractile body, serving as an intermedium both for locomotion and nutrition to all the individuals, so that they are all carried together by the sole movements of the common part, without the particular movements of each occasioning any obstacle, and they are all nourished in a mediate manner, by means of this common portion of which they form a part. The nutriment which favourable circumstances have placed within the reach of one individual, nourishes that individual first, and then by extension, nourishes the common stem, and thus the other polypi, which constitute organic portions of it, receive their share.

It is rather more difficult to conceive the mode of growth in the pennatula, and its mode of reproduction. In all the aggregate polyparia, as the madrepores, the growth proceeds by the extremities, and consequently, very probably, by the adherence of the gemmules, produced by the terminal polypi, to the lodge of the latter. The accidental fall of these gemmules gives birth to new individuals. In the first case, we must consider it rather an accumulation, than a true growth; and, in fact, there is a real death of all that is below the extremities. It cannot be so in the pennatula, which is a terminated and finished whole, so that we must believe that the growth here is really individual, at least in the common part.

As for the composing polypi, each of them, without doubt, must grow. But the truth is, all is conjecture here, nor is our knowledge of the habits of the pennatulæ at all adequate to assist us in throwing light on these very difficult points of physiology.

The pennatulæ, as it seems, live constantly in the high seas, and are always floating. Although they are in a continual movement of systole and diastole, somewhat like the medusæ, which is produced by the contraction of the inflated and posterior part of the animal, and especially by the *winglets,* or polypiferous branches, it is not probable that they possess the least power of directing themselves in the interior of the waters. They are altogether under the influence of the currents, like the medusæ themselves. Still less is it admissible that the composing polypi can conspire together in their movements, to direct themselves in determinate tracks. This is an hypothesis which defies conception, nor is the object of such a faculty at all perceptible. Each polypus acts independently of its neighbour, and the object of the motion of its tentacula, is only to seize the little animals which may come within its reach.

Pennatulæ have been observed in every sea, but their known species are not numerous.

We shall dismiss the consideration of this class, with a notice of the SPONGES *(spongia).*

This is a group of organized bodies, extremely extended in every sea, but particularly in those of warmer climates. They have been known since the remotest antiquity, and, nevertheless, naturalists are far from being agreed with respect to their true character. Some will have them to be vegetables; others a wholly simple animal; and several, polyparia, of which the animal is unknown. All the authors of antiquity allowed them a sensitive life, and admitted that the living sponges seemed to avoid the hand which would touch them, and that they appeared to adhère more closely to the submarine rocks, the greater the efforts that were made to detach them. They considered the sponges to be organized bodies, intermediate to vegetables and animals, such as they subsequently named zoophytes. This opinion was maintained for a long time, and gained the assent of most of the Italian writers, Spallanzani, Olivi, &c. some of whom, however, granted a higher degree of animality to these productions. Rondelet seems to have been the first who utterly refused all sensibility to the sponges, and denied the fact mentioned by Aristotle, above alluded to. From this an hypothesis arose, that they were only vegetables, and it was adopted by Tournefort, by many ancient botanists, and by Linnæus himself, in the first editions of his *Systema Naturæ*. Such, for a time, was also the opinion of Spallanzani, as far as some species were concerned, because he observed in them no signs of contractility. The third opinion, which was Peyssonell's, is that sponges are sorts of polyparia, fabricated by animals found in their excavations. But as this could not be supported, as these animals have no adherence with the sponges, and are often of totally different species, this hypothesis was modified by pronouncing the sponge to be a polyparium, the polypi of which are unknown.

If naturalists still hesitate respecting the nature of the sponges, it is, doubtless, because they have no sufficient notion of their organization. Without touching on the merits of the controversies concerning them, we shall lay before our readers all that is known upon the subject.

Every one agrees that the sponges are generally formed of at least two substances; the first, interior, more or less corneous, fibrous, intercrossing in all ways, and forming a sort of felt-like tissue, more or less compact. It is this which attaches the sponge to the submarine bodies; the second, soft or gelatinous, enveloping the preceding, forms a sort of general stratum, which is the seat of sensibility and life. Olivi tells us that there are species of sponges, which present three very distinct substances. 1. The fibrous matter which forms the support, or as it were, the skeleton of the sponge. 2. The mucous gelatinous substance, surrounding the fibrous parts; and 3. A terrene matter, mingled with the preceding, and forming a sort of cortical substance, which surrounds the gelatine itself. According to Father Vico, however, the organization of the sponges is much more complicated, and he would fain discover in them, a muscular and a nervous system, ovaries, seminiferous vessels, &c. As this opinion, however, seems wholly untenable, it is unnecessary to dwell on the details on which it is founded. Olivi, whose opinion is more probable, regards the fibres, not as very important organs, but as the support, or frame-work of the living body, and the sort of marrow, which is sometimes found in their interior, he supposes to have originated at the period when the animal was young. He thinks that the mucilaginous and cortical envelope really constitutes the animal, but that this animal is amorphous.

Sponges are considered by M. de Blainville, after Pallas, Cavolini, Olivi, Donovan, &c., as organized bodies, intermediate to the two organic kingdoms, without any regular determinate form; presenting an absorbent surface having some obscure traces of feeling, nourished by the molecules from the surrounding medium, pretty nearly like vegetables. Reproduction takes place by a sort of pullulation, or scission of the living or gelatinous matter.

The sponges are always adherent to submarine bodies, of whatsoever nature they may be, at variable, though always considerable depths, and consequently, in places where the sea is tranquil. They are particularly to be found in great abundance in the excavations of rocks. It is certain, however, that some species can exist in places covered and left bare successively by the sea, as is the case in our northern seas. Some of them attain great dimensions, having been found more than three or four feet in height. We are not exactly acquainted with the duration of their life, nor with the degree of rapidity in their growth ; but we must conclude that they grow promptly if it be that they can be fished for, with success, after the second year, in places which had been previously exhausted of them. The sponges are very common in the seas of warmer climates, where they attain the largest dimensions. They are less so in those of temperate regions, and finally, as we approach the ices of the north, they become more and more rare, and smaller, and utterly disappear near the polar circle.

SUPPLEMENT

ON THE

INFUSORIA.

On these curious beings, of which so little is known, and whose place in the living series is so far from being determined, we cannot pretend to offer any thing more than a few general observations.

The denomination of INFUSORIA, was introduced into zoology by Otto-Frederic Müller, to designate a class of animals which are developed in vegetable or animal infusions, and which, from their extreme smallness, have also been named microscopic. All systematic authors since Gmelin, have adopted the division and the name, though some have rather restrained its application, and others, in admitting it, have observed that it was very badly circumscribed. In fact, Müller was guided by no principle in the establishment of this class, and it is therefore probable that it contains an anomalous assemblage of animals of degrees of organization, or types, extremely different, and equally different degrees of development. The only common characters which they possess, if characters they can be called, are their extreme littleness and transparence, which render them appreciable only by the microscope; their constant habitat in a fluid, which is a consequence of their littleness, and their not being developed for the most part, except in vegetable or animal infusions, which, however, is as yet but doubtful. Their general and particular form, the only thing which an observer can seize, confirms the differences in their organization. In fact, there are some whose form is truly binary and symmetrical, not only in their body, but also in the appendages attached to it, and which, besides, are clothed with a true corneous envelope. Some have the body elongated, vermiform, or depressed symmetrically, without any trace of appendages, as the vibrio, &c. Others, on the contrary, have a form evidently radiated, with a mouth or cavity apparent, as most of the vorticellæ which we have seen are transferred by the Baron to the preceding class. Finally, there are some whose body is amorphous, or without any determinate form susceptible of definition, without buccal aperture, or trace of appendages, as the proteus, the volvox, and the monads. Those of the first sort are true animals, and even very elevated in the scale, since we find in them locomotive appendages, very distinct, which have been denominated wheels, filaments, &c.; a tail composed of several articulations, and often terminated by appendages, variable in form and number; a true cephalothoracic buckler, covering a trunk more or less distinct; even a heart, eyes, and ovaries, have been remarked, and consequently we cannot doubt that these animals are provided with an intestinal canal. These animals thus exhibit some relations with those crustacea called *entomostraca*, by Müller, and it is no very improbable conjecture, that some of these infusoria of this first section, may be only degrees of developments of species of entomostraca, well known in the adult state, for some of them are susceptible of very distinct metamorphoses, as has been proved by M. de Jurine, in the nauplia and amynome. As for the second form in the infusoria, which is seen in the vibriones, we may conceive that it might appertain to animals of the class of apod worms, since the body is elongated and symmetrical, without any visible articulations, and certainly without any appendages; but it would be too much to assert this as certain, since observers say nothing of an intestinal canal, nor consequently of mouth or anus. Still the numerous movements of these organized bodies in determinate directions, will not permit us to doubt of their animality. It is the same with the vorticellæ, of which we have spoken elsewhere, and which, though long ranked here, have a great analogy with the hydræ or polypi. There remain then the protei, and the volvoces, which we cannot refer to any known type. They, in fact, are organized bodies, without any determinate form, without any distinct organ, being nothing but a small mass of cellular tissue, in the meshes of which are contained fluids, and which is hardly condensed at the circumference to form an envelope, so that all the functions in these bodies, are reduced to immediate absorption of molecules, already prepared in the ambient fluid, and to exhalation. This may be considered as the term, the last link of animal life, where no organ is distinguishable.

Be this, however, as it may, it clearly appears from what has now been said, that the class of the infusoria is totally inadmissible, because it contains animals of very different types. This, indeed, could not have been recognized previously to the establishment of the principle, that the general form of the body carries with it a determinate degree of organization, for in such little animals it is almost the form alone that can be perceived. Thus we may consider that the genera brachion, urceolaria, cercaria, furcularia, kerone, trichocercus, and himantopus, really belong to the type of articulated animals, and particularly to the class of heteropods, order entomomostraca. Many species of vibriones may be regarded as apod worms, and likewise the genera paramecia, and kolpoda; the rest of the vibriones, the cyclides, and, perhaps, the leucophes, should come near the planaria. The true vorticellæ we have seen to be polypi. Finally, the genera gonium, proteus, volvox and monads, if they can be certainly regarded as animals, ought, in the opinion of M. de Blainville, to form a distinct type, to which he gives the names of *amorpha*, and *agastraria*, which are intended to express that they have no determinate form, and that the external envelope does not turn into or form a stomach, as in all true animals.

Though many first rate naturalists, such as Leuwenhoek, and Spallanzani, have occupied themselves with the infusoria, it is much to be wished that such labours were renewed with juster views, and more scientific exactitude of observation. Many things, no doubt, would be found to require rectification, and the result would be important, not only to zoology, but to general physiology. Many authors, admitting too generally that these animals are born in vegetable or animal infusions, have taken occasion from this position to maintain the doctrine of spontaneous generation, and several other notions, more or less erroneous. From some remarks on the last of the infusoria, the generation from spontaneous scission in the parent, in whose body gemmules might be formed, has been also admitted. This from analogy is more probable; but it would be well, if possible, to ascertain that it were true. The greatest care should be taken in observation to avoid the errors which the instrument we are forced to use may lead us into.

Erratum in title of third order of Polypi.

For polypi polypiferi, *read* polypi *with* polyparia.

ALPHABETICAL LIST

OF THE

FIGURES OF MOLLUSCA*.

PL. FIG.

7. 3. ACASTA spinosula, *Des.*
Hyaline, valves tubercular near the lips.

7. 4. Acasta Montagui, *Leach.*
Ashy white, valves lanceolate.

27. 5. Achatina Mulleri Férussac.
Pale brown, with deeper streaks.

2. 1. Anatifa lævis, *Lam.*
Nearly white.

12. 3. Anatina hispidula, *Cuv.* (Lanterna, *Lam.*)
Ashy, with white spots, shells very thin.

22. 3. Anatina elliptica, *King.*
Nearly white, striated.

32. 13. Ancillaria cinnamomea, *Blain.*
Cinnamon colour.

37. 9. Ancillaria Australis, *Sav.*
Light yellow, with ashy spots.

19. 3. Anodon Georginæ, *Gray.*
Varied, purple red, shell solid, smooth, covered with a thickish olive coloured periosteum.
Rivers of Paraguay.

24. 1. Anodon Susannæ, *Gray.*
Shell thin, covered with pale olive coloured periosteum.
From South America.

24. 2. Anodon tenuis, *Gray.*
Periosteum sea green.

39. 6. Anomia ephippium, *Lin.*
Smooth, silvery, thin, varies in shape with the substance it is attached to.

PL. FIG.

10. 6. Aplidium lobatum, *Sav.*
Pale brown.

39. 12. Arca granosa, *Lam.*
White, with transverse tubercles.

5. 3. Argonauta argo, *Lin.*
(The animal is crythoe antiquorum, and has no muscular attachment to the shell.) Shell white, animal brownish.

35. 1. Arion Empiricorum.
Black, with cross lines.

7. 6. Asemus porosus, *Cuv.*
Purple.

8. 7. Aspergillum vaginiferum, *Lam.*
Ash coloured.

27. 8. Auricula Midæ, *Lam.*
White, covered with brownish periosteum.

39. 2. Avicula heteroptera, *Lam.*
Brown.

7. 1. Balanus ovularis, *Lam.*
White.

7. 2. Balanus sulcatus.
White.

18. Bivalve shells, terminology of.

9. 10. Boltenia ovifera, *Sav.*
Pale brown.

10. 1. Botryllus polycyclus, *Sav.*
Pale ash colour, with green spots.

* Most of the inedited shells figured in this work are from the collection in the British Museum.

PL. FIG.

27. 1. Bulimus Guadalupensis, *Ferus.*
White, with brown bands.

37. 7. Bulimus auris-vulpina, *Gray.*
(Struthiolaria arcuata, *Lam.*)
Pale yellow, with light band.
St. Helena.

37. 8. Bullæa semiplicata, *Gray.*
Pale buff, upper lip and suture callous.

12. 6. Byssomia pholadis, *Cuv.*
Whitish, with a pale thin periosteum; syphon large, pale pink.

6. 8. Calpurnus verrucosus, *Mart.*
(Ovula verrucosa, *Lam.*)
White and pink.

32. 11. Cancellaria asperula, *Des.*
White.

38. 3. Cardium fimbricatum, *Lin.*
Ashy.

39. 10. Cardita caliculata, *Lam.*
White, spotted with black.

32. 8 Cassis decussata, *Lam.*
Brown, with darker bands or spots.

32. 9. Cassidaria echinophora, *Lam.*
Whitish, shell thin.

11. 6. Catillus Cuvierii, *Brongn.*
Fossil.

13. 1. Cerithium læve [1], *Gray.*
White.
New Holland.

14. 1. Cerithium zonate, *Lam.*
Blackish, with white band.

14. 4. Cerithium truncatum, *Lam.*
Yellowish ash.

38. 5. Chama croceata, *Lam.*
Shell orange colour, subspinose.

27. 9. Chrondrus avenaceus, *Cuv.*
Pale brown.

7. 10. and 11. Chthamatus stellatus.
Dirty white.

2. 6. Cineras vittata, *Leach.*
Dark ash, with black bands.

8. 6. Clavagella coronata, *Lam.*
Fossil.

37. 6. Clavatula Griffithii, *Gray.*
Fusiform, solid, ashy, spirally striated, whorls rounded with a subposterior dark band, interrupted by whitish tubercles.

9. 13. Clavellina borealis, *Sav.*
Pale ash.

27. 6. Clausilia inflata, *Lam.*
Pale ash.

3. 8. Cleodora lanceolata.
Hyaline.

3. 1. Clio borealis, *Lin.*
Dirty white.

37. 1. Columbella Tyleræ, *Gray.*
White, with black lines, leaving square white spots; front transversely striated, white and brown.

37. 2. Columbella harpœformis, *Sow.*
Dark ash.

41. 3. Columbella suturalis, *Gray.*
Blue and brown.

32. 7. Concholepas Peruvianus, *Lam.*
Dark brown, inside white.

7. 1. Conia radiata.
Ashy white.

27. 13. Conovulus fasciatus, *Des.*
White, banded with brown.

6. 1. Conus Caledonicus, *Lam.*
Fulvous, with spiral lines.

6. 2. Conus bandanus, *Lam.*
Black, with triangular white spots.

6. 3. Conus tendineus, *Lam.* Blue.

7. 13. Coronula balænaris.
Greenish white.

4. 5. Crania personata, *Lam.*
Fossil.

7. 9. Creusia spinosula.
Hyaline.

11. 3. Crenatula avicularis, *Lam.*
Brown, with narrow pale rays.

3. 9. Creseis virgula, *Rang.*
Hyaline.

41. 1. Cryptostoma Javanica, *Gray.*
White.

3. 10. Cuvieria columnella, *Rang.*
Hyaline.

28. 1. Cyclostoma articulata, *Gray.*
White, spirally grooved with brown dots, umbilicus deeply grooved.

28. 3. Cyclostoma pulchrum, *Gray.*
Pale brown, with interrupted, brown, radiated streaks.

[1] Erratum in the plate, Del. *truncatum*, lege *læve.*

PL. FIG.

28. 4. Cyclostoma Madagasciensis, *Gray*.
Whitish, with narrow brown spiral lines.

28. 5. Cyclostoma auriculare, *Gray*.
White, apex reddish.

41. 2. Cyllene Owenii, *Gray*.
This shell in general appearance is assimilated to *Buccinum*, but it has a groove over the suture, as in *Oliva*, and a sinus in front of the outer lip, as in strombus. Mr. Gray has, therefore, separated it generically.

3. 2. Cymbulia Peronii, *Cuv*.
Blueish white.

9. 11. Cynthia momus, *Sow*.
Rose coloured.

6. 4. Cypræa stolida, *Lam*.
Pale, fulvous, teeth yellow.

6. 5. Cypræa pediculus, *Lam*.
Animal.

20. 2. Cyrena similis, *Gray*.
Periosteum olive green, lateral teeth long.

19. 1. Cytherea Dronia, var.
Pale pinkish white.
Note.—This seems to be Cy. lupinaria, of Lesson.

7. 15. Diadema Coronula Diadema.
Dirty yellow.

32. 10. Dolium perdix, *Lam*.
Brown.

38. 9. Donax hilairea, *Valenc*.
Yellowish grey.

32. 1. Eburna spirata. White.

5. 2. Eledone moschatus; arm of.

4. 8. Etheria elliptica, *Lam*.
Ashy, with yellowish spots.

10. 5. Eucœlium hospitiolum, *Sauv*.
Pale reddish.

3. 12. Eurybia hemispherica. Hyaline.

41. 7. Fasciolaria trapezium, *Lam*.
Brown, with darker spiral lines.

8. 3. Fistulana gregata, *Lam*.
Dirty white.

PL. FIG.

33. 3. Fusus morio, *Lam*.
Purple black, with a posterior pale band.

38. 7. Galathea radiata, *Lam*.
Covered with olive periosteum.

8. 4. Gastrochæna cuneiformis, *Lam*.
Dirty white; shell thin.

11. 4. Gervilia solenoides, *Defrance*.
Fossil.

31. 2. Glauconome Chinensis [1], *Gray*.
White, covered with a pale green periosteum.

12. 4. Glycimeris siliqua, *Lam*.
Dark green.

32. 5. Harpa ventricosa, *Lam*.
Pink, varied with brown and yellow.

27. 7. Helix corocolla, *Lin*.
Animal reddish, with three blue bands along the back.

28. 2. Helix argillacea, *Gray*.
Clay coloured.

36. 1. Helix mora, *Gray*.
Chestnut brown, edge of lips white.

36. 4. Helix Cunninghami, *Gray*.
Pale, with broad bands.
New Holland.

36. 5. Helix Lamarkii.
Pale, with broad brown bands.

36. 2. Helix viridis, *Desh*.
Green, banded with black.

36. 6. Helix Fraseri, *Gray*.
Pale, with many unequal plain bands.
New Holland.

43. 10. Hippocrenes macropterus.
Reddish ash.

3. 6. Hialea globulosa.
Dirty white.

3. 7. Hialea trespinosa, *Leseur*.
Dirty white.

12. 7. Hiatella arctica, *Bosc*.
Whitish.

11. 5. Inoceramus sulcatus, *Cuv*.
Fossil.

3. 5. Limacina helicina, *Cuv*.
Dirty white.

[1] This is named by mistake on the plate Glycemeris Apinensis.

PL. FIG.

4. 4. Orbicula lævigata, *Blain.*
Horn colour.

39. 1. Ostrea cristagalli, *Lam.*
Black.

6. 6. Ovula triticea, *Lam.*
Pale pink.

6. 7. Ovula volva, *Lam.*
Pale red, spirally striated.

1. 5. Paludina Chinensis, *Gray.*
Olive green; mouth with black edge.

1. 6. Paludina pulchra, *Gray.*
Pellucid, with red brown spiral bands.

36. 3. Paludina subcostata, *Gray.*
Olive green.
From China.

12. 5. Panopœa Aldrovandi, hinge of, *Cuv.*

35. 5. Parmacella Olivieri, *Lam.*
Dirty ash.

35. 6. Parmacella palliolum, head of, *Fer.*

31. 6. Pecten purpureus, *Lam.*
Purplish white.

39. 5. Pecten gibbosus, *Lam.*
Reddish brown.

39. 9. Pentadina margaritifera, *Lam.*
Green.

11. 2. Perna ephippium, *Lam.*
Purple.

39. 13. Petricola lucinalis, *Lam.*
White.

9. 12. Phallusia nigra, *Sav.*
Black.

8. 1. Pholas striata, (papyracea.)
White.

3. 11. Phyche globulosa, animal.

27. 4. Physa Novæ Hollandiæ.
Pale brown.

38. 11. Pinna angustana, *Lam.*
Yellowish ash.

39. 8. Placuna placenta, *Lam.*
Silvery White.

23. 1. Pleurostoma grandis, *Gray.*
White with red spots.

23. 2. Pleurostoma carinata, *Gray.*
White, whorles keeled.

33. 1. Pleurostoma Babylonia, *Lam.*
White, with black spots.

39. 4. Plicatula cristata, *Lam.*
Dirty white.

3. 3. Pneumodermon diaphanum, *Quoy and Gay.*
Dirty white.

PL. FIG.

3. 4. Pneumodermon Peronii, *Cuv.*
Dirty white.

2. 2. Polliceps cornucopia, *Lam.*
White.

2. 3. Polliceps mitella, *Lam.*
Dirty white.

2. 4. Polliceps scalpellum.
Dirty white.

10. 4. Polyclinum constellatum, *Sow.*
Bluish ash colour, with yellow spots.

32. 12. Potamis fragilis, *Brogn.*
Fossil.

12. 10. Psammothea candida, *Lam.*
White.

3. Psyche globulosa.
Hyaline.

33. 6. Pterocera scorpio, *Lam.*
Brown, with reddish brown streaks.

11. 7. Pulvinites Adansonii, *Dep.*
Fossil.

27. 3. Pupa striatella, *Fer.*
Ash coloured, striated.

3. 13. Pyrgo lævis, *Dep.*
Hyaline.

7. 7,8. Pyrgona cancellata.

10. 2. Pyrosoma rufum, *Quoy and G.*
Straw colour.

10. 3. Pyrosoma giganteum, Anatomy of.

25. 3,4. Pyrula Mawæ, *Gray.*
Shell deformed. White.
From China.

33. 5. Pyrula perversa, *Lam.*
Light ash-coloured, with brown streaks.

37. 4. Pyrula striata, *Sav.*
Pale yellow, with deeper spots.

32. 2. Ricinula arachnoides, *Lam.*
White, with black tubercles.

33. 8. Rostellaria pes-pelicani, *Lam.*
Brownish ash.

9. 2. Salpa cristata (scutigera) *Cuv.*
Ashy.

9. 3. Salpa infundibuliformis, *Quoy.*
White, with ashy lines.

9. 4. Salpa tricuspis, *Quoy.*
White, with ashy lines.

9. 5. Salpa longicauda, *Quoy.*
Ash coloured.

9. 6. Salpa fusiformis, *Quoy.*
Ash coloured.

PL. FIG.

9. 7. Salpa zonaria, *Quoy.*
White, with ashy shadow.

9. 8. Salpa cylindrica, *Cuv.*
White, with ashy shadow.

9. 9. Salpa pyramidalis, *Quoy.*
White with ashy shadow.
None of these salpæ have a shell.

12. 9. Sanguinolaria livida, *Lam.*
Livid flesh colour.

27. 11. Scarabus imbrium
Pale brown.

5. 4. Sepia officinalis, *Lin.*
Purplish.

12. 8. Solen vagina (truncata *Wood*)
Yellowish.

31. 1. Solen novaculina, *Gray.*
White, periosteum dirty olive.

31. 3. Solen Sayii.
Pale flesh-coloured.

31. 4. Solen tenuis, *Gray.*
Yellowish white.

39. 11. Spondylus Americanus, *Lam.*
Reddish White.

4. 3. Spirifer trigonalis, *Sav.*
Fossil.

5. 8. Spirula australis, *Peron.*
Animal reddish.

25. 5. Strombus deformis, *Gray.*
White.

25. 6. Strombus Campbellii, *Gray.*
Brown, with obscure bands.

33. 2. Strombus papilio, *Lam.*
Black brown.

27. 12. Succinea rubescens, *Desm.*
Reddish.

19. 2. Tellina Guildfordiæ, *Gray.*
White, inside bright yellow.

38. 2. Tellina linguafelis, *Lam.*
White, with reddish rays.

6. 9. Terebellum subulatum, *Lam.*
White, mottled with brown.

23. 3. Terebra subulata, *Lam.*
Reddish white, with brown spots. The hinder part of the whorles rounded into acute keels.

23. 5. Terebra Africana, *Gray.*
Pale ash, with central brown band, and streaked with brown.

32. 3. Terebra muscaria, *Lam.*
White.

4. 2. Terebratula Gaudichaudii.
Colour of horn.

PL. FIG.

8. 5. Teredina personata, *Lam.*
Fossil.

8. 2. Teredo navalis, *Lin.*
Dirty white.

35. 4. Testacellus haliotideus,
Reddish ash-coloured.

2. 7. Tetralesmis hirsutus, *Cuv.*
Brown.

9. 1. Thalia cristata, *Cuv.*
Ash-coloured.

38. 1. Tridacna gigas, *Lam.*
White.

39. 7. Trigonia pectinata, *Lam.*
Pearly white, varying with orange, reddish, and purple.

2. 8. Triton (alepis) fasciculatus,
Yellow.

41. 4. Triton Nassoides.
Spanish white.

23. 4. Triton iostoma, *Gray.*
White, streaked with brown, mouth violet.

25. 1. Triton turbinelloides, *Gray.*
Reddish yellow.

25. 2. Triton elegans, *Gray.*
White, varied with brown.

37. 5. Triton vexillum, *Gray.*
Pale reddish, with bright orange spiral bands.

1. 1. Trochus bicarinatus, *Gray.*
White, streaked with brown.

1. 7. Trochus Cunninghami, *Gray.*
Pale buff, with deeper spots.

7. 14. Tubicinella balænarum
Yellowish white.

41. 5. Turbinella ceratus.
White, with brown spots.

33. 9. Turbinella pyrum, *Lam.*
White, spotted brown.

30. 3. Turbinella tubercularis, *Sav.*
White, with brown patches.

13. 5. Turritella suturnalis, *Sav.*
Brown, with a white spiral line.

20. Unio Childreni (unio chinensis) *Gray.*
Dark brownish periosteum, teeth small, compressed; from South America.

20. 3. Unio Smithii, *Gray.*
Periosteum dark olive, disk whitish, varied with olive.

21. 1. Unio Leaii, *Gray.*
Dark ash periosteum; from China.

PL.	FIG.	
21.	2.	Unio Douglasiæ, *Gray*. Periosteum olive, inside yellowish.
21.	3.	Unio Grayii, *Lea*. Dark olive.
24.	2.	Unio tenuis, *Gray*. Sea green, inclining to pink toward the apex.
15.		Univalve spiral shells.
16.		Univalve non-spiral shells.
35.	7.	Vaginula Taunaysii, *Ferr.* Dark green, underneath yellowish.
38.	4.	Venus cancellata. Pale brown, radiated.
38.	12.	Venus exoleta, *Lam.* White.
38.	10.	Venus Danmonensis (Crassina). Periosteum pale brown.
31.	5.	Villorita cyprinoides, *Gray* (Cyrena cyprinoides, *Wood*). Olive green.
35.	3.	Vitrina pellucida, *Lam.* Animal ashy, shell green.
6.	10.	Voluta nivosa, *Lam.* Brown, with white streaked pale spiral bands.
6.	11.	Voluta Æthiopica, *Lam.* Pale brown.
26		Voluta Broderipii, *Gray*. Pale yellow, with two bands of angular lines, spire nearly flat.
29.		Voluta Miltonis, *Gray*. Pale ash, marbled with brown spots. From New Holland.
30.	1.	Voluta rudis, *Gray*. Clay coloured, mixed with white.
30.	4.	Voluta pallida, *Gray* (Vol. Grayii, *Sow.*) Pale buff, with two darker bands.
34.		Voluta Georginæ, *Gray*. Brown, varied in degree. From Swan River.
40.	1.	Voluta Turneri. Brownish, with reddish irregular lines.
40.	3.	Voluta papillosa. Brown, with deeper patches, and a light transverse band.
40.	4.	Voluta gracilis. Dark brown, with zigzag lines.
6.	14.	Volvaria pallida, *Lam.* (Marginella pallida, *Gray*). Palebrown,obscurelybanded.

END OF VOL. XII.